ADVANCES IN CHEMICAL PHYSICS

VOLUME 138

ADVANCES IN
CHEMICAL PHYSICS

VOLUME 138

Series Editor

STUART A. RICE

Department of Chemistry
and
The James Franck Institute
The University of Chicago
Chicago, Illinois

WILEY-INTERSCIENCE
A JOHN WILEY & SONS, INC. PUBLICATION

For general information on our other products and services or for technical support, please contact our Customer Care Department within the United States at (800) 762-2974, outside the United States at (317) 572-3993 or fax (317) 572-4002.

Wiley also publishes its books in a variety of electronic formats. Some content that appears in print may not be available in electronic formats. For more information about Wiley products, visit our web site at www.wiley.com.

Library of Congress Catalog Number: 58-9935

ISBN: 978-0-471-68234-9

Printed in the United States of America

10 9 8 7 6 5 4 3 2 1

CONTRIBUTORS TO VOLUME 138

STUART C. ALTHORPE, Department of Chemistry, University of Cambridge, Cambridge, CB2 1EW, UK

GABRIEL G. BALINT-KURTI, School of Chemistry, University of Bristol, Bristol BS8 1TS, UK

DAVID S. BOUCHER, Department of Chemistry, Washington University in St. Louis, One Brookings Drive, CB 1134, St. Louis, MO 63130 USA

ALEX BROWN, Department of Chemistry, University of Alberta, Edmonton, AB, T6G 2G2, Canada

JUAN CARLOS JUANES-MARCOS, Leiden Institute of Chemistry, Gorlaeus Laboratories, Leiden University, P.O. Box 9502, 2300 RA Leiden, The Netherlands

RICHARD A. LOOMIS, Department of Chemistry, Washington University in St. Louis, One Brookings Drive, CB 1134, St. Louis, MO 63130 USA

R. B. METZ, Department of Chemistry, University of Massachusetts Amherst, Amherst, MA 01003 USA

HIROKI NAKAMURA, Institute for Molecular Science National Institutes of Natural Sciences, Myodaiji, Okazaki 444-8585, Japan

DAVID L. OSBORN, Combustion Research Facility, Sandia National Laboratories, Livermore, CA 94551-0969 USA

IVAN POWIS, School of Chemistry, University of Nottingham, Nottingham NG7 2RD, UK

ECKART WREDE, Department of Chemistry, University of Durham, South Road, Durham, DH1 3LE, UK

SHIYANG ZOU, School of Chemistry, University of Bristol, Bristol BS8 1TS, UK

INTRODUCTION

Few of us can any longer keep up with the flood of scientific literature, even in specialized subfields. Any attempt to do more and be broadly educated with respect to a large domain of science has the appearance of tilting at windmills. Yet the synthesis of ideas drawn from different subjects into new, powerful, general concepts is as valuable as ever, and the desire to remain educated persists in all scientists. This series, *Advances in Chemical Physics*, is devoted to helping the reader obtain general information about a wide variety of topics in chemical physics, a field that we interpret very broadly. Our intent is to have experts present comprehensive analyses of subjects of interest and to encourage the expression of individual points of view. We hope that this approach to the presentation of an overview of a subject will both stimulate new research and serve as a personalized learning text for beginners in a field.

STUART A. RICE

CONTENTS

THE INFLUENCE OF THE GEOMETRIC PHASE ON REACTION DYNAMICS

STUART C. ALTHORPE

Department of Chemistry, University of Cambridge, Cambridge, CB2 1EW, UK

JUAN CARLOS JUANES-MARCOS

Leiden Institute of Chemistry, Gorlaeus Laboratories, Leiden University, P.O. Box 9502, 2300 RA Leiden, The Netherlands

ECKART WREDE

Department of Chemistry, University of Durham, South Road, Durham, DH1 3LE, UK

CONTENTS

Advances in Chemical Physics, Volume 138, edited by Stuart A. Rice
Copyright © 2008 John Wiley & Sons, Inc.

1

I.   INTRODUCTION

Research over the past 15 years has established that conical intersections (CIs) are much more common than previously thought, and that they play a central role in photochemistry [1, 2]. The main role of a CI is to act as a funnel, transferring population between the upper and lower adiabatic electronic states. This effect has been studied extensively, and is well known to be caused by the derivative coupling terms between the two electronic surfaces [3], which are singular at the CI. However, an accompanying effect, which has received less attention, is the so-called geometric (or Berry) phase (GP), in which the adiabatic electronic wave function changes sign upon following a closed loop around the CI [4–9]. This effect is particularly interesting when the system is confined to the lower adiabatic surface, since it is then the only nonadiabatic effect produced by the CI.

The GP affects the nuclear dynamics by introducing a corresponding sign change in the continuity boundary condition, which cancels out the sign change in the electronic wave function (in order to keep the total wave function single valued). In a model particle-on-a-ring system, with the CI at the center of the ring, the GP boundary condition changes the allowed values of M in the nuclear wave function, $\exp(iM\phi)$, from integer to half-integer values. The GP brings about analogous changes in the quantum numbers and energy levels of more realistic systems, and some of these have been predicted and observed in a variety of Jahn–Teller molecules [10–13]. One can say with some confidence, therefore, that the effect of the GP on the nuclear wave function of a bound-state system is well understood.

Until very recently, however, the same could not be said for reactive systems, which we define to be systems in which the nuclear wave function satisfies scattering boundary conditions. It was understood that, as in a bound system, the nuclear wave function of a reactive system must *encircle* the CI if nontrivial GP effects are to appear in any observables [6]. Mead showed how to predict such effects in the special case that the encirclement is produced by the requirements of particle-exchange symmetry [14]. However, little was known about the effect of the GP when the encirclement is produced by reaction paths that loop around the CI.

Very recently, this state of affairs has changed, and there is now a good general understanding of GP effects in gas-phase reactions. This has come about mainly through detailed reactive-scattering studies, both theoretical [15–37] and experimental [30–39], on the prototype hydrogen-exchange reaction $(H + H_2 \rightarrow H_2 + H)$. This is the simplest reaction to possess a CI, and at high energies ($> 1.8eV$ above the potential minimum) it is just possible for some reaction paths to encircle the CI. Reactive-scattering calculations are difficult, and the first calculations that included the GP boundary condition were reported by Kuppermann and co-workers in the early 1990s [15–19]. These calculations predicted large GP effects at high energies, but unfortunately these predictions were not reproduced by later calculations [20–29], nor, most crucially, by experiment [30–39]. Instead, the experiments found no evidence at all of GP effects in the $H + H_2$ reaction. Detailed scattering data from the experiments agreed quantitatively with theoretical predictions that omitted the GP boundary condition.

This negative result seemed to imply that $H + H_2$ could not be used to investigate the effect of the GP because the nuclear wave function does not encircle the CI. However, a series of calculations by Kendrick [20–22] yielded a surprising result, which we will refer to in this chapter as the "cancellation puzzle". This is that GP effects appear in the scattering observables at specific values of the total angular momentum quantum number J, but cancel on summing over J to give the full reactive scattering wave function (describing a rectilinear collision of the reagents). This is puzzling because the sum over J is a unitary transformation in the external, angular, degrees of freedom, which describe the scattering of the products, and there is no direct relation linking this space to the GP boundary condition (which acts on the internal degrees of freedom describing motion around the CI).

This chapter surveys the work we did [25–29] to solve the cancellation puzzle, and the general explanation of GP effects in reactive systems that came out of it [27,28]. The latter uses ideas that were introduced in the late 1960s in Feynman path-integral [40] work on the analogous Aharonov–Bohm effect [41–45]. These early papers seem to have passed unnoticed in the chemical physics community, perhaps because they are written in the language of path-integral theory and algebraic topology. However, the central result of this work is surprisingly simple, and can be derived without path-integrals [27,28]. It is that the nuclear wave function has two components, and that the sole effect of the GP is to change their relative sign. One component contains all the Feynman paths that loop an even number of times around the CI; the other contains all the paths that loop an odd number of times.

In Section II, we introduce this central result, deriving it first without path integrals, by using a diagrammatic representation of the nuclear wave function; we then give a heuristic summary of the ideas behind the early Aharonov–Bohm papers, and show how they can be combined with the diagrammatic approach. In Section III, we describe in detail the solution to the cancellation puzzle in $H + H_2$, which demonstrates how to explain GP effects in a reaction in terms of the even- and odd-looping Feynman paths. In Section IV, we discuss some further aspects of the topology, explaining the effect of particle-exchange symmetry, and the difference between GP effects in bound and reactive systems. Section V concludes the chapter.

II. UNWINDING THE NUCLEAR WAVE FUNCTION

A. Topology and Encirclement

We will consider a system with N nuclear degrees of freedom, which possesses one CI seam [1,2] of dimension $N - 2$. The topology of the nuclear space can then be represented schematically as shown in Fig. 1 [28]. The line at the center represents every CI point in the seam. Each circular cut through the cylinder represents the two degrees of freedom in the nuclear "branching space", in which the adiabatic potential energy surfaces have the familiar double-cone shape, centred about the CI point. We assume that the seam line extends throughout the entire region of energetically accessible nuclear coordinates space. We also assume that the system is confined to the lower (adiabatic) electronic state, because it has insufficient energy to approach the region of strong coupling with the upper state close to the conical intersection. The CI seam line is therefore surrounded by a tube of inaccessible coordinate space.

We then define an internal coordinate ϕ such that $\phi = 0 \rightarrow 2\pi$ denotes a a path that has described one complete loop around the CI in the nuclear branching space. Other than this, we need specify no further details about ϕ. We do not even need to specify whether the complete set of nuclear coordinates give a direct product representation of the space. It is sufficient that ϕ permits us to count how many times a closed loop has wound around the CI. Using this definition of ϕ, we can express the effect of the GP on the

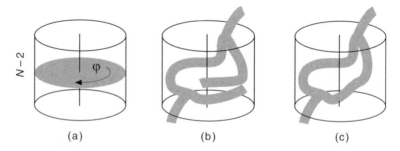

(a)   (b)   (c)

Figure 1. (a) Diagram illustrating the topology of the N-dimensional nuclear coordinate space of a reactive system with a CI. The vertical line represents the $(N - 2)$-dimensional space occupied by the CI seam. The gray disk represents a two-dimensional (2D) branching-space cut through one point on the seam, with the angle ϕ describing internal rotation around the CI. (b) A nuclear wave function that wraps around the CI, but is not a torus, and thus exhibits only trivial GP effects. (c) A torus-shaped nuclear wave function encircling the CI. The "arms" are to be understood as extending to infinity, and are the portions of the wave function in the reagent and product channels.

adiabatic ground-state electronic wave function $\Phi(\phi)$ and the nuclear wave function $\Psi(\phi)$ as

$$\Phi(\phi + 2n\pi) = (-1)^n \Phi(\phi) \tag{1}$$
$$\Psi(\phi + 2n\pi) = (-1)^n \Psi(\phi) \tag{2}$$

The dependence on the other $N - 1$ nuclear degrees of freedom has been suppressed.

The effects of the GP are therefore the differences between the nuclear dynamics described by the wave function

$$\Psi_G(\phi) = (-1)^n \Psi_G(\phi + 2n\pi) \tag{3}$$

which correctly includes the GP boundary condition, and the wave function

$$\Psi_N(\phi) = \Psi_N(\phi + 2n\pi) \tag{4}$$

which ignores it (and is therefore physically incorrect). It is well known in the literature that the GP will only produce a nontrivial effect on the dynamics when $\Psi_G(\phi)$ *encircles* the CI. Otherwise the effect is simply a change in the phase of $\Psi_N(\phi)$, which has no effect on any observables. Hence, throughout this chapter, we are seeking to explain how the dynamics described by $\Psi_G(\phi)$ differs from the dynamics described by $\Psi_N(\phi)$, when these wave functions encircle the CI.

It is worth clarifying what is meant by encirclement. As already mentioned, the nuclear coordinates need not form a direct product, and in fact the notion of taking a cut through the nuclear coordinate space, in order to see whether $\Psi_G(\phi)$ encircles the CI in this cut, is not useful. Figure 1(b) shows a nuclear wave function which, if a certain choice of nuclear coordinates were used, could easily be made to "encircle" the CI if a suitable 2D cut were taken. However, this particular wave function would not show nontrivial GP effects, because it does not encircle the CI: it has unconnected "ends". For nontrivial GP effects to appear, $|\Psi_G(\phi)|^2$ must have the form of a *torus* in the nuclear coordinate space, as shown in Fig. 1(c). If one were to take a series of branching-space cuts through this wave function, none of them would encircle the CI, and hence one might get the mistaken impression that this wave function would only show a trivial phase change upon inclusion of the GP boundary condition. However, the wave function of Fig. 1(c) would definitely show strong, nontrivial GP effects. There are various ways in which one can prove this and we will mention one below. It is important to emphasise that it is $|\Psi_G(\phi)|^2$, which has the form of a torus and not the wave function $\Psi_G(\phi)$.

B. Symmetry Approach

To explain the effect of the GP on the nuclear dynamics [i.e., to explain the difference between the dynamics described by an encircling $\Psi_G(\phi)$ and an encircling $\Psi_N(\phi)$], we need to compare the topology of $\Psi_G(\phi)$ with the topology of $\Psi_N(\phi)$. In Section II. C, we review how this can be done using the homotopy of the Feynman paths [41–45] that make up these wave functions. But first, to demonstrate the simplicity of the problem, we use the diagrammatic approach developed in Refs. [27 and 28].

We represent the internal coordinate space occupied by the nuclear wave function as shown in Fig. 2. The gray area represents the energetically accessible region of the potential energy surface; the conical intersection is the point at the center; the arms represent the reagent entrance and product exit channels. To simplify the discussion, we place a restriction on ϕ (which will be relaxed later), stating that ϕ tends to a constant value as the system moves down the entrance or exit channel toward an asymptotic separation of the reagents or products. This places no restriction on the generality of the diagram, other than that the conical intersection should be located in the "strong-interaction region" of the potential energy surface, where all the nuclei are close together. Note that, although we have restricted the number of product channels to one, the diagram is immediately generalizable to systems with multiple product channels. We also assume that the reaction is bimolecular (leaving unimolecular reactions until Section II.D), which means that it is initiated at the asymptotic limit of the reagent channel, at the value of ϕ that is reached in this limit. We will define this to be $\phi = 0$.

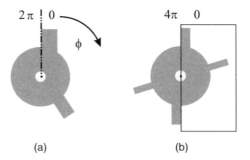

(a)   (b)

Figure 2.   (a) Schematic picture of the potential surface of a reactive system, indicating that there is an energetically accessible (gray) "tube" through the potential surface, permitting encirclement of the CI (dot at center). The "arms" are the reagent and product channels. (b) The same surface, represented in the $0 \rightarrow 4\pi$ cover space. The rectangle represents a $0 \rightarrow 2\pi$ sector that can be cut out of the double space so as to map back onto the single space, where the $\phi = 0$ and $\phi = 2\pi$ "edges" are joined together at the cut line (chains).

Figure 2(b) represents the potential surface of the identical system, mapped onto the double-cover space [28]. The latter is obtained simply by "unwinding" the encirclement angle ϕ, from $0 \to 2\pi$ to $0 \to 4\pi$, such that two (internal) rotations around the CI are represented as one in the page. The potential is therefore symmetric under the operation $\hat{R}_{2\pi}$ defined as an internal rotation by 2π in the double space. To map back onto the single space, one cuts out a 2π-wide sector from the double space. This is taken to be the $0 \to 2\pi$ sector in Fig. 2(b), but any 2π-wide sector would be acceptable. Which particular sector has been taken is represented by a cut line in the single space, so in Fig. 2(b) the cut line passes between $\phi = 0$ and 2π. Since the single space is the physical space, any observable obtained from the total (electronic + nuclear) wave function in this space must be independent of the position of the cut line.

To construct a diagrammatic representation of the wave function, we start in the double space, as shown in Fig. 3a. The arrow at the top indicates that the incoming boundary condition is applied here, and the arrows at each of the other channels indicate outgoing boundary conditions. Note that we are treating the second appearance of the reagent channel (at $\phi = 2\pi$) as though it were a product channel, and are treating the second appearance of the product channel (in the $2\pi \to 4\pi$ sector) as though it were physically distinct from the first appearance of this channel, which is indicated by the use of wavy lines. Consequently, the wave function Ψ_e is neither symmetric nor antisymmetric under $\phi \to \phi + 2\pi$, which means it cannot be mapped back onto the physical space independently of the position of the cut line. In other words Ψ_e is the wave function of a completely artificial system.

To construct wave functions that can be mapped back onto the physical space, one needs to take symmetric and antisymmetric linear combinations of $\Psi_e(\phi)$ and $\Psi_o(\phi) = \Psi_e(\phi + 2\pi)$, and these are illustrated in Fig. 3b. It is then

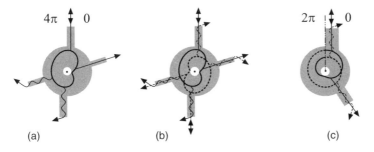

(a)   (b)   (c)

Figure 3. (a) The unsymmetrised nuclear wave function Ψ_e (solid line) in the double space. The arrows indicate the application of incoming and outgoing scattering boundary conditions, (b) The symmetrized linear combinations of Ψ_e (solid) and Ψ_o (dashed), which yield $\Psi_{N/G} = 1/\sqrt{2}[\Psi_e \pm \Psi_o]$. (c) The same functions mapped back onto the single space.

clear that these functions can mapped onto the physical space (Fig. 3c), and that they correspond to Ψ_N and Ψ_G, respectively. Thus we may write,

$$
\begin{aligned}
\Psi_G &= \frac{1}{\sqrt{2}}[\Psi_e + \Psi_o] \\
\Psi_N &= \frac{1}{\sqrt{2}}[\Psi_e - \Psi_o]
\end{aligned}
\tag{5}
$$

This equation is the main result needed to explain the effect of the GP on the nuclear dynamics of a chemical reaction. Clearly, the sole effect of the GP is to change the relative sign of Ψ_e and Ψ_o. Within each of these functions the dynamics is completely unaffected by the GP. We emphasize that, despite remaining unnoticed for so long in the chemical physics community, Eq. (5) is exact.

If we can compute Ψ_G and Ψ_N numerically (as described below), it is therefore trivial to extract Ψ_e and Ψ_o by evaluating

$$
\begin{aligned}
\Psi_e &= \frac{1}{\sqrt{2}}[\Psi_N + \Psi_G] \\
\Psi_o &= \frac{1}{\sqrt{2}}[\Psi_N - \Psi_G]
\end{aligned}
\tag{6}
$$

Once one has extracted Ψ_e and Ψ_o, an explanation of the GP effect on the nuclear dynamics will follow immediately. The dynamics in Ψ_e is decoupled from the dynamics in Ψ_o, and thus any observable will show GP effects only if the corresponding operator samples Ψ_e and Ψ_o in a region of space where these functions overlap. In a nonencircling nuclear wave function, Ψ_e and Ψ_o never overlap, and this gives us a diagrammatic proof (Fig. 4) of the well-known result that a nonencircling wave functions shows no nontrivial GP effects.

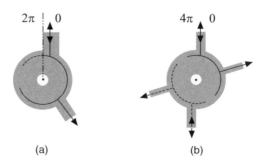

(a)   (b)

Figure 4. (a) Single- and (b) double-space representations of Ψ_e (solid) and Ψ_o (dashed) for a system that does not encircle the CI.

C. Feynman Path Integral Approach

We now explain the physical significance of the two components Ψ_e and Ψ_o in terms of the path integral theory developed in Refs. [41–45]. This theory was developed originally to treat the Aharonov–Bohm system, in which an electron encircles, but does not touch, a magnetic solenoid. The vector potential of the solenoid has an effect that is exactly equivalent to the application of the GP boundary condition, and scattering boundary conditions are applied at long range. The Aharonov–Bohm system is therefore exactly analogous to a nuclear wave function in a reactive system that encircles a CI.

To avoid discouraging the reader, we point out that only a few, basic concepts of path integrals are required. We review these here in a heuristic manner, beginning with the celebrated result of Feynman and Hibbs [40], which is that the time-evolution operator or Kernel, $K = \exp(-i\hat{H}t/\hbar)$, can be constructed using

$$K(x, x_0|t) = \int \mathcal{D}x(t)e^{iS(x,x_0)/\hbar} \qquad (7)$$

Here, $\mathcal{D}x(t)$ represents the sum over all possible paths connecting the points x and x_0 in the time interval t, and S is the classical action evaluated along each of these individual paths. It is useful to point out two properties of this expression: (1) the overall sign of the Kernel is arbitrary, because S is only defined up to an overall constant (because S is the time integral over the Lagrangian, and the latter is only defined up to a total derivative in t [46]); (2) each path has equal weight, so the relative contribution of a given path to the sum is determined by the extent to which it is canceled out by its immediate neighbors.

Any prediction expressed in the language of path integrals must have an equivalent formulation in the language of wave functions. Point (1) is equivalent to saying that a wave function is only specified up to an overall phase factor. Point (2) can be thought of as saying that, when computing $K(x, x_0|t)$, all possible paths between x and x_0 in time t are coupled. If we start with one particular path between x and x_0, then we need to know all of its immediate neighbors, in order to assess the extent to which this path is canceled out by them. These neighbouring paths are obtained by all possible tiny distortions that can be applied to the first path. We then need to know all of the immediate neighbors of each of the latter paths (in order to assess the extent to which each of these is canceled out), and then we need to find out the immediate neighbors of the new paths, and so on. In other words, if we start with one particular path between x and x_0, then this path is coupled (in the sense just described) to all the other paths into which it can be continuously deformed. This is equivalent to saying that one cannot accurately compute just part of a wave function; one must compute all of it, since all parts of the function are coupled by the Hamiltonian operator.

In their work on the Aharonov–Bohm system, Schulman, deWitt, and co-workers [41–44] found that points (1) and (2) must be modified when applying path integral theory in a *multiply connected* space. The term multiply connected simply means that the space contains an inaccessible region or obstacle, which gets in the way, such that a given path between x and x_0 cannot be continuously deformed into all other possible paths. It can only be deformed into a subset of such paths. This subset defines a *homotopy class:* paths that belong to different homotopy classes are called different *homotopes*. The concept that there exist different classes of paths, such that a path that belongs to one class cannot be continuously deformed into a path that belongs to another, is called *homotopy*.

The nuclear coordinate space shown in Fig. 1 is a multiply connected space, because there is an energetically inaccessible "tube" of space surrounding the CI seam. An explanation of the homotopy of such a space will be found in any elementary text on topology [47]. Let us take first a system with only two nuclear degrees of freedom, so that the CI is just a point at the center of the branching space, and there are no other degrees of freedom. The homotopy of a given path within this space is simply the number of entire loops it follows around the CI. We can thus classify each homotopic class according to a winding number n, as defined in Fig. 5. Note that the sign of n indicates the *sense* of the path, and that it is useful to adopt the convention that even n refer to paths that make an even number of clockwise loops or an odd number of counterclockwise loops; and odd n vice versa. It is easy to prove that the set of all these homotopic classes forms an infinite group, which is called the "Fundamental Group" of a circle [47].

The same classification into winding numbers can be used in a system with N nuclear degrees of freedom, in which the CI seam is an $(N-2)$-dimensional hyperline as in Fig. 1. For example, if we take $N = 3$, then the seam is a line; the

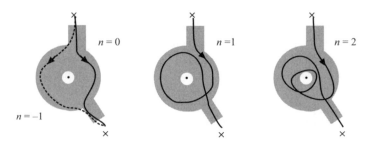

Figure 5. Examples of Feynman paths belonging to different homotopy classes, illustrating how the winding number n is defined.

homotopy of this system is just the same as for the $N = 2$ system, since the number of loops made around the line can be represented by a winding number defined exactly as for the 2D case. Although it is difficult to visualize, the generalization continues to all higher N in the same way, so that one can always classify a path by its winding number around the $(N - 2)$-dimensional CI hyperline. In many systems, each class of paths designated by the winding number n will in fact include more than one homotopy class, because it will be possible to further classify the paths according to their winding about other energetically inaccessible regions in the potential surface, which may exist in addition to the tube around the CI seam. However, to understand the GP we do not need to consider these classes, and so, for shorthand, we will use the terms "homotopy class" and "winding number" interchangeably.

Retracing the argument used to justify point (2), it is clear that, in a multiply connected space, a given path is only coupled to those paths into which it can be continuously deformed. By definition, these are all the paths that belong to the same homotopy class. Paths belonging to different homotopy classes are thus decoupled from one another [41–45]. For a reactive system with a CI that has the space of Fig. 1, this means that a path with a given winding number n is coupled to all paths with the same n, but is decoupled from paths with different n. As a result, the Kernel separates into [41–45]

$$K(\mathbf{x}, \mathbf{x}_0|t) = \sum_{n=-\infty}^{\infty} e^{in\alpha} K_n(\mathbf{x}, \mathbf{x}_0|t) \tag{8}$$

where

$$K_n(\mathbf{x}, \mathbf{x}_0|t) = \int \mathcal{D}_n \mathbf{x}(t) e^{iS(\mathbf{x}, \mathbf{x}_0)/\hbar} \tag{9}$$

and $\mathcal{D}_n \mathbf{x}$ denotes the sum over all paths linking $\mathbf{x}_0$ to $\mathbf{x}$ that have winding number n.

Each K_n in Eq. (8) has a different overall phase, which arises because a different Lagrangian can be used for each value of n. However, there is a strong constraint on these phases, which arises because the set of Kernels K_n must form an irreducible representation of the Fundamental Group of the circle. As a result, the phases have the form $e^{in\alpha}$ [given in Eq. (8)], so that there is only one parameter α that can be varied. To determine possible values of α, let us consider the operation $\phi \rightarrow \phi + 2\pi$ on K_n. This operation is equivalent to rotating the end points of all the paths around the CI by 2π, thus increasing the winding number of each path from n to $n + 1$. As a result, $K_n \rightarrow K_{n+1}$, which means that, overall, $K \rightarrow \exp(i\alpha)K$. In other words, specifying α is equivalent to specifying the $\phi \rightarrow \phi + 2\pi$ boundary condition that is to be satisfied by the

Kernel. Thus, we can obtain Kernels corresponding to GP and non-GP boundary conditions by choosing $\alpha = \pi$ and $\alpha = 0$, which gives

$$
\begin{aligned}
K_G(x, x_0|t) &= K_e(x, x_0|t) - K_o(x, x_0|t) \\
K_N(x, x_0|t) &= K_e(x, x_0|t) + K_o(x, x_0|t)
\end{aligned}
\tag{10}
$$

where $K_e = \sum K_n$, with the sum running over all even n, and K_o is similarly defined for odd n.

To put this result in context, one should imagine a crude semi classical calculation, in which one propagates Newtonian trajectories, each of which is given a phase $\exp(iS/\hbar)$. One could implement the GP boundary condition by counting the number of loops n made by each trajectory around the CI, and adding an extra $n\pi$ to the associated phase. To our knowledge, no such calculation has been reported, almost certainly because it would be difficult to disentangle genuine GP effects from errors in the approximation. However, Eq. (10) tells us that such an intuitive approach can be applied to the Feynman paths, and thus implemented rigorously, without approximation.

D. Consistency between the Symmetry and Feynman Approaches

The separation of the Feynman paths in Eq. (10) is equivalent to the splitting of the wave function into Ψ_e and Ψ_o in Eq. (6). To demonstrate this, we connect the Kernel to the wave function using [48],

$$
\Psi(x) = \frac{1}{A(E)} \int dx_0 \int_0^\infty dt e^{iEt/\hbar} K(x, x_0|t)\chi(x_0)
\tag{11}
$$

where $\chi(x_0)$ is an initial wave packet, which contains a spread of energies $A(E)$. At time $t = 0$, $\chi(x_0)$ is localized in the reagent channel, at a sufficiently large reagent separation that the interaction potential can be neglected. The function $\Psi(x)$ given by Eq. (11) is the time-independent wave function, with incoming boundary conditions in the reagent channel, as represented schematically in Fig. 3. It follows immediately from Eq. (11) that K_e generates $\Psi_e(\phi)$, and K_o generates $\Psi_o(\phi)$. Hence, unwinding the nuclear wave function according to Eq. (6) is equivalent to separating the even n Feynman paths, which are contained in $\Psi_e(\phi)$, from the odd n paths, which are contained in $\Psi_o(\phi)$.

Some care must be taken when applying the Feynman interpretation to Eq. (6), as the Feynman interpretation must be consistent with the position of the cut line (used to map from the double to the single space). For example, Fig. 6a and b shows two different choices of cut line. It is clear that the relative sign of $\Psi_e(\phi)$ and $\Psi_o(\phi)$, and hence all the GP effects, are independent of the position of the cut line. However the overall phase of $\Psi_G(\phi)$ does depend on the cut line. This phase is important, because it must cancel out a corresponding

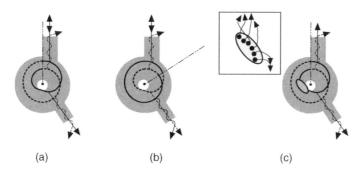

Figure 6. Diagram showing how the winding number n of the Feynman paths should be defined with respect to the cut line. In (a), the cut line (chains) is placed between $\phi = -\epsilon$ and $2\pi - \epsilon$; in (b), between $\phi = \pi/4$ and $-7\pi/4$. In (c), the wave function describes a unimolecular reaction, in which the initial state occupies the (gray shaded) area shown. Feynman paths originate from all points within this area (inset); their winding number n is defined with respect to the common cut line.

phase in the electronic wave function $\Phi(\phi)$, to give a total wave function $\Psi_G(\phi)\Phi(\phi)$, which is independent of the position of the cut line. Because of this, one needs to define the winding number n with respect to the cut line and *not* with respect to the $(\phi = 0)$ point at which the nuclear wave function enters the encirclement region. Thus in Fig. 6b, a path that starts at $\phi = 0$ and terminates at $\phi = \pi/2$, and has made no loops around the CI, is classified as an $n = 0$ path. However, a path that starts at $\phi = 0$ and terminates just short of the cut line at, say, $\phi = \pi/6$, and has also made no loops around the CI, is an $n = -1$ path. A path that enters at $\phi = 0$ and makes one clockwise loop around the CI will be an $n = 0$ path if it terminates just short of the cut line, and will only become an $n = 1$ path once it has passed the cut line, and so on. By classifying the Feynman paths with respect to the cut line in this way we ensure that the overall phase of $\Psi_G(\phi)$ has the correct dependence on ϕ needed to cancel the corresponding dependence of the phase of $\Phi(\phi)$.

In Fig. 3, we placed the cut line between $\phi = -\epsilon$ and $\phi = 2\pi - \epsilon$, where ϵ is an arbitrarily small number. This choice will often be the most convenient cut line, because n then exactly describes the number of complete loops that the system has made around the CI since entering the encirclement region. Thus the paths that scatter inelastically will each have described an (internal) rotation of exactly $\phi = 2n\pi$.

However, it will not always be possible to fix the cut line at the same value of ϕ as the entry points, for the reason that the system does not enter the encirclement region at one unique value of ϕ. Up till now, we have assumed (see Section II.B) that the reaction is bimolecular, that it can only encircle the CI when the nuclei are all close together, and that the reagents and products are distinguishable. These conditions are what are required to guarantee that the system starts at one unique value of ϕ, which we have taken to be $\phi = 0$. We can now relax these conditions, and consider unimolecular reactions, and reactions that can encircle the CI at large separations of the reagents or products. We consider the case of bimolecular scattering with identical reagents and products in Section IV. A.

We can represent a unimolecular reaction using the diagram of Fig. 6c. The gray blob indicates the initial state of the system. For example, it could be the Frank–Condon region accessed in a photodissociation experiment [49]. All the Feynman paths that contribute to the nuclear wave function will originate in the initial state. Hence, the paths will have a spread of start points, distributed over the range of ϕ for which the initial state is nonnegligible. Clearly, the symmetry argument of Section II.B applies immediately to this system, so we may unwind the wave function, and extract Ψ_e and Ψ_o using Eq. (6). We can then interpret these functions as containing the even n and odd n Feynman paths, respectively, where n is defined with respect to a fixed cut line. Note that, when we discuss, say, the even n Feynman paths, we are not referring to paths that all necessarily complete an even number of loops around the CI, since the paths may have started on different sides of the cut line (if the latter passes through the initial state), or on different sides of the end point. The reader may verify that, in either of these cases, the even n paths will contain a mixture of paths that have looped an even and an odd number of times around the CI.

Hence, when applied to a unimolecular reaction, Eq. (6) does not give such a neat separation into even- and odd-looping Feynman paths. However, the separation that it does give (into even and odd n, each of which contains a mixture of even- and odd-looping paths) is the one that is necessary to explain the effect of the GP, since these are the two contributions to Ψ whose relative sign is changed by the GP. Clearly, if we were to compute directly the Kernels, we could then separate out the odd- and even-looping paths, because we would know the starting point of each path. In the wave function, however, we neither know the starting points of the individual paths, nor do we need to in order to explain the effect of the GP.

Similar arguments to those just given apply to bimolecular reactions in which the CI can be encircled when the reagents are still well separated from one another. For such systems, one cannot define ϕ such that it tends to a unique value as the system travels out along the reagent channel. The incoming boundary condition must then be applied across a range of ϕ, which is analogous to the range of ϕ contained in the initial state of the unimolecular reaction. Applying Eq. (6) will then separate out the even and odd n paths with respect to a fixed cut line, which paths may contain a mixture of odd- and even-looping paths (as in the unimolecular case).

III.   APPLICATION TO THE HYDROGEN-EXCHANGE REACTION

The above theory is completely general, but was developed in response to a specific challenge, which was to explain the "cancellation puzzle" posed by the results of Kendrick's calculations [20–22] on the hydrogen-exchange reaction. This application of the theory gives a very good illustration of the effect of the GP on a gas-phase reaction, and we discuss it here in some detail. Note that the only familiarity with reactive scattering assumed is a basic knowledge of quantum scattering theory (e.g., given in standard introductory texts on quantum mechanics).

A.   Reaction Paths and Potential Energy Surface

The first thing to be done when applying the theory is to identify the e and o reaction paths. One can then proceed to calculate Ψ_G and Ψ_N, and then to extract Ψ_e and Ψ_o using Eq. (6). In $H + H_2$, the form of the potential energy surface is very well characterized [50–53], and the form of the CI is a standard example of an $E \times e$ Jahn–Teller intersection.

Figure 7 shows a schematic representation of the $H + H_2$ potential energy surface [29], plotted using the hyperspherical coordinate scheme of Kuppermann [54]. In this section, we will treat the three hydrogen nuclei as distinguishable particles, ignoring the requirement that the nuclear wave function be antisymmetric under exchange of two ${}^1H$ nuclei. The exchange symmetry is

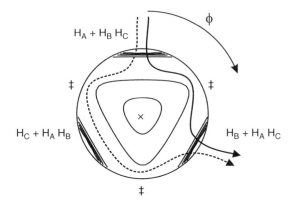

Figure 7. Schematic representation of the 1-TS (solid) and 2-TS (dashed) (where TS = transition state) reaction paths in the reaction $H_A + H_B H_C \rightarrow H_A H_C + H_B$. The H_3 potential energy surface is represented using the hyperspherical coordinate system of Kuppermann [54], in which the equilateral-triangle geometry of the CI is in the center ($\times$), and the linear transition states ($\ddagger$) are on the perimeter of the circle; the hyperradius $\rho = 3.9$ a.u. The angle ϕ is the internal angular coordinate that describes motion around the CI.

easy to incorporate by taking appropriate linear combinations of the unsymmetrised (distinguishable particle) wave functions, and we explain how to do this in Section IV.A. Hence, we consider here only the effects of the GP on the unsymmetrised wave functions, since these are the effects caused by reaction paths that encircle the CI, which give rise to the cancellation puzzle.

Hence, we consider wave functions in which the reaction starts at an asymptotic separation of one uniquely specified arrangement of the atoms (A + BC), and analyze the cross-sections produced by reactive scattering into one of the product channels (AC + B). It is well known [55,56] that the dominant H + H$_2$ reaction path passes over one transition state (1-TS), as illustrated schematically in Fig. 7. Since GP effects are found in the reaction probabilities at sufficiently high energies [20–22, 25–27] (>1.8 eV above the potential minimum), the wave function must encircle the CI at these energies, and thus also contain reaction paths that pass over two transition states (2-TS). For the AC + B products, the 1-TS paths make less than one full revolution, loop in a clockwise sense around the CI (see Fig. 7), and are assigned a winding number $n = 0$ (following the convention of Section II.C). The 2-TS paths also make less than one full revolution, but loop in an anticlockwise sense, and are assigned $n = -1$ [28]. This means that the 1-TS (Feynman) paths are contained in Ψ_e and the 2-TS paths in Ψ_o. In principle, there are also paths with higher winding numbers present in both Ψ_e and Ψ_o, but these can be ignored (since reaction paths passing over three or more TS are highly unlikely in H + H$_2$). In this chapter, we will therefore use the terms Ψ_e paths and 1-TS paths (and Ψ_o paths and 2-TS paths) interchangeably.

B. Application of Topology to Reactive Scattering

Here we summarize what the reader will need to know about reactive scattering in order to understand the application of the topological ideas of Section II to the hydrogen-exchange reaction. The only thing we assume is that the reader is familiar with the concept of scattering wave functions and boundary conditions. Generalizing these concepts to reactive scattering is conceptually straightforward (although technically difficult [57–62], but we will not need to discuss the technicalities here).

We consider a nuclear wave function describing collisions of type A + BC(n) → AC(n') + B, where $n = \{v, j, k\}$ are the vibrational v and rotational j quantum numbers of the reagents (with k the projection of j on the reagent velocity vector of the reagents), and $n' = \{v', j', k'\}$ are similarly defined for the products. The wave function is expanded in the terms of the total angular momentum eigenfunctions $D^J_{kk'}(\theta, \eta, \chi)$ [63], and takes the form [57–61]

$$\Psi_n^{[\lambda]}(R, r, \gamma; \theta, \eta, \chi) = \frac{1}{2k_n R} \sum_{Jk'} (2J + 1) D^J_{kk'}(\theta, \eta, \chi) F^{[\lambda]}_{Jnk'}(R, r, \gamma) \qquad (12)$$

where $\hbar k_n$ is the magnitude of the A—BC approach momentum. The label λ will be taken to be each of G, N, e, or o (see below), which indicate whether the wave function is Ψ_G, Ψ_N, Ψ_e, or Ψ_o. The internal degrees of freedom (in which the GP boundary condition is applied in Ψ_G) are described by the coordinates (R, r, γ), which can be defined in either the reagent or product arrangements; in the product arrangement, r is the AC bond length, R is the length of the vector joining B to the AC center of mass, and γ is the angle between this vector and the AC bond. The external, spatial degrees of freedom are described by the Euler angles (θ, η, χ).

The most important observable is the angular distribution of the scattered products with respect to the initial approach direction of the reagents, which is called the state-to-state *differential cross-section* (DCS). The DCS can be written [57–61]

$$\frac{d\sigma_{n'\leftarrow n}^{[\lambda]}}{d\Omega}(\theta, E) = \frac{1}{2j+1} |f_{n'\leftarrow n}^{[\lambda]}(\theta, E)|^2 \tag{13}$$

where $f_{n'\leftarrow n}^{[\lambda]}(\theta, E)$ is the scattering amplitude, obtained by taking the asymptotic limit of $\Psi_n^{[\lambda]}(R, r, \gamma; \theta, \eta, \chi)$ as $R \to \infty$. To obtain an expression for $f_{n'\leftarrow n}^{[\lambda]}(\theta, E)$, we take the $R \to \infty$ limits of the components $F_{Jnk'}^{[\lambda]}(R, r, \gamma)$, which are

$$F_{Jnk'}^{[\lambda]}(R, r, \gamma) \to \sum_{v'j'} \sqrt{\frac{k_{n'}}{k_n}} \Theta_{j'k'}(\gamma) \psi_{v'}(r) e^{ik_{n'}R} S_{n'\leftarrow n}^{[\lambda]}(J, E) \tag{14}$$

where $\Theta_{j'k'}(\gamma)$ and $\psi_{v'}(r)$ are the rotational and vibrational wave functions of AC. The aim of the calculation is thus to determine the matrix of coefficients $S_{n'\leftarrow n}^{[\lambda]}(J, E)$, which is called the reactive scattering S matrix. Using this expression, we obtain

$$f_{n'\leftarrow n}^{[\lambda]}(\theta, E) = \frac{1}{2ik_{vj}} \sum_J F(J)(2J + 1) d_{k'k}^J(\pi - \theta) S_{n'\leftarrow n}^{[\lambda]}(J, E) \tag{15}$$

where $d_{k'k}^J(\pi - \theta)$ is a reduced Wigner rotation matrix [63]. We have included a filter $F(J)$ in Eq. (15), which allows us to calculate separate DCS corresponding to different ranges of J. There is a rough correspondence between J and the classical impact parameter b, such that low values of J correspond to low-impact (i.e., head on) collisions of the reagents, and high values of J correspond to high-impact (i.e., glancing) collisions.

In addition to the DCS, we also need to consider the state-to-state integral cross-section (ICS), which is a measure of the total amount of scattered AC product in quantum state n', and is given by

$$\sigma_{n'\leftarrow n}^{[\lambda]}(E) = \frac{2\pi}{2j+1} \int_0^\pi |f_{n'\leftarrow n}^{[\lambda]}(\theta, E)|^2 \sin\theta \, d\theta, \tag{16}$$

We will also need to consider the reaction probability, which is a measure of the amount of scattered product at a given value J, defined by

$$P^{[\lambda]}_{n' \leftarrow n}(J, E) = |S^{[\lambda]}_{n' \leftarrow n}(J, E)|^2 \qquad (17)$$

In general, it is difficult to map contributions from different reaction paths onto the DCS. However, Eq. (6) tells us that, in a reaction with a CI, one can easily map the contributions from the e and o (Feynman) paths onto the DCS. Since Eq. (6) applies to the entire wave function, we can apply it to the asymptotic limit of the wave function in Eq. (14), and thus to $S^{[G]}(E)$ and $S^{[N]}(E)$, to obtain

$$
\begin{aligned}
S^{[e]}(E) &= \frac{1}{\sqrt{2}} [S^{[N]}(E) + S^{[G]}(E)] \\
S^{[o]}(E) &= \frac{1}{\sqrt{2}} [S^{[N]}(E) - S^{[G]}(E)]
\end{aligned}
\qquad (18)
$$

These equations are all that we need to explain the effect of the GP on scattering cross-sections, such as the DCS and ICS. They allow us to compute separate e and o cross-sections using Eq. (15), which show the scattering produced by the e and o reaction paths in isolation. They tell us that we can only expect GP effects if $f^{[e]}_{n' \leftarrow n}(\theta, E)$ and $f^{[o]}_{n' \leftarrow n}(\theta, E)$ overlap.

In the case of the $H + H_2$ reaction, Eq. (18) specializes to

$$
\begin{aligned}
S^{[1-TS]}(E) &= S^{[e]}(E) \\
S^{[2-TS]}(E) &= S^{[o]}(E)
\end{aligned}
\qquad (19)
$$

Hence, simply by adding and subtracting the computed $S^{[G]}(E)$ and $S^{[N]}(E)$, we can identify the contributions from the 1-TS and 2-TS reaction paths in the DCS and ICS, and thus explain the effects of the GP on the $H + H_2$ reaction.

C. Details of the Calculation: The Use of Vector Potentials

Before discussing the results of applying Eq. (19), we explain how the GP boundary condition is implemented numerically in the calculations of $S^{[G]}(E)$. This is the most technical part of the chapter, and the material here is not needed to understand the sections that follow.

There are three ways of implementing the GP boundary condition. These are (1) to expand the wave function in terms of basis functions that themselves satisfy the GP boundary condition [16]; (2) to use the vector-potential approach of Mead and Truhlar [6,64]; and (3) to convert to an approximately diabatic representation [3, 52, 65, 66], where the effect of the GP is included exactly through the adiabatic–diabatic mixing angle. Of these, (1) is probably the most elegant method, but it can be implemented efficiently only if the encirclement angle ϕ is a simple function of the coordinates used to represent the Hamiltonian. This is not the case for the coordinates (R, r, γ), which we use in our calculations. Approach (3) is in general numerically the most robust way to include nonadiabatic effects, but it requires that one carry out the calculation on two coupled surfaces, which is clearly inefficient if the system is confined to the lower adiabatic surface. Approach (2) is both numerically robust and uses just the one (lower adiabatic) surface, and is hence the approach we used in the $H + H_2$ calculations [26].

In the vector potential approach [6], the (real) electronic wave function $\Phi(\phi)$ is multiplied by a complex phase factor $f(\phi)$, defined such that

$$f(\phi + 2\pi)\Phi(\phi + 2\pi) = f(\phi)\Phi(\phi) \tag{20}$$

A simple choice of phase factor is one that has the form

$$f(\phi) = e^{i\frac{l}{2}\phi} \tag{21}$$

where l must be chosen odd to ensure that Eq. (20) is satisfied. All odd values of l will correctly incorporate the GP, and give physically equivalent wave functions, which will differ only in an overall phase factor. Similarly, all even values of l will give the non-GP wave function.

When $f(\phi)$ takes the form of Eq. (21), the nuclear Laplacian operator is modified according to

$$-\nabla^2 \rightarrow (-i\nabla - A) \cdot (-i\nabla - A) \tag{22}$$

where the *vector potential* A is given by

$$A = -\frac{l}{2}\nabla\phi \tag{23}$$

Hence, the method of Mead and Truhlar [6] yields a single-valued nuclear wave function by adding a vector potential A to the kinetic energy operator. Different values of odd (or even) l yield physically equivalent results, since they yield $\Psi(\phi)$ that are identical to within an integer number of factors of $\exp(i\phi)$. By analogy with electromagnetic vector potentials, one can say that different odd (or even) l are related by a gauge transformation [6, 7].

To implement the vector potential in the Jacobi coordinate system (R, r, γ), one proceeds as follows. The Jacobi kinetic energy operator splits into three parts [61]:

$$\hat{T} = \hat{T}_R + \hat{T}_r + \hat{T}_{\mathrm{ang}} \tag{24}$$

where each contains a derivative term in just one of the Jacobi coordinates (R, r, γ), and hence contains one component of the vector potential,

$$A_a(R, r, \gamma) = -\frac{l}{2}\frac{\partial\phi(R, r, \gamma)}{\partial a} \tag{25}$$

where a denotes, respectively, R, r, and γ. Note that ϕ is a function of all three of the coordinates.

The method used to propagate solutions to the Schrödinger equation [61] requires $\hat{T}$ to be represented on a grid of points distributed in (R, r, γ), which we will denote using the labels $|klm>$. The first term in $\hat{T}$ is given by

$$\hat{T}_R = -\frac{\hbar^2}{2\mu_R}\frac{\partial^2}{\partial R^2} \tag{26}$$

(where μ_R is the reduced mass associated with R). Application of Eq. (22) changes the derivative operator according to

$$
\begin{aligned}
-\frac{\partial^2}{\partial R^2} &\rightarrow \left(-i\frac{\partial}{\partial R} - A_R\right)\left(-i\frac{\partial}{\partial R} - A_R\right) \\
&\rightarrow -\frac{\partial^2}{\partial R^2} + A_R^2 + i\left(\frac{\partial}{\partial R}A_R + A_R\frac{\partial}{\partial R}\right)
\end{aligned}
\tag{27}
$$

This operator is diagonal in all but the R grid basis functions (denoted $|k\rangle$), and its matrix elements change according to

$$
\begin{aligned}
\langle k|\hat{T}_R|k'\rangle \rightarrow \langle k|\hat{T}_R|k'\rangle + \frac{\hbar^2}{2\mu_R}\Big\{ &\delta_{kk'}A_R(R_k, r_l, \gamma_m)^2 \\
&+ i\left\langle k\left|\frac{\partial}{\partial R}\right|k'\right\rangle[A_R(R_k, r_l, \gamma_m) + A_R(R_{k'}, r_l, \gamma_m)]\Big\}
\end{aligned}
\tag{28}
$$

where R_k denotes the value of R at the kth grid point; r_l and γ_m are the r and γ grid points (see below). Note that this expression was derived by keeping the operator in the symmetric form of Eq. (27), and acting outward with the first derivative operators, on the bra and the ket. This approach (as opposed to taking the second derivative of the ket [20,21]) yields a grid matrix which is exactly Hermitian.

The second term $\hat{T}_r$ has exactly the same form as $\hat{T}_R$ (with r in place of R) and produces an exactly analogous change in the matrix elements between the r-grid basis functions $|l\rangle$.

The most complicated changes are those produced in the third term $\hat{T}_{\text{ang}}$. This operator can be split into three terms [67]

$$\hat{T}_{\text{ang}} = \hat{T}_{\text{ang}}^{(1)} + \hat{T}_{\text{ang}}^{(2)} + \hat{T}_{\text{ang}}^{(3)} \tag{29}$$

which are given by

$$\hat{T}_{\text{ang}}^{(1)} = \frac{\hat{J}^2 - 2\hat{J}_z^2}{2\mu_R R^2}$$

$$\hat{T}_{\text{ang}}^{(2)} = \left(\frac{1}{2\mu_R R^2} + \frac{1}{2\mu_r r^2}\right)\hat{j}^2 \tag{30}$$

$$\hat{T}_{\text{ang}}^{(3)} = -\frac{\hat{J}\hat{j} + \hat{j}\hat{J}}{2\mu_R R^2}$$

The term $\hat{T}_{\text{ang}}^{(1)}$ contains the total angular momentum operators $\hat{J}^2$ and $\hat{J}_z^2$. These do not operate on the internal degrees of freedom, and are thus not changed by Eq. (22). The term $\hat{T}_{\text{ang}}^{(2)}$ contains the BC angular momentum operator j^2, which involves a γ-derivative operator. The change brought about in this operator by Eq. (22) is similar to Eq. (27). The matrix elements of $\hat{T}_{\text{ang}}^{(2)}$ are diagonal in all but the γ grid basis functions $|m\rangle$, and change according to

$$\langle m|\hat{T}_{\text{ang}}^{(2)}|m'\rangle \rightarrow \langle m|\hat{T}_{\text{ang}}^{(2)}|m'\rangle + \hbar^2\left(\frac{1}{2\mu_R R_k^2} + \frac{1}{2\mu_r r_l^2}\right)\Big\{\delta_{mm'}A_\gamma(R_k, r_l, \gamma_m)^2$$

$$+ i\left\langle m\left|\frac{\partial}{\partial\gamma}\right|m'\right\rangle[A_\gamma(R_k, r_l, \gamma_m) + A_\gamma(R_k, r_l, \gamma_{m'})]\Big\} \tag{31}$$

The operator $\hat{T}_{\text{ang}}^{(3)}$ contains the cross-terms that give rise to the Coriolis coupling that mixes states with different Ω (the projection of the total angular momentum quantum number J onto the intermolecular axis). This term contains first derivative operators in γ. On application of Eq. (22), these operators change the matrix elements over $\hat{T}_{\text{ang}}^{(3)}$ according to

$$\langle mJ\Omega|\hat{T}_{\text{ang}}^{(3)}|m'J\Omega'\rangle \rightarrow \langle mJ\Omega|\hat{T}_{\text{ang}}^{(3)}|m'J\Omega'\rangle$$

$$+ \frac{i\hbar^2}{2\mu_R R_k^2}\Big\{\delta_{\Omega\Omega'+1}C_{J\Omega'}^+\langle m\Omega|m'\Omega'\rangle A_\gamma(R_k, r_l, \gamma_{m'}) \tag{32}$$

$$- \delta_{\Omega'\Omega+1}C_{J\Omega}^+\langle m'\Omega'|m\Omega\rangle A_\gamma(R_k, r_l, \gamma_m)\Big\}$$

where

$$C_{ab}^{\pm} = \sqrt{a(a+1) - b(b\pm 1)} \tag{33}$$

To apply the above equations to $H + H_2$, we need an expression for the vector potential $\mathbf{A}(R, r, \gamma)$, which can be obtained from Eq. (25) once the angle ϕ has been specified. As mentioned above, we are free to define ϕ in any way, provided that $\phi = 0 \rightarrow 2\pi$ describes a closed path around the CI. We chose the form,

$$\phi(R, r, \gamma) = \tan^{-1}\left(\frac{d^2R^2 - r^2/d^2}{2Rr\cos\gamma}\right) \tag{34}$$

where d is a dimensionless scaling factor defined in [26]. In the potential cut shown in Fig. 7, this definition of ϕ corresponds to the circular polar angle describing an internal revolution about the CI.

D. Solving the Cancellation Puzzle

We are now in a position to discuss the cancellation puzzle in $H + H_2$. We start by considering the state-to-state reaction probabilities $P_{n'\leftarrow n}^{[\lambda]}(J, E)$, computed according to Eq. (17), with the filter $F(J) = 1$. Representative results, taken from ref. [26], are shown in Fig. 8. These results [26] reproduce those obtained earlier by Kendrick [21], and show that there are noticeable GP effects in some of the state-to-state reaction probabilities, which indicate that a small proportion of the wave function encircles the CI. A curious feature of these

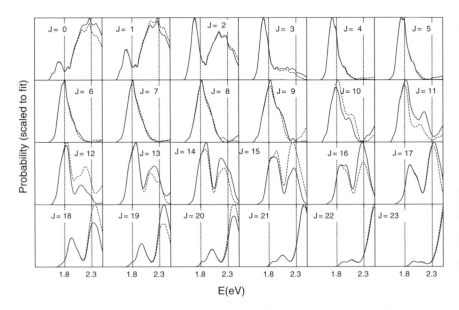

Figure 8. State-to-state reaction probabilities, for $H + H_2(100) \rightarrow H_2(250) + H$, computed using GP (solid lines) and non-GP (dashed lines) boundary conditions.

results is the apparent alternation in sign of the difference between the GP and non-GP probabilities,

$$\Delta P_{n' \leftarrow n}(J, E) = P^{[G]}_{n' \leftarrow n}(J, E) - P^{[N]}_{n' \leftarrow n}(J, E) \tag{35}$$

as a function of J.

One would expect that effects of similar magnitude to those shown in Fig. 8 should also appear in the corresponding state-to-state differential and integral cross-sections. However, this is not the case. As already mentioned, there is a considerable amount of cancellation of GP effects in these quantities, which we refer to as the cancellation puzzle. The unexpected cancellations appear in the state-to-state DCS at low impact parameters (i.e., low values of J), and in the state-to-state ICS (including all impact parameters). We now discuss each of these cancellations in turn.

1. Low Impact-Parameter Cross-Sections

The cancellation in GP effects in the state-to-state DCS are found [20–22, 26, 27, 29] at low impact parameters, when $F(J)$ in Eq. (15) is chosen to include only contributions for which $J \leq 9$. It is well known [55,56] that most of the reactive scattering in this regime consists of head-on collisions, in which the reaction proceeds mainly by the H atom striking the H_2 diatom at geometries that are close to linear. Most of the products are then formed by direct recoil in the backward $(\theta = 180°)$ region, this being typical behavior for a hydrogen-abstraction reaction.

Figure 9 shows the low impact DCS obtained for the same initial and final states $(250 \leftarrow 100)$, and at the same energy (2.3 eV above the potential minimum),

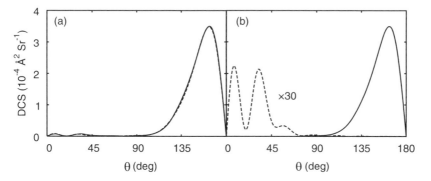

Figure 9. Low-impact parameter DCS at 2.3 eV, for $H + H_2(100) \rightarrow H_2(250) + H$, describing the scattering of (a) Ψ_G (solid lines) and Ψ_N (dashed lines), and (b) Ψ_e (solid lines) and Ψ_o (dashed lines).

as the reaction probabilities of Fig. 8. These DCS are clearly consistent with the abstraction mechanism just mentioned. Most of the scattering is concentrated in the backward direction, but there is a tiny component in the forward ($\theta = 0$) direction. The striking feature is that the GP and non-GP DCS are in perfect agreement. The noticeable GP effects in the reaction probabilities of Fig. 8 appear to have canceled completely.

This observation is the first part of the cancellation puzzle [20, 21, 27, 29]. We know from Section III.B that we should be able to solve it directly by applying Eq. (19), which will separate out the contributions to the DCS made by the 1-TS and 2-TS reaction paths. That this is true is shown by Fig. 9(b). It is apparent that the main backward concentration of the scattering comes entirely from the 1-TS paths. This is not a surprise, since, by definition, the direct abstraction mechanism mentioned only involves one TS. What is perhaps surprising is that the small lumps in the forward direction, which might have been mistaken for numerical noise, are in fact the products of the 2-TS paths. Since the 1-TS and 2-TS paths scatter their products into completely different regions of space, there is no interference between the amplitudes $f_{n' \leftarrow n}^{[e]}(\theta)$ and $f_{n' \leftarrow n}^{[o]}(\theta)$, and hence no GP effects.

We note that the particular $(2, 5, 0 \leftarrow 1, 0, 0)$ state-to-state DCS that we have chosen happens to have a perfectly clean separation between the 1-TS and 2-TS scattering, which is why the cancellation in GP effects is perfect. Most of the other low impact DCS, however, have a small amount of overlap between the 1-TS and 2-TS scattering, which means that the GP effects almost cancel out in the cross-sections, but that tiny, genuine GP effects remains. Examples of such cross-sections are given in [29].

The fact that the 1-TS and 2-TS paths scatter mainly in opposite directions is also the reason for the alternation in sign of $\Delta P_{n' \leftarrow n}(J, E)$ pointed out above. It is important to realize that this alternation is not exact. For example, in Fig. 8, we see that $\Delta P_{n' \leftarrow n}(J, E)$ has the same sign for $J = 13$ and 14, and that, at 2.3 eV, the alternation is broken by $\Delta P_{n' \leftarrow n}(2, E) = 0$.

We can explain the approximate alternation of $\Delta P_{n' \leftarrow n}(J, E) = 0$ by substituting Eq. (19) into Eq. (35), to obtain

$$\Delta P_{n' \leftarrow n}(J, E) = -2F(J)^2 \mathrm{Re} \left[S_{n' \leftarrow n}^{[1-TS]}(J, E)^* S_{n' \leftarrow n}^{[2-TS]}(J, E) \right] \qquad (36)$$

We then invert Eq. (15) by integrating over θ, which yields the following expression for the S-matrix elements,

$$S_{n' \leftarrow n}^{[\lambda]}(J, E) = \frac{i k_{vj}}{F(J)} \int_0^\pi f_{n' \leftarrow n}^{[\lambda]}(\theta, E) d_{k'k}^J (\pi - \theta) \sin \theta \, d\theta \qquad (37)$$

Equations (36) and (37), together with the property $d_{k'0}^{J}(\pi - \theta) = (-1)^{J}d_{k'0}^{J}(\theta)$ [63], show that if the scattering amplitudes satisfied

$$f_{n' \leftarrow n}^{[1-TS]}(\theta, E) \propto f_{n' \leftarrow n}^{[2-TS]}(\pi - \theta, E) \tag{38}$$

then the sign of $\Delta P_{n' \leftarrow n}(J, E)$ would follow $(-1)^{J}$ exactly. In the DCS of Fig. 9, there is a major component in the 1-TS and 2-TS amplitudes that satisfies Eq. (38), as well as a minor component that does not. Hence, overall the sign of $\Delta P_{n' \leftarrow n}(J, E)$ displays an approximate alternation with J.

2.   Full Cross-Sections

The second part of the cancellation puzzle concerns the full state-to-state DCS and ICS (i.e., including all the impact parameters). In this case, the GP effects do not cancel in the DCS [26, 27, 29], as is shown in Fig. 10. Instead, they shift the phase of the fine oscillations that are superimposed on the main DCS envelope. Following the above, this indicates that the 1-TS and 2-TS paths scatter into overlapping regions of space, so that the GP produces an effect by changing the sign of the interference between $f_{n' \leftarrow n}^{[1-TS]}(\theta)$ and $f_{n' \leftarrow n}^{[2-TS]}(\theta)$. This is confirmed by Fig. 10b, which shows that the 1-TS and 2-TS DCS do indeed overlap.

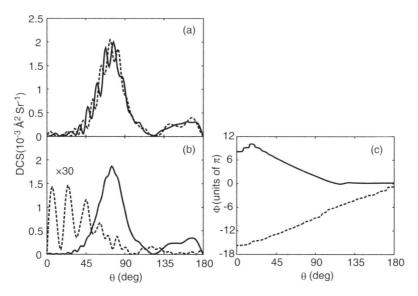

Figure 10.   Full DCS (i.e., including all impact parameters), for $H + H_2(100) \rightarrow H_2(250) + H$, describing the scattering of (a) Ψ_G (solid lines) and Ψ_N (dashed lines), and (b) Ψ_e (solid lines) and Ψ_o (dashed lines). (c) The phases of the corresponding e and o scattering amplitudes.

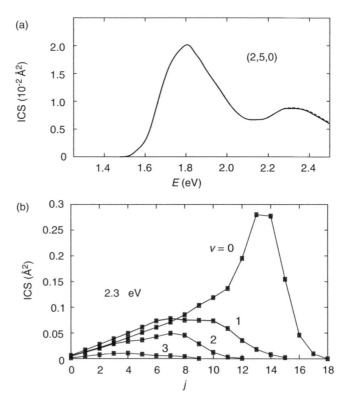

Figure 11. The ICS for $H + H_2$, computed using GP (solid lines) and non-GP (dashed lines) boundary conditions, for (a) $H + H_2(100) \rightarrow H_2(250) + H$ over a range of energies, and (b) $H + H_2(100) \rightarrow H_2(v,j) + H$ at 2.3 eV.

The surprizing result is that these GP effects cancel out completely when the DCS is integrated over θ to yield the ICS [via Eq. (16)]. This cancellation is shown in Fig. 11(a) for the $(2, 5, 0 \leftarrow 1, 0, 0)$ ICS over a range of energies, and in Fig. 11(b) for all the nonzero state-to-state ICS at $E = 2.3$ eV. In general, one expects small differences in a DCS to average out on integrating over θ, but a complete cancellation of the differences, which is found to hold for all final states and collision energies tested [26], suggests that there is a systematic difference between the 1-TS and 2-TS scattering dynamics that is causing the cancellation.

This last point is the second part of the cancellation puzzle, and is soon explained by plotting the phases $\Phi^{[e]}_{n' \leftarrow n}(\theta, E)$ and $\Phi^{[o]}_{n' \leftarrow n}(\theta, E)$ of the scattering amplitudes $f^{[e]}_{n' \leftarrow n}(\theta, E)$ and $f^{[o]}_{n' \leftarrow n}(\theta, E)$ (Fig. 10c). It is clear that these phases depend in opposite senses on θ, and the same trend is observed for most of the other final states considered [29]. As a result the integrand in

$$\sigma_{n'\leftarrow n}^{[G]}(E) - \sigma_{n'\leftarrow n}^{[N]}(E) = -\frac{4\pi}{2j+1}\operatorname{Re}\int_0^\pi f_{n'\leftarrow n}^{[e]}(\theta,E)f_{n'\leftarrow n}^{[o]*}(\theta,E)\sin\theta\,d\theta. \quad (39)$$

is highly oscillatory (with period $\sim 12°$), and thus integrates to a very small value (although not to zero).

From semiclassical scattering theory [68,69], it is known that a *negative* dependence of $\Phi_{n'\leftarrow n}(\theta,E)$ on θ indicates scattering into *positive* deflection angles, and vice versa. The terms nearside and farside are sometimes used to describe these two types of scattering (see Fig. 12). Hence, the reason that GP effects cancel in the state-to-state ICS is that the 1-TS and 2-TS paths scatter in opposite senses (with respect to the center-of-mass). There is thus a mapping between the sense in which the reaction paths loop around the CI (clockwise for 1-TS, counterclockwise for 2-TS), and the sense in which the products scatter into space.

3.   The 2-TS Mechanism

To complete the explanation of why GP effects cancel in the ICS, we need to explain why the 2-TS paths scatter into negative deflection angles. (It is well known that the 1-TS paths scatter into positive deflection angles via a direct recoil mechanism [55, 56].) We can explain this by following classical trajectories, which gives us the opportunity to illustrate a further useful consequence of the theory of Section II.

This is that, once we have separated the nuclear wave function into Ψ_e and Ψ_o using Eq. (6), we are free to model the dynamics of each component separately using classical trajectories, secure in the knowledge that we have removed the

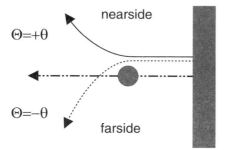

Figure 12.   Diagram illustrating the difference between nearside scattering into positive deflection angles Θ, and farside scattering into negative Θ. The arrow (chains) represents the initial approach direction of the reagents in center-of-mass frame; the gray rectangle represents the spread of impact parameters in the initial plane wave. Most of the 1-TS paths scatter into positive Θ, and most of the 2-TS paths into negative Θ.

effects of the GP. Hence, although we cannot use classical mechanics to predict the effect of the GP, which is entirely a quantum effect, we can use it to model the dynamics of the e and o paths separately. This allows us to predict the extent to which these paths will overlap, and hence estimate the likely magnitude of the GP effects. Of course, we must make allowances for the other types of quantum effects found in reaction dynamics, such as tunneling, zero-point energy, reactive resonances, and threshold effects. The best way to do this is to make detailed comparisons with quantum scattering data (e.g., the state-to-state product distributions), using the quasiclassical trajectory (QCT) approach [70–72].

Hence, in $H + H_2$ we were able to use QCT to model the dynamics of the 1-TS and 2-TS reaction paths separately [29]. The main feature of the quantum calculations that the QCT calculations must reproduce is the scattering of the 1-TS paths into positive deflection angles, and the 2-TS paths into negative deflection angles (since this is what causes the cancellation of GP effects in the ICS). Figure 13 shows that the scattering of the classical 1-TS and 2-TS paths agrees strikingly in this regard, suggesting that, at the very least the classical trajectories are able to give a good overall explanation of why the 2-TS paths scatter into negative deflection angles.

Figure 14 shows a representative 2-TS trajectory, which demonstrates that the 2-TS paths follow a direct S-bend insertion mechanism. The trajectory passes through the middle of the molecule, and avoids the CI; this forces the products to scatter into negative deflection angles. The 2-TS QCT total reaction

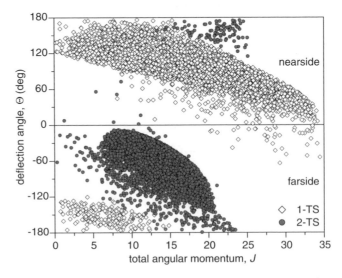

Figure 13. Density plot of the correlation between the deflection angle, Θ, and the total angular momentum, J, for 1-TS (open diamonds) and 2-TS (circles) trajectories at 2.3-eV total energy. Note, only 5000 trajectories are plotted for each type for clarity.

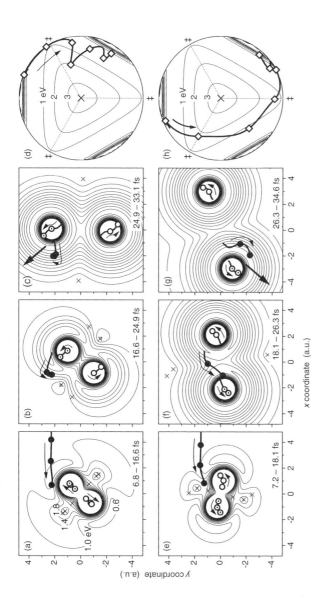

Figure 14. Classical trajectories for the $H + H_2(v = 1, j = 0)$ reaction representing a 1-TS (a–d) and a 2-TS reaction path (e–h). Both trajectories lead to $H_2(v' = 2, j' = 5, \kappa' = 0)$ products and the same scattering angle, $\theta = 50°$. (a–c) 1-TS trajectory in Cartesian coordinates. The positions of the atoms (H_A, solid circles; H_B, open circles; H_C, dotted circles) are plotted at constant time intervals of 4.1 fs on top of snapshots of the potential energy surface in a space-fixed frame centered at the reactant $H_B H_C$ molecule. The location of the conical intersection is indicated by crosses ($\times$). (d) 1-TS trajectory in hyperspherical coordinates (cf. Fig. 1) showing the different $H + H_2$ arrangements (open diamonds) at the same time intervals as panels (a–c); the potential energy contours are for a fixed hyperradius of $\rho = 4.0$ a.u. (e–h) As above for the 2-TS trajectory. Note that the 1-TS trajectory is deflected to the nearside, whereas the 2-TS trajectory proceeds via an insertion mechanism and is deflected to the farside (deflection angle $\Theta = +50°$).

29

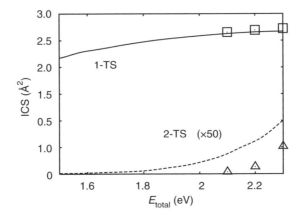

Figure 15. Quantum (lines) and QCT (points) total cross-sections for the $H + H_2$ 1-TS and 2-TS paths.

cross-section (Fig. 15) is roughly one-half of the quantum result, suggesting that the reaction is enhanced by quantum tunnelling. This seems reasonable for such a constrained reaction path, and in [29] we use the product rotational distributions (not shown here) to argue that the insertion is facilitated by tunneling through the side of the lower cone of the CI.

IV. FURTHER ASPECTS OF TOPOLOGY

A. Including Particle-Exchange Symmetry

So far, we have treated the atoms as distinguishable particles, both in the general theory of Section II and in the application to $H + H_2$ in Section III. Here, we explain how to incorporate the effects of particle exchange symmetry. First, we discuss how the symmetry of the system maps from the physical onto the double space, and then explain what effect the GP has on wave functions of reactions that (like $H + H_2$) have identical reagents and products.

1. Symmetry in Double Space

A useful property of the double space is that it clarifies the treatment of symmetry [28]. In the single space, the symmetry of Ψ_G can appear confusing, because it depends on the position of the cut line. One way to avoid this confusion is to consider the symmetry of the total (electronic + nuclear) wave function $\Psi\Phi$, which is of course independent of the position of the cut line [6, 7]. Another way is to map Ψ_G onto the double space.

In Section II, we explained that Ψ_N and Ψ_G are respectively symmetric and antisymmetric under the operator $\hat{R}_{2\pi}$ in the double space. More generally, if the

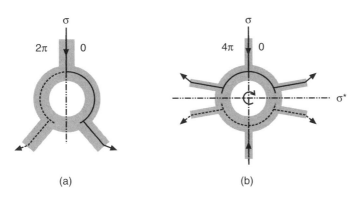

Figure 16. Relation between symmetry in (a) the single space and (b) the double space of a system whose molecular symmetry group has a σ plane in the single space, and is isomorphic with C_{2v} in the double space.

molecular symmetry group in the single space is G, then the symmetry group in the corresponding double space is the direct product double group $G \otimes R$, where $R = \{E, \hat{R}_{2\pi}\}$ (and E is the identity operator). The properties of direct product double groups are well known from molecular spectroscopy [73]. A double group $G \otimes R$ has exactly twice as many symmetry operations as the corresponding single group G, though not necessarily twice as many classes and irreducible representations (irreps). All irreps in the double group will be either completely symmetric or antisymmetric under $\hat{R}_{2\pi}$. Clearly, the antisymmetric (symmetric) irreps constitute all the allowed symmetries of the $\Psi_G(\Psi_N)$ states.

For example, consider the system shown in Fig. 16, in which the (single-space) molecular symmetry group is $\{E, \sigma\}$, where σ is a mirror plane of symmetry running from top to bottom of the figure. The symmetry group in the double space is then $\{E, \sigma\} \otimes \{E, \hat{R}_{2\pi}\}$, which is isomorphic with C_{2v}, with a second plane of symmetry $\sigma^* = \hat{R}_{2\pi}\sigma$. Clearly, there are two Ψ_N irreps (symmetric under $\hat{R}_{2\pi}$) and two Ψ_G irreps (antisymmetric under $\hat{R}_{2\pi}$). The Ψ_N irreps have the same symmetry under σ and σ^*; the Ψ_G irreps have opposite symmetries. Of the latter, let us take the irrep that is symmetric under σ and antisymmetric under σ^*, which describes a reaction in which the reagents are prepared in a state that is symmetric with respect to σ. In the single space, this function is antisymmetric under σ when the cut line is placed at $\phi = 0$, symmetric when it is placed at $\phi = -\pi$, and unsymmetric when it is placed at, say, $\phi = \pi/4$. Use of the double space removes this ambiguity.

2. Identical Reagents and Products

When the reagents and products are identical, then the system enters the encirclement region at several different values of ϕ, and must therefore be

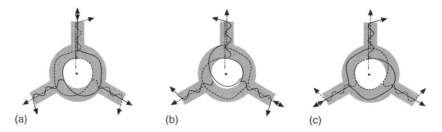

Figure 17. The unsymmetrized functions (a) $\Psi_G(\phi)$, (b) $-\Psi_G(\phi - 2\pi/3)$, and (c) $\Psi_G(\phi - 4\pi/3)$ for a system, such as H_3, which has three identical reagent and product channels. Superposing these components gives the wave function $\Psi_G^{\text{sym}}(\phi)$ of Eq. (40) which is fully symmetric under cyclic permutation of identical nuclei.

treated analogously to a unimolecular system [28] (Section II.D). Figure 17a represents the nuclear wave function of the $H + H_2$ reaction (in the single space), which was considered earlier by Mead under the assumption that it did not encircle the CI [14]. In Fig. 17a, Ψ_G (or Ψ_N, depending on which relative phase of Ψ_e and Ψ_o is assumed in the diagram) is drawn as though the reagents and products of the reaction were distinguishable (as they would be for say $D + H_2$). Thus all the Feynman paths that enter the encirclement region start at one unique value of ϕ.

To treat the reagents and products as indistinguishable, one must make the total (electronic + nuclear) wave function symmetric under a cyclic exchange of nuclei, which is equivalent to making it symmetric under rotations $\hat{R}_{2\pi/3}, \hat{R}_{4\pi/3}$, about the threefold axis of symmetry. Mead showed that, because the electronic wave function Φ is antisymmetric under $\hat{R}_{2\pi/3}$, then Ψ_G must be symmetrized according to

$$\Psi_G^{\text{sym}}(\phi) = 1/\sqrt{3}[\Psi_G(\phi) - \Psi_G(\phi - 2\pi/3) \\ + \Psi_G(\phi - 4\pi/3)] \tag{40}$$

By assuming that the system does not encircle the CI, Mead showed [14] that this equation implies that the GP changes the relative sign of the inelastic and reactive contributions in the scattering amplitude.

It is straightforward to combine Eq. (40) with the arguments of Section II, in order to extend Mead's result to systems that encircle the CI. One has simply to substitute Eq. (5) into each term of Eq. (40), which yields

$$\Psi_G^{\text{sym}}(\phi) = 1/\sqrt{6}[\Psi_e(\phi) - \Psi_o(\phi) \\ - \Psi_e(\phi + 2\pi/3) + \Psi_o(\phi + 2\pi/3) \\ + \Psi_e(\phi + 4\pi/3) - \Psi_o(\phi + 4\pi/3)] \tag{41}$$

We can represent this function in the single space, provided we use a *common cut line* for all three components. This is shown schematically in Fig. 17. Use of the common cut line is equivalent to taking the linear combinations in the double space, then cutting a 2π-wide section out of the entire $\Psi_G^{sym}(\phi)$. The winding numbers n of the Feynman paths that enter the three equivalent reagent channels must all be defined with respect to the common cut line, since they are analogous to paths starting at different points in the initial state of a unimolecular reaction (Section II.D).

We can now extend Mead's argument in order to find out the relative sign of the inelastic and reactive contributions to an encircling nuclear wave function. The symmetrized wave function of Eq. (41) can be represented graphically by combining the three functions of Figs. 17a–c, using the convention that the dashed lines have the opposite sign to the solid lines. Four types of Feynman path contribute to Ψ_G^{sym} in a given reagent–product channel. We will call these the direct inelastic, looping inelastic, direct reactive, and looping reactive. To work out whether a contribution (in a given reagent–product channel) is direct or looping, one should identify the shortest route back to the point at which the path entered the encirclement region. If this route passes by one or more exit channels then the path is looping; if it does not, then the path is direct. It is then clear that the GP changes the sign of the direct reactive with respect to the direct inelastic contribution, and that it leaves unchanged the sign of the looping reactive with respect to the direct inelastic. The first of these observations is Mead's result [14]; the second is the required generalization of Mead's result to an encircling nuclear wave function. Although we have considered here a reaction that has the same particle-exchange symmetry as the $H + H_2$ reaction, the arguments above can clearly be generalized to treat reactions of any symmetry.

B.   Complete Unwinding of the Nuclear Wave Function

One can think of the mapping of the nuclear wave function onto the double space in Eq. (5) as a partial unwinding of the wave function [28]. This amount of unwinding is sufficient to explain completely the effect of the GP. However, it is interesting to consider unwinding the wave function further. If the range of n that contribute significantly to Ψ_G is finite, then Eq. (8) implies that, in principle, one can unwind Ψ_G completely, separating the contributions from individual values of n.

First, let us consider a system in which n is restricted to $n = 0$ and $n = -1$. The Kernels K_n are therefore negligibly small for $n < -1$ or $n > 0$. In such a system, Ψ_e contains only the $n = 0$ paths, and Ψ_o only the $n = -1$ paths. Hence, mapping onto the double space, to generate Ψ_e using Eq. (6), is sufficient to unwind completely the nuclear wave function (Figs. 18a and b). In the double space (Fig. 18b) the $n = 0$ and $n = -1$ paths are the branches of Ψ_e accessed by

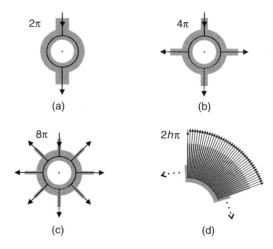

Figure 18. Complete unwinding of an encircling nuclear wave function Ψ_G by mapping onto higher cover spaces. (a) The function Ψ_G in the single space; (b) Ψ_e in the double space; (c) Ψ_4 in the quadruple space; (d) schematic picture of Ψ_h in a $2h\pi$ cover space. In each case, Ψ_G will be completely unwound if it contains contributions from Feynman paths belonging to (b) 2, (c) 4, and (d) h different winding-number classes.

rotating clockwise and counterclockwise from the entry point at $\phi = 0$. There is a gap between these two branches, such that Ψ_e in the double space is like a nonencircling Ψ_G in the single space (Fig. 4).

When higher n Feynman paths contribute to the wave function, one has simply to apply repeatedly the single- to double-space mapping, until the nuclear wave function is completely unwound (in the sense just defined). Thus, if the wave function contains only $n = -2, -1, 0, 1$ paths, then we need to compute a function Ψ_e' in the double space that satisfies the boundary condition $\Psi_e'(\phi) = -\Psi_e'(\phi + 4\pi)$. Adding this function to Ψ_e [which satisfies $\Psi_e(\phi) = \Psi_e(\phi + 4\pi)$] then gives a new function, $\Psi_4(\phi)$, which occupies the quadruple space $\phi = 0 \rightarrow 8\pi$ (see Fig. 18c). This new quadruple-space wave function will be completely unwound, such that there is a gap between its clockwise and counterclockwise branches. The $n = -2, -1, 0, 1$ contributions will lie in the $-4\pi \rightarrow -2\pi, -2\pi \rightarrow 0, 0 \rightarrow 2\pi$, and $2\pi \rightarrow 4\pi$ sectors, respectively. If desired, we can convert $\Psi_4(\phi)$ back to Ψ_G by taking the combinations,

$$\Psi_G(\phi) = \frac{1}{2}[\Psi_4(\phi) - \Psi_4(\phi + 2\pi)$$
$$+ \Psi_4(\phi + 4\pi) - \Psi_4(\phi + 6\pi)] \tag{42}$$

and then cutting a 2π-wide sector out of the quadruple space, to map back onto the single space.

Clearly, the above procedure can be continued (in principle) as many times as required. Thus, if the wave function includes $n = -4 \cdots 3$ paths, we have simply to define the function $\Psi_4'(\phi) = -\Psi_4'(\phi + 8\pi)$, and then map onto the $\phi = 0 \rightarrow 16\pi$ cover space, which will unwind the function completely. In general, if there are h homotopy classes of Feynman paths that contribute to the Kernel, then one can unwind Ψ_G by computing the unsymmetrised wave function Ψ_h in the $0 \rightarrow 2h\pi$ cover space. The symmetry group of the latter will be a direct product of the symmetry group in the single space and the group $\{E, \hat{R}_{2\pi}, \hat{R}_{4\pi}, \ldots, \hat{R}_{2(h-1)\pi}\}$.

This approach is applicable even to a system that supports long-lived scattering resonances that correspond classically to periodic orbits [74] looping around the CI. Clearly, such systems can support paths for which $-\infty \leq n \leq \infty$, meaning that if one wants to compute Ψ_G *exactly* then it will never be possible to unwind it completely. However, if one wants to compute Ψ_G to within a given accuracy, then the number of homotopic classes h will be finite because the time-dependent wave function $\Psi_G(t)$ decays exponentially from within the encirclement region as a function of t. Hence, for a reactive system, there must be a value of h, such that mapping onto a $\phi = 0 \rightarrow 2h\pi$ cover space completely unwinds Ψ_G, to within a specified accuracy—meaning that there will be a region of ϕ in the $2h_\pi$ cover space, over which $|\Psi h(\phi)|^2$ is negligibly small. This gap region will contain contributions from Feynman paths with $|n| > h$, which could themselves be unwound (if higher accuracy were later required) by mapping $\Psi_h(\phi)$ onto a yet higher cover space.

In a numerical calculation, the number of times that one can unwind Ψ_G will be limited by the maximum size of cover space that can be treated computationally. An efficient way to unwind onto an $2h\pi$ cover space will be to compute the h single-space wave functions that satisfy the boundary conditions

$$\Psi_n(\phi + 2m\pi) = e^{i2nm\pi/h}\Psi_n(\phi) \qquad (n = 0 \ldots h - 1) \qquad (43)$$

The wave function Ψh in the $2h_\pi$ cover space is then given by

$$\Psi_h(\phi) = \frac{1}{\sqrt{h}}\sum_{n=0}^{h-1}\Psi_n(\phi) \qquad (44)$$

To compute each of the $\Psi_n(\phi)$, one can generalize the methods used to compute Ψ_G. Hence, the most elegant method would be to use basis functions that satisfy the boundary conditions of Eq. (43), if this were practical to implement. A more general method would be to extend the Mead–Truhlar vector-potential approach [6]. This approach would involve carrying out h calculations, each including a vector potential of the form

$$A = -i\frac{n}{h}\nabla\phi \tag{45}$$

where $n = 0 \ldots h - 1$.

To clarify, the complete unwinding of the wave function is not required to explain the effect of the GP. The latter affects only the sign of the odd n Feynman paths with respect to the even n paths, and is thus explained completely once one has unwound these two classes of path by mapping onto the double space. The complete unwinding explains the interference *within* the even n and odd n contributions, by unwinding each of them further, into the contributions from individual values of n.

C. Difference between Bound and Scattering Systems

This chapter has focused on reactive systems, in which the nuclear wave function satisfies scattering boundary conditions, applied at the asymptotic limits of reagent and product channels. It turns out that these boundary conditions are what make it possible to unwind the nuclear wave function from around the CI, and that it is impossible to unwind a bound-state wave function.

To see why this is so, let us attempt to apply the procedure of Section II.B to a bound-state wave function. This is illustrated schematically in Fig. 19. It is clear immediately that we cannot construct an unsymmetric Ψ_e in the double space, because each bound-state eigenfunction must be an irreducible representation of the double-space symmetry group. Thus a bound-state function in the double space is necessarily symmetric or antisymmetric under $\hat{R}_{2\pi}$, and is thus either a Ψ_G or a Ψ_N function. For a Ψ_G function, we have $\Psi_N = 0$ (since Ψ_G and Ψ_N cannot form a degenerate pair), which implies [from Eq. (6)] that

$$\Psi_e = -\Psi_o = \Psi_G \tag{46}$$

Similarly, for a Ψ_N function,

$$\Psi_e = \Psi_o = \Psi_N \tag{47}$$

In other words, if we map a bound-state wave function onto the double-cover space using Eq. (6), we simply duplicate the function, because the contribution from the even n Feynman paths is exactly equal to (or equal and opposite to) the contribution from the odd n paths.

If we continue mapping onto successively higher cover spaces, following the procedure of Section IV.B, then the effect is the same. Instead of completely unwinding the nuclear wave function, and producing a gap (where $|\Psi_h(\phi)|^2$ is

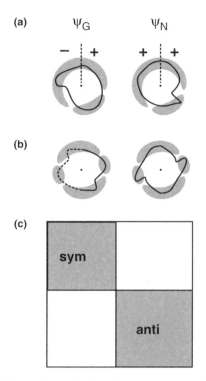

Figure 19. Relation between Ψ_G and Ψ_N for a bound-state system. The functions in the single space (a) can be mapped onto the double space (b) where they have opposite symmetries under $\hat{R}_{2\pi}$, and belong to different symmetry blocks of the double-space Hamiltonian matrix (c). Unlike reactive wave functions, bound-state functions cannot be unwound from around the CI.

negligibly small), the mapping onto the $2h\pi$ space generates a periodic repetition of the original single-space function. Hence, the technique of Section IV.B, in which one can unwind a reactive wave function completely from around the CI (to within a specified accuracy) does not work for a bound-state function. Heuristically, one may think of this as arising because, in the bound-state system, the Feynman paths (like the classical paths) can describe an infinite number of loops around the CI [28].

An encircling reactive wave function is thus topologically different from an encircling bound-state wave function. This is the reason why, in Section II.A, we said that, when the wave function encircles the CI, it is $|\Psi(\phi)|^2$ that has the form of a torus, rather than $\Psi(\phi)$. A reactive wave function $\Psi(\phi)$ is not a torus—it is essentially a coil, since it can be unwound. A bound-state function $\Psi_G(\phi)$, on the other hand, is a torus (with a twist), because if one imagines calculating it by propagating a function around the CI, then the two ends of the function must match up. In a reactive wave function, no such matching is ever required, and we may loop each end of the function around the CI as many times as we please, before allowing each end to pass out one of the entrance or exit channels (where it is then matched up to the asymptotic scattering functions). We therefore suggest that the term encirclement often used in the GP literature should be qualified as weak encirclement, when the system is reactive, and strong encirclement when it is bound.

Mapping onto the double space will therefore reveal nothing new about the effect of the GP on a bound-state system, since this is purely a boundary-condition effect. However, it does gives us an alternative representation of the GP and non-GP wave functions, which may sometimes be clearer than the equivalent single-space representation (in which one deduces the symmetry Ψ from the total wave function $\Phi\Psi$). For example, in the double space, it is very clear that the GP will cause all the bound states to be doubly degenerate when the (single-space) molecular symmetry group is isomorphic with C_{2v} (because the double-space group is then isomorphic with C_{4v}). Similarly, the double-space picture is analogous to a double-well system, with periodic boundary conditions, and this may also sometimes be useful in rationalizing the effect of the GP on the spectrum. We note that use of a double space has proved very useful in the analogous field of Möbius molecules, where the electronic wave function satisfies what is in effect a GP boundary condition, on account of a twist in the nuclear structure [75].

Of course, the distinction between reactive- and bound-state wave functions becomes blurred when one considers very long-lived reactive resonances, of the sort considered in Section IV.B, which contain Feynman paths that loop many times around the CI. Such a resonance, which will have a very narrow energy width, will behave almost like a bound-state wave function when mapped onto the double space, since $|\Psi_e|$ will be almost equal to $|\Psi_o|$. The effect of the GP boundary condition would be therefore simply to shift the energies and permitted nodal structures of the resonances, as in a bound-state function. For short-lived resonances, however, $|\Psi_e|$ and $|\Psi_o|$ will differ, since they will describe the different decay dynamics produced by the even and odd n Feynman paths; separating them will therefore reveal how this dynamics is changed by the GP. The same is true for resonances which are long lived, but which are trapped in a region of space that does not encircle the CI, so that the decay dynamics involves just a few Feynman loops around the CI.

V.   OUTLOOK AND CONCLUSIONS

The central theme of this chapter is that the effect of the GP on the dynamics of a chemical reaction is very simple, thanks to the topological property of homotopy.

Feynman paths that loop a different number of times around the CI are decoupled from one another in the sum-over-paths, with the result that the nuclear wave function can be split into separate contributions from the even- and odd-looping paths. The sole effect of the GP is to change the relative sign of these two components.

Although the basic physics behind this idea is not new (it was applied to the Aharonov—Bohm effect many years ago [42, 43, 45]), its application to chemical reaction dynamics was made only very recently, in our application to solve the cancellation puzzle in $H + H_2$ [27–29]. This application has been discussed here in some detail, since it illustrates how to use homotopy to explain GP effects in chemical reactions. If one can compute the GP and non-GP nuclear wave functions, then adding and subtracting these functions yields the even- and odd-looping components. If one cannot compute the wave functions, one can still estimate the likely magnitude of GP effects by modeling the dynamics of the even- and odd-looping reaction paths using classical trajectories.

Hence, the GP has a much milder effect on reactive systems than on bound-state systems. This difference has been overlooked in the past, but becomes apparent on noting that an encircling bound-state function contains Feynman paths that loop an infinite number of times around the CI [28]. Consequently, the encirclement of a bound-state wave function is much stronger than that of a reactive wave function; the bound wave function cannot be unwound from around the CI, whereas the reactive wave function can. One consequence of this is that the separation into even- and odd-looping paths yields no information about the dynamics of a bound state system, in which these two contributions are necessarily equal and opposite [28].

The $H + H_2$ example gives clues as to whether GP effects are likely to be important in other reactions. The GP effects cancel in the integral cross-section of $H + H_2$ because the even- and odd-looping paths scatter into different regions of angular phase space. It seems reasonable to assume that many other direct reactions will scatter in a similar manner (since looping different numbers of times around the CI necessarily entails very different types of reaction mechanism), and therefore show similar cancellations. We can also conclude that there will be no GP effects in reactions that involve phase averaging (e.g., statistical capture reactions [76]). For similar reasons, GP effects are very unlikely to affect the outcome of reactions in liquids. The best experiments to see GP effects in reactions are thus likely to be ultrafast control experiments, in which an encircling wave function is probed at short times [77].

Acknowledgments

This research was funded by a grant from the UK Engineering and Physical Sciences Research Council (EPSRC), and by the award of a Royal Society University Research Fellowship to SCA.

References

1. W. Domcke, D. R. Yarkony and H. Köppel (eds.), *Conical Intersections: Electronic Structure, Dynamics and Spectroscopy*, World Scientific, New Jersey, 2003.
2. G. A. Worth and L. S. Cederbaum, *Annu. Rev. Phys. Chem.* **55**, 127 (2004).
3. C. A. Mead and D. G. Truhlar, *J. Chem. Phys.* **77**, 6090 (1982).
4. G. Herzberg and H. C. Longuet-Higgins, *Discuss. Faraday Soc.* **35**, 77 (1963).
5. H. C. Longuet-Higgins, *Proc. R. Soc. A* **344**, 147 (1975).
6. C. A. Mead and D. G. Truhlar, *J. Chem. Phys.* **70**, 2284 (1979).
7. C. A. Mead, *Rev. Mod. Phys.* **64**, 51 (1992).
8. M. S. Child, *Adv. Chem. Phys.* **124**, 1 (2002).
9. M. V. Berry, *Proc. R. Soc. A* **392**, 45 (1984).
10. B. K. Kendrick, *Phys. Rev. Lett.* **79**, 2431 (1997).
11. B. E. Applegate, T. A. Barckholtz, and T. A. Miller, *Chem. Soc. Rev.* **32**, 38 (2003).
12. C. A. Mead, *Chem. Phys.* **49**, 23 (1980).
13. D. Babikov, B. K. Kendrick, P. Zhang, and K. Morokuma, *J. Chem. Phys.* **122**, 044315 (2005).
14. C. A. Mead, *J. Chem. Phys.* **72**, 3839 (1980).
15. B. Lepetit and A. Kuppermann, *Chem. Phys. Lett.* **166**, 581 (1990).
16. Y. M. Wu, A. Kuppermann, and B. Lepetit, *Chem. Phys. Lett.* **186**, 319 (1991).
17. A. Kuppermann and Y.-S. M. Wu, *Chem. Phys. Lett.* **205**, 577 (1993).
18. A. Kuppermann and Y.-S. M. Wu, *Chem. Phys. Lett.* **241**, 229 (1995).
19. A. Kuppermann and Y.-S. M. Wu, *Chem. Phys. Lett.* **349**, 537 (2001).
20. B. K. Kendrick, *J. Chem. Phys.* **112**, 5679 (2000).
21. B. K. Kendrick, *J. Phys. Chem. A* **107**, 6739 (2003).
22. B. K. Kendrick, *J. Chem. Phys.* **118**, 10502 (2003).
23. A. J. C. Varandas and L. P. Viegas, *Chem. Phys. Lett.* **367**, 625 (2003).
24. J. C. Juanes-Marcos and S. C. Althorpe, *Chem. Phys. Lett.* **381**, 743 (2003).
25. J. C. Juanes-Marcos and S. C. Althorpe, *Faraday Discuss.* **127**, 115 (2004).
26. J. C. Juanes-Marcos and S. C. Althorpe, *J. Chem. Phys.* **122**, 204324 (2005).
27. J. C. Juanes-Marcos, S. C. Althorpe, and E. Wrede, *Science* **309**, 1227 (2005).
28. S. C. Althorpe, *J. Chem. Phys.* **124**, 084105 (2006).
29. J. C. Juanes-Marcos, S. C. Althorpe, and E. Wrede, *J. Chem. Phys.* **126**, 044317 (2007).
30. E. Wrede, L. Schnieder, K. H. Welge, F. J. Aoiz, L. Bañares, and V. J. Herrero, *Chem. Phys. Lett.* **265**, 129 (1997).
31. E. Wrede, L. Schnieder, K. H. Welge, F. J. Aoiz, L. Bañares, J. F. Castillo, B. Martínez-Haya, and V. J. Herrero, *J. Chem. Phys.* **110**, 9971 (1999).
32. F. Fernández-Alonso, B. D. Bean, J. D. Ayers, A. E. Pomerantz, R. N. Zare, L. Bañares, and F. J. Aoiz, *Angew. Chem. Int. Ed. Engl.* **39**, 2748 (2000).
33. F. Fernández-Alonso, B. D. Bean, R. N. Zare, F. J. Aoiz, L. Bañares, and J. F. Castillo, *J. Chem. Phys.* **115**, 4534 (2001).
34. S. C. Althorpe, F. Fernández-Alonso, B. D. Bean, J. D. Ayers, A. E. Pomerantz, R. N. Zare, and E. Wrede, *Nature* (London) **416**, 67 (2002).

35. A. E. Pomerantz, F. Ausfelder, R. N. Zare, S. C. Althorpe, F. J. Aoiz, L. Bañares, and J. F. Castillo, *J. Chem. Phys.* **120**, 3244 (2004).

36. F. Ausfelder, A. E. Pomerantz, R. N. Zare, S. C. Althorpe, F. J. Aoiz, L. Bañares, and J. F. Castillo, *J. Chem. Phys.* **120**, 3255 (2004).

37. S. A. Harich, D. Dai, C. C. Wang, X. Yang, S. D. Chao, and R. T. Skodje, *Nature (London)* **419**, 281 (2002).

38. E. Wrede and L. Schnieder, *J. Chem. Phys.* **107**, 786 (1997).

39. E. Wrede, *Energy-Dependent Investigation of the Hydrogen-Exchange Reaction* Ph.D. thesis, University of Bielefeld, Germany, 1998.

40. R. P. Feynman and A. R. Hibbs, *Quantum Mechanics and Path Integrals*, McGraw-Hill, New York, 1965.

41. L. S. Schulman, *Phys. Rev.* **176**, 1558 (1968).

42. L. S. Schulman, *J. Math. Phys.* **12**, 304 (1971).

43. L. S. Schulman, *Techniques and Applications of Path Integration*, John Wiley & Sons Inc., New York, 1981.

44. M. G. G. Laidlaw and C. M. Morette DeWitt, *Phys. Rev. D* **3**, 1375 (1971).

45. G. Morandi and E. Menossi, *Eur. J. Phys.* **5**, 49 (1984).

46. H. Goldstein, *Classical Mechanics*, Addison-Wesley, Reading MA, 1980.

47. J. Stillwell, *Classical Topology and Combinatorial Group Theory*, Springer, New York, 1993.

48. D. J. Kouri and D. K. Hoffman, *Few Body Systems* **18**, 203 (1995).

49. R. Schinke, *Photodissociation Dynamics*, Cambridge University Press, Cambridge, 1993.

50. A. J. C. Varandas, F. B. Brown, C. A. Mead, D. G. Truhlar, and N. C. Blais, *J. Chem. Phys.* **86**, 6258 (1987).

51. A. I. Boothroyd, W. J. Keogh, P. G. Martin, and M. R. Peterson, *J. Chem. Phys.* **104**, 7139 (1996).

52. S. Mahapatra, H. Köppel, and L. S. Cederbaum, *J. Phys. Chem. A* **105**, 2321 (2001).

53. C. R. Evenhuis, X. Lin, D. H. Zhang, D. Yarkony, and M. A. Collins, *J. Chem. Phys.* **123**, 134110 (2005).

54. A. Kuppermann, *Chem. Phys. Lett.* **32**, 374 (1975).

55. F. Fernández-Alonso and R. N. Zare, *Annu. Rev. Phys. Chem.* **53**, 67 (2002).

56. F. J. Aoiz, L. Bañares, and V. J. Herrero, *Int. Rev. Phys. Chem.* **24**, 119 (2005).

57. G. C. Schatz and A. Kuppermann, *J. Chem. Phys.* **65**, 4642 (1975).

58. W. H. Miller, *J. Chem. Phys.* **50**, 407 (1969).

59. R. T. Pack and G. A. Parker, *J. Chem. Phys.* **87**, 3888 (1987).

60. D. Neuhauser, M. Baer, R. S. Judson, and D. J. Kouri, *J. Chem. Phys.* **90**, 5882 (1989).

61. S. C. Althorpe, *J. Chem. Phys.* **114**, 1601 (2001).

62. S. C. Althorpe and D. C. Clary, *Annu. Rev. Phys. Chem.* **54**, 493 (2003).

63. R. N. Zare, *Angular Momentum*, John Wiley & Sons Inc., New York, 1988.

64. B. Kendrick and R. T. Pack, *J. Chem. Phys.* **106**, 3519 (1997).

65. H. Nakamura and D. G. Truhlar, *J. Chem. Phys.* **22**, 10353 (2001).

66. H. Köppel, *Faraday Discuss.* **127**, 35 (2004).

67. J. Tennyson and B. T. Sutcliffe, *J. Chem. Phys.* **77**, 4061 (1982).

68. M. S. Child, *Molecular Collision Theory*, Academic Press, London, 1974.

69. A. J. Dobbyn, P. McCabe, J. N. L. Connor, and J. F. Castillo, *Phys. Chem. Chem. Phys.* **1**, 1115 (1999).

70. D. G. Truhlar and J. T. Muckerman in *Atom–Molecule Collision Theory*, R. B. Bernstein (ed.), Plenum, New York 1979.

71. F. J. Aoiz, V. J. Herrero, and V. Sáez Rábanos, *J. Chem. Phys.* **94**, 7991 (1991).

72. L. Bonnet and J.-C. Rayez, *Chem. Phys. Lett.* **277**, 183 (1997).

73. G. Herzberg, *Electronic Spectra and Electronic Structure of Polyatomic Molecules*, Van Nostrand Reinhold, New York, 1966.

74. M. S. Child, *Semiclassical Mechanics with Molecular Applications*, Clarendon Press, Oxford, UK, 1991.

75. P. W. Fowler, *Phys. Chem. Chem. Phys.* **4**, 2878 (2002).

76. E. J. Rackham, T. Gonzalez-Lezana, and D. E. Manolopoulos, *J. Chem. Phys.* **24**, 12895 (2003).

77. M. Abe, Y. Ohtsuki, Y. Fujimura, Z. Lan, and W. Domcke, *J. Chem. Phys.* **124**, 224316 (2006).

OPTIMAL CONTROL THEORY FOR MANIPULATING MOLECULAR PROCESSES

GABRIEL G. BALINT-KURTI and SHIYANG ZOU

School of Chemistry, University of Bristol, Bristol BS8 1TS, UK

ALEX BROWN

Department of Chemistry, University of Alberta, Edmonton, AB, T6G 2G2, Canada

CONTENTS

Advances in Chemical Physics, Volume 138, edited by Stuart A. Rice

I. INTRODUCTION

Optimal Control Theory (OCT) is a broad subject with applications in very many fields. In this chapter, we consider the application of OCT in the area of designing laser pulses for manipulating molecular processes. This application of OCT falls within the area of coherent control, which is the science of using shaped or tailored laser pulses to determine or steer the outcome of a chemical process. The term "coherent" implies that the phase of the light plays an important role and that the same control could not be achieved through manipulation of just the amplitude of the light intensity. Two excellent books, [1, 2], the proceedings of two conferences, [3, 4], special issues of two journals [5, 6], and many review articles [7–25], have already been published on this subject, as well as a host of research papers. The current review will deal only with the theoretical aspects of the subject and presents, in a coherent fashion, many of the details of the OCT formulation of the coherent control problem in quantum dynamics. In particular, we focus on the equations necessary for determining the optimal field and the numerical methods for the solution of the resultant equations, including methods for applying constraints on the field. We also discuss an approach, with illustrative examples, for including effects beyond the dipole approximation and the use of analytic methods for guiding or interpreting OCT solutions. This chapter will not directly address any experimental problems. The interested reader is referred to the several excellent reviews focussing on the experimental aspects of control. [21–25]. Having said this, our general approach is that the theoretical treatment should be accurate and capable of correctly predicting and interpreting real experiments.

There are two general theoretical approaches to laser control: (1) a small number of interfering optical pathways are chosen based on physical intuition or (2) many interfering pathways are created using algorithmic methods. In the former, the mechanisms leading to control need to be identified *a priori*. In the latter, the control mechanisms are often unidentified due to the number of interfering pathways and the complexity of the laser fields. Approach (1) can be further broken down into two general subgroups: phase control methods as pioneered by Brumer and Shapiro [26] and the "pump–dump" method of Tannor and Rice [7, 27–29]. While all coherent-control scenarios require the existence of multiple paths from the initial to final state, the phase-control approach is a weak field one utilizing (usually) two paths. Control is achieved by manipulating the relative phases and amplitudes of the two pathways. On the other hand, the pump–dump approach envisages a pump pulse (often of femtosecond duration) that excites a system to an electronically excited state. After a time delay, during which there is free evolution of the excited-state wave packet on the potential energy surface, a dump–pulse deexcites the system back to the ground-state surface. Control is achieved by varying the properties of the initial excitation pulse, the time delay, and the properties of the dump pulse. Tannor and Rice demonstrated that these laser parameters could be optimized to maximize control objectives. The algorithmic approach is due to the extensive work of Rabitz and co-workers [30, 31] and is based on optimal control theory, which has been widely used in engineering applications. The theory is centered around the definition of an "objective functional", which has its maximum value when the desired transformation is successfully achieved by the laser pulse under consideration. Much of the following chapter will be directed at discussion of this method and its application. In related work, Rabitz and co-workers advocated the use of a "closed-loop" or feedback process [8, 32] in which the results of an experiment are fed back to a pulse design procedure. The design of the laser pulse is progressively modified so as to produce an optimal outcome. This approach has led to several successful experimental applications [33–40] and can also, in principle, be used as a theoretical procedure for the design of optimal laser pulses.

In Section II, the basic equations of OCT are developed using the methods of variational calculus. Methods for solving the resulting equations are discussed in Section III. Section IV is devoted to a discussion of the Electric Nuclear Born–Oppenheimer (ENBO) approximation [41, 42]. This approximation provides a practical way of including polarization effects in coherent control calculations of molecular dynamics. In general, such effects are important as high electric fields often occur in the laser pulses used experimentally or predicted theoretically for such processes. The limits of validity of the ENBO approximation are also discussed in this section.

All of the methods for designing laser pulses to achieve a desired control of a molecular dynamical process require the solution of the time-dependent Schrödinger equation for the system interacting with the radiation field. Normally, this equation must be solved many times within an iterative loop. Different possible approaches to the solution of these equations are discussed in Section V.

Optimal control theory, as discussed in Sections II–IV, involves the algorithmic design of laser pulses to achieve a specified control objective. However, through the application of certain approximations, analytic methods can be formulated and then utilized within the optimal control theory framework to predict and interpret the laser fields required. These analytic approaches will be discussed in Section VI.

Section VII presents a brief summary of this chapter. Technical aspects of the necessary derivations are, in general, presented in the appendices.

II.   OPTIMAL CONTROL THEORY

A.   Basic Formalism

Optimal control theory provides a formal framework for designing laser pulses to achieve different objectives. The theory operates by defining an objective (or cost) functional, J, which measures the success of a particular laser pulse in achieving the assigned objective. The details of the objective functional depend on the desired purpose of the laser pulse. The theory aims to maximize the objective functional so as to find the optimal laser pulse for achieving the desired objective. Thus normally a target wave function, Φ, will be defined and the aim of the optimal control procedure will be to design a laser pulse that will force the system from its initial state [described by a wave function $\psi(t = 0) = \phi_i$] into the desired final state at the end of the laser pulse [i.e., $\psi(t = T) = \Phi$, where the pulse duration lasts from $t = 0$ to $t = T$]. The principal term in the objective functional will therefore be $|\langle \psi(t = T)|\Phi\rangle|^2$, the square of the overlap between the system wave function at the end of the pulse and the target wave function. Following the work of Shi and Rabitz [43]. The cost functional may be defined in the form:

$$J = |\langle\psi(T)|\Phi\rangle|^2 - \int_0^T f[\epsilon(t)]dt - 2\Re\left\{\int_0^T dt\, \langle\chi(t)\middle|\left[\frac{\partial}{\partial t} + \frac{i}{\hbar}\hat{H}(\epsilon(t))\right]\middle|\psi(t)\rangle\right\}$$

(1)

where Φ is the target wave function, $\chi(t)$ is an undetermined Lagrange multiplier, which ensures that the time-dependent Schrödinger equation is obeyed at all times and $\epsilon(t)$ is the electric field at time t. In the third term, the notation $\hat{H}(\epsilon(t))$ indicates that the Hamiltonian depends on time through the variation of the electric field of the laser with time.

The second term in Eq. (1) is a penalty term representing constraints on the control field $\epsilon(t)$ via a functional f and is extremely important in determining the outcome of the optimization. The most common penalty term, and that first introduced by Rabitz and co-workers [41], is

$$f[\epsilon(t)] = \beta\epsilon(t)^2$$

(2)

where β is a constant penalty parameter set to provide constraints on the total laser fluence, $\int_0^T \epsilon(t)^2 dt$. If it is desired to constrain the total pulse energy, the penalty term may be written in the form $f[\epsilon(t)] = \beta[\epsilon(t)^2 - E]$ where E is a constant and β is a Lagrange multiplier rather than a constant [1, 29]. An alternative penalty term for the laser fluence has also been utilized:

$$f[\epsilon(t)] = \beta\epsilon(t)^4$$

(3)

where β is a constant. This alternate form was chosen to account for effects of molecular polarizability (see discussion in Appendix A).

Other forms of the objective functional have also been utilized. Zhu and Rabitz have formulated expressions [44, 45] for the more general problem of achieving quantum control over the expectation value of a positive definite operator, that is, a physical observable. Ohtsuki et al. [46] have extended the formalism to apply to multiple targets beyond population transfer to a single state, that is, beyond $|\langle\psi(t=T)|\Phi\rangle|^2$. Tesch and de Vivie-Riedle generalized the optimal control algorithm in order to find the optimized field for simultaneously steering a set of initial states to a set of final states [47]. Such a generalization has direct applicability to molecular quantum computing [47–54]. Xu and co-workers formulated the optimal control problem for dissipative non-Markovian systems [55]. Optimal control schemes for time-dependent targets are presented in Refs. [56–59].

In Appendix A, we follow the derivation of Shi and Rabitz and carry out the functional variation of the objective functional [Eq. (1)] so as to obtain the equations that must be obeyed by the wave function $(\psi(t))$, the undetermined Lagrange multiplier $(\chi(t))$, and the electric field $(\epsilon(t))$. Since the results discussed in Section IV.B focus on controlled excitation of H_2, where molecular polarizability must be considered, the penalty term given by Eq. (3) is used and the equations that must be obeyed by these functions are (see Appendix A for a detailed derivation):

$$i\hbar\frac{\partial}{\partial t}\psi(t) = \hat{H}(\epsilon(t))\psi(t); \qquad \psi(0) = \phi_i \tag{4.a}$$

$$i\hbar\frac{\partial}{\partial t}\chi(t) = \hat{H}(\epsilon(t))\chi(t); \qquad \chi(T) = \langle\Phi|\psi(T)\rangle\Phi \tag{4.b}$$

$$4\beta[\epsilon(t)]^3 = \frac{2}{\hbar}\text{Im}\left\langle\chi(t)\left|\frac{\partial\hat{H}(\epsilon(t))}{\partial\epsilon(t)}\right|\psi(t)\right\rangle \tag{4.c}$$

Equation (4.a) states that the wave function $\psi(t)$ must obey the time-dependent Schrödinger equation with initial condition $\psi(t=0) = \phi_i$. Equation (4.b) states that the undetermined Lagrange multiplier, $\chi(t)$, must obey the time-dependent Schrödinger equation with the boundary condition that $\chi(T) = \langle\Phi|\psi(T)\rangle\Phi$ at the end of the pulse, that is at $t = T$. As this boundary condition is given at the end of the pulse, we must integrate the Schrödinger equation backward in time to find $\chi(t)$. The final of the three equations, Eq. (4.c), is really an equation for the time-dependent electric field, $\epsilon(t)$.

Equation (4.c) is discussed in Appendix A. For a symmetric molecule that does not possess a dipole moment and interacts with the electric field of the laser pulse through its polarizability, the choice of the penalty function for the fluence $[-\beta \int_0^T dt\, \epsilon(t)^4]$ in Eq. (1) is the appropriate choice. For such a system, the customary choice of penalty function $(-\beta \int_0^T dt\, \epsilon(t)^2)$ does not lead to an equation for determining the electric field, see Appendix A [Eq. (A-20)] for details.

B. Additional Restrictions on Laser Pulse

In early work in the optimal control theory design of laser fields to achieve desired transformations, the optimal control equations were solved directly, without constraints other than those imposed implicitly by the inclusion of a penalty term on the laser fluence [see Eq. (1)]. This inevitably led to laser fields that suddenly increased from very small to large values near the start of the laser pulse. However, physically realistic laser fields should turn-on and -off smoothly. Therefore, during the optimization the field is not allowed to vary freely but is rather expressed in the form [60]:

$$\epsilon(t) = s(t)\, \epsilon_0(t) \tag{5}$$

where $s(t)$ is a pulse envelope function. The pulse envelope, $s(t)$, is kept fixed thus forcing the laser field to go to zero at the beginning and the end of the pulse and only the residual field $\epsilon_0(t)$ is varied so as to maximize the objective functional [41, 42, 60–62]. Two different forms of the envelope are often considered: a sine-squared pulse or a Gaussian pulse. The $\sin^2$ pulse has the form:

$$s(t) = \sin^2(n\, \pi t/T) \tag{6}$$

where the pulse lasts from $t = 0$ to $t = T$ and n corresponds to the number of maxima in the pulse envelope (normally chosen to be 1). The choice of a Gaussian function centered at the mid-point of the pulse [62] has the slight disadvantage that the field does not go strictly to zero before and after the pulse. However, it has the advantage that it corresponds more closely to an experimentally realizable form.

Another restriction we may often wish to place on the laser pulse is to limit the frequency range of the electric field in the pulse. One method that has been used to accomplish this is simply to eliminate frequency components of the field that lie outside a specified range [63]. Another possibility is to use a frequency filter, such as the twentieth-order Butterworth bandpass filter [64], which is a smoother way of imposing basically the same restrictions [41, 42]. In order to impose such restrictions on the frequency content of the pulse, the time-dependent electric field of the laser pulse must be Fourier transformed so as to obtain its frequency spectrum. After the frequency spectrum of the laser pulse has been passed through the filter, it is back transformed to yield back a modified time-dependent electric laser field (see Section III.A below for further details).

As has often been noted [60] the designed laser pulses, which result from an optimal control theory calculation, are generally quite complex and this is an impediment to their experimental realization. It may therefore be desirable to simplify the frequency spectrum of the pulse, which may be done by reducing or eliminating the contributions from frequencies that occur only with a small amplitude. This procedure has been called "sifting" [42]. If $d(\omega)$ is the Fourier transform of the electric field increment to be added to the old or previous electric field at some particular iteration, then we wish to multiply the largest value of $d(\omega)$ by 1.0 and all other components by smaller numbers to reduce their significance. This may be done in many ways, but one sifting function that has been applied is

$$d_{new}(\omega) = \frac{1}{2}\left\{1 + \tanh\left[\frac{90.0}{d_{max}}\left(|d_{old}(\omega)| - \frac{d_{max}}{3}\right)\right]\right\}d_{old}(\omega) \qquad (7)$$

where $d_{old}(\omega)$ is the Fourier transform of the electric field increment to be added to the old or previous electric field, and d_{max} is the maximum amplitude for all ω of this function.

Other methods for imposing constraints on the frequency components of the optimized pulses have also been considered in the literature. A subspace projection method was developed by de Vivie-Riedle and co-workers to reduce the spectral complexity of the final field from the OCT algorithm [65]. They have also designed an alternate form of the objective functional that allows for large values of the penalty parameter β, [see Eq. (2)], which leads to spectrally simple pulses. Two experimental methods implemented within the closed-loop algorithm for reducing spectral complexity are also worth noting: a simple reduction in the number of adjustable parameters [66] and control pulse cleaning [67], where genetic pressure is applied on spectral components to reduce complexity. The imposition of the constraint on frequency content is again discussed in Section III as it is imposed in conjunction with the iterative solution of the OCT equations [Eq. (4)].

The main mechanism for restricting the magnitude of the electric field of the laser pulse is through the second term in Eq. (1), which is a penalty term specifically to address this point. In some cases, it may be desirable to further guarantee that the field does not exceed some specified limits. In this case, various restrictions may be placed on the variation of the electric field strength during the iterative solution of Eq. (4). One method for imposing these restrictions is discussed briefly in Section III [see Eqs. (13) and (14) and the accompanying discussion]. Shen and Rabitz [68] showed that explicit restrictions on the maximum field amplitude can be applied through alternate definitions of the cost functional, Eq. (1). More recently, Farnum and Mazziotti [69] introduced a trigonometric mapping into the standard optimal control scheme to restrict the maximum field strength explicitly rather than indirectly through limiting the field energy with a penalty term of the form given by Eq. (2).

An alternate method for introducing pulse restrictions has been introduced by one of us (AB) recently within an iterative scheme for solving the optimal control equations [70]. The idea is that a new reference field $\epsilon(t)$ is constructed based on the field from the previous iteration after the application of a filter function F to ensure the fulfilment of some predesigned temporal and spectral properties. Therefore, a penalty term of the form

$$f[\epsilon(t)] = \beta(\epsilon(t) - \epsilon_{ref}(t))^2 \tag{8}$$

is utilized where $\epsilon_{ref}(t)$ is the filtered field obtained in a previous iteration (see Section III.B for a discussion of iterative methods for solving the optimal control problem). The utility of the method has been demonstrated by applying both constraints on the spectral bandwidth and on the maximum field amplitude [70], although more sophisticated filters could also be applied. A similar method was also introduced by Werschnik and Gross [71].

Rather than applying constraints on an optimized pulse, one can *a priori* choose a fixed form for the laser field described by multiple parameters, for example, amplitudes, frequencies, and phases, and then optimize this set of variables for achieving the desired objective. Then one must perform a multi-parameter optimization or search, which is the basis of the closed-loop experiments[8, 22, 23, 32, 39, 40]. Several different approaches have been utilized for finding the optimized fields: conjugate gradient methods [72]; simulated annealing [73]; and, now most widely used, genetic algorithms [74–77]. An interesting discussion comparing pulses determined using optimal control algorithms versus parameter space searching is given in Ref. [78]. We will not discuss OCT methods using such fixed-form fields further in this chapter.

C. Photodissociation

The problem of controlling the outcome of photodissociation processes has been considered by many authors [63, 79–87]. The basic theory is derived in detail in Appendix B. Our set objective in this application is to maximize the flux of dissociation products in a chosen exit channel or final quantum state. The theory differs from that set out in Appendix A in that the final state is a continuum or dissociative state and that there is a continuous range of possible energies (i.e., quantum states) available to the system. The equations derived for this case are [see Appendix B Eq. (B.18)]:

$$i\hbar \frac{\partial}{\partial t}\psi(t) = \hat{H}\psi(t), \qquad \psi(0) = \phi_i \tag{9a}$$

$$i\hbar \frac{\partial}{\partial t}\chi(t) = \hat{H}\chi(t) - i\hbar \Lambda \hat{F}\psi(t), \qquad \chi(T) = 0 \tag{9b}$$

$$\epsilon(t) = -\frac{1}{\hbar\beta}\mathrm{Im}\,\langle\chi(t)|\mu|\psi(t)\rangle \tag{9c}$$

Here, we have considered the case where the interaction of the system with the light is mediated through a dipole operator. Several electronic states of the system may be involved and the wave functions are generalized to column vectors each of whose components correspond to a nuclear wave function associated with a different electronic state of the system. The matrices $\hat{H}$, Λ, $\hat{F}$, and μ similarly have diagonal terms that are associated with a single electronic state and off-diagonal terms that couple different electronic states. Λ is a diagonal matrix whose elements allow us to choose which dissociation channels we wish to target for optimization. As with all numerical photodissociation calculations, the wave function must be absorbed at the edge of the finite grid. This is normally achieved through the use of a complex absorbing potential [88–91].

III. METHODS FOR SOLVING THE OPTIMAL CONTROL PROBLEM

The solution of the coupled equations, Eq. (4.a–4.c), must of necessity be performed in an iterative manner, as knowledge of the electric field, $\epsilon(t)$, is needed to solve the time-dependent Schrödinger equations and to determine $\psi(t)$ and $\chi(t)$, but it is $\epsilon(t)$ which we vary in order to maximize J. Another way of viewing the problem is that the field strength $\epsilon(t)$, which maximizes J, depends on $\psi(t)$ and $\chi(t)$ through Eq. (4.c). The most straightforward approach to the iterative solution of Eqs. (4.a–c) is to first guess an initial time-dependent laser field. Having specified an initial guess to the time-dependent electric field of the laser pulse, the time-dependent Schrödinger equation, Eq. (4.a) must be solved with the initial boundary condition that the wave function be equal to the specified initial wave function of the system, ϕ_i, before the laser pulse. After solving this equation, we will know the wave function of the system at the end of the laser pulse ($\psi(T)$) and we will be in a position to calculate the space integral $\langle\Phi|\psi(T)\rangle$, where Φ is the wave function of our target state. We can now solve the equation for the undetermined Lagrange multiplier, $\chi(t)$, that is, Eq. (4.b). This equation must be solved backward in time, starting with the boundary condition at $t = T$ of $\chi(T) = \langle\Phi|\psi(T)\rangle\Phi$. Having obtained both $\psi(t)$ and $\chi(t)$ we can now calculate a new time-dependent laser field using Eq. (4.c) [see also Eq. (A-16) and the associated discussion]. This now constitutes a self-consistent loop. Having obtained a new electric field, we repeat the whole procedure again starting with the solution of Eq. (4.a) for $\psi(t)$ and this is repeated until the changes in the objective functional J (arising from the variation of the time-dependent electric field) are smaller than some preassigned threshold. While this approach is straightforward, in general it converges very slowly, if at all, to the optimal solution [92, 93]. In Section III.B we discuss various more recently devised iterative methods that exhibit improved convergence properties [56, 81, 92–94].

An alternative to such an iterative approach is to attempt to optimze the objective functional J directly. Such techniques work best if the derivative of the functional with respect to the variable is known. In our case, we split the time duration of the pulse into small increments, δt. The field strength is taken as constant during each of these increments. In Appendix C, we derive an expression for the derivative of the objective functional with respect to the electric field strength during the ith time interval [41, 42]. With a knowledge of $\psi(t)$ and $\chi(t)$ we can evaluate the derivative of the objective functional in the ith time interval. Knowing this derivative we can apply standard optimization techniques, such as steepest descent or conjugate gradient [95] to find the field parameters that maximize the objective functional [96–99].

A.   Conjugate Gradient Method

With the objective functional, J, being defined as in Eq. (1) and using the penalty term $f[\epsilon(t)] = \beta\epsilon^4(t)$, the gradient of the objective functional with respect to variation of ϵ_0 at time t is given by (see Appendix C):

$$g^k(t) \equiv \frac{\delta J^k}{\delta\epsilon_0^k(t)} = -s(t)\left[4\,\beta\left[\epsilon^k(t)\right]^3 - \frac{2}{\hbar}\,\mathrm{Im}\left\langle\chi(t)\left|\frac{\partial\hat{H}(R,\epsilon^k(t))}{\partial\epsilon^k(t)}\right|\psi(t)\right\rangle\right] \quad (10)$$

where the superscript k indicates the iteration number in the optimization cycle. Using Eq. (10), the Polak–Ribière–Polyak [100] search direction can be calculated as:

$$d^k(t_i) = g^k(t_i) + \zeta^k d^{k-1}(t_i) \quad (11)$$

where

$$\zeta^k = \frac{\sum_j g^k(t_j)^T\left(g^k(t_j) - g^{k-1}(t_j)\right)}{\sum_j g^{k-1}(t_j)^T g^{k-1}(t_j)} \quad (12)$$

$k = 2, 3, \ldots, d^1(t_i) = g^1(t_i)$, ζ^k is the conjugate gradient update parameter and the summation is over all time intervals. A line search is then performed along this direction to determine the maximum value of the objective functional.

There are different variants of the conjugate gradient method each of which corresponds to a different choice of the update parameter ζ^k. Some of these different methods and their convergence properties are discussed in Appendix D. The time has been discretized into N time steps $(t_i = i \times \delta t$ where $i = 0, 1, \cdots, N - 1)$ and the parameter space that is being searched in order to maximize the value of the objective functional is composed of the values of the electric field strength in each of the time intervals.

Although the overall cost of the conjugate gradient algorithm may be higher than that of some of the iterative algorithms described in Section III.B, the algorithm allows us easily to restrict the spectral and temporal structure of optimal pulses and enables us to incorporate the exact form of the laser–molecule interactions.

We now discuss the imposition of an additional restriction on the magnitude of the electric field strength and also the restriction of the frequency content of the laser pulse. These items are discussed here (rather than in Section II.B) as they are generally implemented during the line search stage of the conjugate gradient maximization of the objective functional, although they may be imposed on the electric field at other stages of the optimization process.

The penalty function for the laser fluence, which gives rise to the term involving β in Eq. (10), acts to limit the magnitude of the electric field to within physically acceptable limits. In order to further limit the field strength the search direction $d^k(t_i)$ is projected as follows [101]:

$$d_p^k(t_i) = P(\epsilon_0^k(t_i) + d^k(t_i)) - \epsilon_0^k(t_i) \tag{13}$$

Here, the projector $P(x)$ is defined as:

$$
\begin{aligned}
P(x) &= \text{sign}(x)\, x_{\lim} \quad &&\text{if} \quad |x| > x_{\lim} \\
P(x) &= x \quad &&\text{if} \quad x \in [-x_{\lim}, x_{\lim}]
\end{aligned}
\tag{14}
$$

where $x_{\lim}$ is some number $\epsilon_{\max}$.

Straightforward application of OCT as described above often results in a quite complicated pulse shapes and may especially introduce some high frequency components, which are difficult to realize experimentally, into the pulse. It is thus highly desirable to find an optimized pulse with spectral components within a predefined frequency range. With this end in view the projected search direction is subjected to a spectral filter

$$\tilde{d}_p^k(t) = \int h(\omega) F_\omega [d_p^k(t)] e^{-i\omega t} d\omega \tag{15}$$

where $F_\omega[d_p^k(t)]$ is the Fourier component at frequency ω and $h(\omega)$ is a spectral filtering function. One possible form of filter function is the twentieth-order Butterworth bandpass filter [64], in which an upper (ω_h) and a lower (ω_ℓ) bounds of the frequency window are defined

$$h(\omega) = \left\{ \left[1 + \left(\frac{\omega_\ell}{\omega} \right)^{40} \right] \left[1 + \left(\frac{\omega}{\omega_h} \right)^{40} \right] \right\}^{-1/2} \tag{16}$$

in order to restrict the frequency components of the electric field to a predefined range [63]. The time-dependent electric field for the next iteration in the optimization cycle is given by

$$\epsilon^{k+1}(t_i) = \epsilon^k(t_i) + \lambda s(t_i) d_p^k(t_i) \tag{17}$$

where $d_p^k(t_i)$ is the projected and frequency filtered search direction and λ is determined by the line search.

B. Iterative Methods

One of the most used iterative algorithms was suggested by Rabitz and others [44, 45, 92] and may be considered to be an extension of an algorithm due to Tannor et al. [81, 93]. Both formulations share a common property: that at each iteration they are guaranteed to increase the magnitude of the objective functional. Later Maday and Turinici [94] presented a unified framework for the monotonically convergent algorithms that contains, as particular cases, the two classical methods cited above. This class of algorithms was reported to converge much faster than gradient-type methods and also to be relatively insensitive to the initially guessed pulse.

To improve the convergence of the gradient-type method, Tannor et al. [81, 93] suggested employing the Krotov iteration method [102]. In formulating their method, they utilize a penalty function of the form $f[\epsilon(t)] = \beta\epsilon^2(t)$. In Tannor's Krotov method, the kth iteration step of the solution process is given by

$$i\hbar \frac{\partial}{\partial t} \psi^k(t) = [\hat{H}_0 - \epsilon^k(t)\mu] \psi^k(t), \qquad \psi^k(0) = \phi_i \tag{18.a}$$

$$\epsilon^k(t) = -\frac{1}{\beta\hbar} \text{Im}\langle\chi^{k-1}(t)|\mu|\psi^k(t)\rangle \tag{18.b}$$

$$i\hbar \frac{\partial}{\partial t} \chi^k(t) = [\hat{H}_0 - \mu\epsilon^k(t)] \chi^k(t), \qquad \chi^k(T) = \langle\Phi|\psi^k(T)\rangle\Phi \tag{18.c}$$

The coupled Eqs. (18.a and b) may be solved by propagating the nonlinear equation,

$$ i\hbar \frac{\partial}{\partial t} \psi^k(t) = \left[\hat{H}_0 + \frac{1}{\beta\hbar} \text{Im}\langle \chi^{k-1}(t)|\mu|\psi^k(t)\rangle \mu \right] \psi^k(t), \qquad \psi^k(t=0) = \phi_i $$

(19)

This equation is obtained by substituting Eq. (18.b) into Eq. (18.a). The difference between the Krotov method and the straightforward iterative approach lies in the details of how the electric field term, $\epsilon^k(t) = -\frac{1}{\beta\hbar}\text{Im}\langle\chi^{k-1}(t)|\mu|\psi^k(t)\rangle$, is computed at each iteration. This is discussed in more detail in Appendix E. The key aspect of the method resides in the fact that the electric field is updated as soon as possible in the iterative process for the solution of the time-dependent Schrödinger equation Eq. (18.a) or (19). Thus, if a first-order method is used for the solution of these equations, then the electric field, $\epsilon^k(t=t_i)$ needed to propagate the solution from $t = t_i$ to $t = t_{i+1}$ is computed using $\chi^{k-1}(t_i)$ and $\psi^k(t_i)$ (which has just been computed in the preceding time step). In the more straightforward method, the electric field would have been precomputed using $\psi^{k-1}(t)$, which is known from the preceding iteration.

It was reported that the convergence of the Krotov iteration method [81, 93] was four or five times faster than that of the gradient-type methods. The formulation of Rabitz and others, [44, 45, 92], designed to improve the convergence of the above algorithm, introduces a further nonlinear propagation step into the adjoint equation (i.e., the equation for the undetermined Lagrange multiplier $\chi(t)$) and is expressed as

$$ i\hbar \frac{\partial}{\partial t} \psi^k(t) = (\hat{H}_0 - \epsilon^k(t)\mu)\psi^k(t), \qquad \psi^k(0) = \phi_i \qquad (20.a) $$

$$ \epsilon^k(t) = -\frac{1}{\beta\hbar}\text{Im}\langle\chi^{k-1}(t)|\mu|\psi^k(t)\rangle \qquad (20.b) $$

$$ i\hbar \frac{\partial}{\partial t} \chi^k(t) = (\hat{H}_0 - \mu\tilde{\epsilon}^k(t))\chi^k(t), \qquad \chi^k(T) = \langle\Phi|\psi^k(T)\rangle\Phi \qquad (20.c) $$

$$ \tilde{\epsilon}^k(t) = -\frac{1}{\beta\hbar}\text{Im}\langle\chi^k(t)|\mu|\psi^k(t)\rangle \qquad (20.d) $$

The similarity between the two algorithms inspired the work of Maday and Turinici [94]. They clarified the relationships between them and presented an unified formulation. In the formulation of Maday and Turinici [94], the iteration scheme is written as:

$$ i\hbar \frac{\partial}{\partial t} \psi^k(t) = (\hat{H}_0 - \epsilon^k(t)\mu)\psi^k(t), \qquad \psi^k(0) = \phi_i \qquad (21.a) $$

$$ \epsilon^k(t) = (1-\delta)\tilde{\epsilon}^{k-1}(t) - \frac{\delta}{\beta\hbar}\text{Im}\langle\chi^{k-1}(t)|\mu|\psi^k(t)\rangle \qquad (21.b) $$

$$i\hbar \frac{\partial}{\partial t} \chi^k(t) = (\hat{H}_0 - \mu \tilde{\epsilon}^k(t)) \chi^k(t), \qquad \chi^k(T) = \langle \Phi | \psi^k(T) \rangle \Phi \qquad (21.c)$$

$$\tilde{\epsilon}^k(t) = (1 - \eta)\epsilon^k(t) - \frac{\eta}{\beta\hbar} \text{Im} \langle \chi^k(t) | \mu | \psi^k(t) \rangle \qquad (21.d)$$

It can be seen that the algorithm of Rabitz et al. [44, 45, 92] corresponds to $(\delta = 1, \eta = 1)$ and that the algorithm of Tannor et al. [81, 93] is given by $(\delta = 1, \eta = 0)$. In Appendix F, we follow the proof used in Ref. [94] and show that the iteration procedure laid out in Eqs. (21.a–d) is guaranteed to converge.

It was observed that the Krotov method ($\delta = 1$ and $\eta = 0$) usually achieves its convergence limit [moderately accurate $J(\epsilon^{k+1}) - J(\epsilon^k) \le 10^{-4}$ or 10^{-5}] with a smaller number of iteration steps in comparison with the algorithm of Rabitz et al. [44, 45, 92] ($\delta = 1$ and $\eta = 1$) (see Refs. [94, 56, and 103]). On the other hand, the algorithm of Rabitz et al. [44, 45, 92] may achieve high accuracy, for example, $J(\epsilon^{k+1}) - J(\epsilon^k) \le 10^{-10}$. It was also observed that the case of ($\delta = 0.5$ and $\eta = 0$) exhibited faster convergence than the Krotov method in certain cases. The optimal choice of δ and η for achieving fast convergence depends on the molecular system and also the stage of optimization (the choice may be changed during the course of an optimisation). A procedure was established for the choice of "good" parameters [56, 103].

Note that the iterative algorithms given above rely on the fact that (1) the interaction of the molecule with the laser is treated within the electric dipole approximation; and (2) that the penalty term is defined as $f[\epsilon(t)] = \beta\epsilon^2(t)$. Recently, Salomon et al. [104] extended the monotonically convergent algorithms to incorporate dipole polarization. For the high field strengths, which occur in many experiments, significant high order interaction terms in the laser field–molecule interaction may need to be considered. In this case, the laser–molecule interaction may not be known analytically, and it may not be possible to establish an iterative scheme. In such a case, a gradient-type method may have to be used.

IV.   THE ELECTRIC NUCLEAR BORN–OPPENHEIMER APPROXIMATION

Many of the initial theoretical models used to validate the concept of coherent control and optimal control have been based on the interaction of the electric field of the laser light with a molecular dipole moment [43, 60, 105]. This represents just the first, or lowest, term in the expression for the interaction of an electric field with a molecule. Many of the successful optimal control experiments have used electric fields that are capable of ionizing the molecules and involve the use of electric field strengths that lead to major distortions of the molecular electronic structure. With this in mind, there has been discussion in the literature of methods suitable for modeling the interaction of molecules with such strong laser fields [41, 42, 106] and also with time-dependent strong laser fields [107–113]. In this section, we discuss an approach that takes account of the distortion of the molecular electron density by the electric field of the laser light.

The approach has been called the ENBO approximation [41, 42]. It uses the concept of an electronically adiabatic electronic state [106, 114, 115]. The assumption is made that the electrons can react instantaneously to the "slow" changes of the electric field of the laser light, as well as to the changes in nuclear geometry. This is therefore a generalization of the Born–Oppenheimer approximation [116] to include the electric field strength and direction, as well as the nuclear coordinates, as parametric, slowly varying variables in the electronic wave function. By using this approach, the electronic wave function and the electronic energy depend on the fixed values of the nuclear geometry and the static electric field strength and direction for which they have been evaluated. The evaluation of these quantities requires no new electronic structure codes, as nearly all available codes can compute the electronic wave function, at several sophisticated levels of theory, in the presence of a static electric field [117–121]. The electronic energy, which results from solving the electric field-dependent electronic Schrödinger equation, may be written in the form $V(\boldsymbol{R}, \epsilon)$. This electronic energy is then taken to act as the potential that governs the motion of the nuclei in the subsequent dynamics.

As with all adiabatic theories, the electronically adiabatic wave functions may be used as a basis set to describe the exact wave function of the system even in situations where transitions between different adiabatic electronic states occur. This provides an accurate way of testing the validity of using the ENBO approximation, in which it is assumed that the system will remain throughout in its lowest adiabatic electronic state. The ENBO approximation has been applied to several calculations involving vibrational and rotational excitation of the H_2 [41, 42, 61, 62] and HF molecules [122]. As dihydrogen (H_2) is a homonuclear diatomic molecule it does not, by symmetry, possess any dipole moment. So previous treatments, which took account only of the electric field–dipole moment interaction term, are inapplicable in this case. For an electric field strength of 2.06×10^8 V cm^{-1} (5.6×10^{13} W cm^{-2}) and a frequency of 2.63×10^{14} s^{-1} it has been shown [41] that a laser pulse of 1.55 ps will excite at most 0.02% of the H_2 molecules out of their lowest adiabatic electronic state. For frequencies and field strengths below these values, the ENBO approximation may therefore be used in its simplest form (i.e., without accounting for electronically nonadiabatic processes). These findings agree qualitatively with other estimates based on theoretically computed ionization rates. Based on the findings of Usachenko and Chu [123] a laser pulse of 1.5 ps duration and having an intensity of 5.6×10^{13} W cm^{-2} with a frequency of 5.8×10^{14} s^{-1} would ionize 0.6% of a sample of H_2 and up to 3% of a sample of N_2 molecules.

Below we give a brief theoretical overview of the ENBO approximation. The derivation follows closely that given in Ref. [41] and serves to make clear the approximations inherent in the method.

A.   Theoretical Formulation

Treating the radiation in the semiclassical dipole approximation, the Hamiltonian operator in the presence of an electric field $\epsilon(t)$ may be written in atomic units as:

$$\hat{H}_{\text{total}}(\mathbf{r}, \mathbf{R}, \epsilon(t)) = \hat{T}_{\text{nu}}(\mathbf{R}) + \hat{H}_{\text{el}}(\mathbf{r}; \mathbf{R}, \epsilon(t)) \tag{21}$$

where $\mathbf{r}$ represents the electronic coordinates and $\mathbf{R}$ the nuclear coordinates.

The electronic Hamiltonian is

$$\hat{H}_{\text{el}}(\mathbf{r}; \mathbf{R}, \epsilon(t)) = -\frac{\hbar^2}{2m_e} \sum_i \nabla_{\mathbf{r}_i}^2 - \sum_{\alpha i} \frac{Z_\alpha e}{4\pi\epsilon_0 R_{\alpha i}} + \sum_{i>j} \frac{e^2}{4\pi\epsilon_0 r_{ij}}$$

$$+ \left\{ \sum_i \mathbf{r}_i - \sum_\alpha Z_\alpha \mathbf{R}_\alpha \right\} \cdot \epsilon(t) + \sum_{\alpha < \beta} \frac{Z_\alpha Z_\beta}{4\pi\epsilon_0 R_{\alpha\beta}} \tag{22}$$

and the nuclear kinetic energy operator is

$$\hat{T}_{\text{nu}}(\mathbf{R}) = -\frac{\hbar^2}{2} \sum_\alpha \frac{1}{M_\alpha} \nabla_{\mathbf{R}_\alpha}^2 \tag{23}$$

where $R_{\alpha i} = |\mathbf{R}_\alpha - \mathbf{r}_i|$, $r_{ij} = |\mathbf{r}_j - \mathbf{r}_i|$, Z_α are the nuclear charges and M_α the nuclear masses.

This treatment differs from the usual approach to molecule–radiation interaction through the inclusion of the contribution from the electric field from the beginning and by not treating it as a perturbation to the field free situation. The notation $\hat{H}_{\text{el}}(\mathbf{r}; \mathbf{R}, \epsilon(t))$ makes the parametric dependence of the electronic Hamiltonian on the nuclear coordinates and on the electric field explicit.

The full time-dependent Schrödinger equation may be written as

$$i\hbar \frac{\partial}{\partial t} \Psi(\mathbf{r}, \mathbf{R}, t; \epsilon(t)) = \hat{H}_{\text{total}}(\mathbf{r}, \mathbf{R}, \epsilon(t)) \Psi(\mathbf{r}, \mathbf{R}, t; \epsilon(t)) \tag{24}$$

The adiabatic electronic potential energy surfaces (a function of both nuclear geometry and electric field) are obtained by solving the following electronic eigenvalue equation

$$\hat{H}_{\text{el}}(\mathbf{r}; \mathbf{R}, \epsilon(t)) \, \phi_k(\mathbf{r}; \mathbf{R}, \epsilon(t)) = E_k(\mathbf{R}, \epsilon(t)) \, \phi_k(\mathbf{r}; \mathbf{R}, \epsilon(t)) \tag{25}$$

where $E_k(\mathbf{R}, \epsilon(t))$ are the desired adiabatic electronic potential energy surfaces. The notation $E_k(\mathbf{R}, \epsilon(t))$ makes clear that the surface depends parametrically on both the nuclear coordinates and on the electric field. It depends indirectly on time through the electric field.

The time-dependent wave function, $\Psi(\mathbf{r}, \mathbf{R}, t; \epsilon(t))$, can be expanded in terms of the field-dependent adiabatic electronic eigenfunctions of Eq. (25):

$$\Psi(\mathbf{r}, \mathbf{R}, t; \epsilon(t)) = \sum_k \psi_k(\mathbf{R}, t)\, \phi_k(\mathbf{r}; \mathbf{R}, \epsilon(t)) \tag{26}$$

Substituting the expansion in Eq. (26) into the time-dependent Schrödinger equation Eq. (24) we obtain

$$i\hbar \sum_k \left\{ \frac{\partial \psi_k(\mathbf{R}, t)}{\partial t} + \psi_k(\mathbf{R}, t)\frac{d\epsilon(t)}{dt} \cdot \nabla_\epsilon \right\} \phi_k(\mathbf{r}; \mathbf{R}, \epsilon(t))$$
$$= \sum_k \{ E_k(\mathbf{R}, \epsilon(t))\psi_k(\mathbf{R}, t) + [\hat{T}_{\mathrm{nu}}(\mathbf{R})\psi_k(\mathbf{R}, t)] \tag{27}$$
$$+ \psi_k(\mathbf{R}, t)\hat{T}_{\mathrm{nu}}(\mathbf{R})\} \phi_k(\mathbf{r}; \mathbf{R}, \epsilon(t))$$

Multiplying on the left by $\phi_\ell^*(\mathbf{r}; \mathbf{R}, \epsilon(t))$, integrating over the electronic coordinates and using the orthonormality of the adiabatic electronic wave functions of Eq. (25) we obtain

$$i\hbar \frac{\partial \psi_\ell(\mathbf{R}, t)}{\partial t} + i\hbar \frac{d\epsilon(t)}{dt} \cdot \sum_k \tau_{\ell k}^{(3)}(\mathbf{R}, \epsilon(t))\psi_k(\mathbf{R}, t)$$
$$= \{\hat{T}_{\mathrm{nu}}(\mathbf{R}) + E_\ell(\mathbf{R}, \epsilon(t))\}\psi_\ell(\mathbf{R}, t) \tag{28}$$
$$- \frac{\hbar^2}{2} \sum_k \sum_\alpha \frac{1}{M_\alpha} \{2\tau_{\ell k}^{(1)}(\mathbf{R}_\alpha, \epsilon(t)) \cdot [\nabla_{\mathbf{R}_\alpha} \psi_k(\mathbf{R}, t)]$$
$$+ \tau_{\ell k}^{(2)}(\mathbf{R}_\alpha, \epsilon(t))\psi_k(\mathbf{R}, t)\}$$

where

$$\tau_{\ell k}^{(1)}(\mathbf{R}_\alpha, \epsilon(t)) = \int d\mathbf{r}\, \phi_\ell^*(\mathbf{r}; \mathbf{R}, \epsilon(t))\, \nabla_{\mathbf{R}_\alpha} \phi_k(\mathbf{r}; \mathbf{R}, \epsilon(t)) \tag{29}$$

and

$$\tau_{\ell k}^{(2)}(\mathbf{R}_\alpha, \epsilon(t)) = \int d\mathbf{r}\, \phi_\ell^*(\mathbf{r}; \mathbf{R}, \epsilon(t))\, \nabla_{\mathbf{R}_\alpha}^2 \phi_k(\mathbf{r}; \mathbf{R}, \epsilon(t)) \tag{30}$$

are the usual field-free nonadiabatic coupling matrices of the first (vector) and second (scalar) kind, respectively, [124] whereas

$$\tau_{\ell k}^{(3)}(\mathbf{R}, \epsilon(t)) = \int d\mathbf{r} \, \phi_\ell^*(\mathbf{r}; \mathbf{R}, \epsilon(t)) \, \nabla_\epsilon \phi_k(\mathbf{r}; \mathbf{R}, \epsilon(t)) \tag{31}$$

is the nonadiabatic (vector) coupling matrix of the third kind due to the external electric field $\epsilon(t)$.

Equation (28) is the set of exact coupled differential equations that must be solved for the nuclear wave functions in the presence of the time-varying electric field. In the spirit of the Born–Oppenheimer approximation, the ENBO approximation assumes that the electronic wave functions can respond immediately to changes in the nuclear geometry and to changes in the electric field and that we can consequently ignore the coupling terms containing $\tau_{\ell k}^{(1)}(\mathbf{R}_\alpha, \epsilon(t))$, $\tau_{\ell k}^{(2)}(\mathbf{R}_\alpha, \epsilon(t))$, and $\tau_{\ell k}^{(3)}(\mathbf{R}, \epsilon(t))$. The equation for the nuclear wave function then reduces to

$$i\hbar \, \frac{\partial \psi(\mathbf{R}, t)}{\partial t} = \{\hat{T}_{\mathrm{nu}}(\mathbf{R}) + V(\mathbf{R}, \epsilon(t))\} \psi(\mathbf{R}, t) \tag{32}$$

where, for simplicity, the subscript on the nuclear wave function has been omitted. It is assumed that the electronic state corresponds to the lowest adiabatic electronic state and the ground state electronic potential energy surface has been renamed as $V(\mathbf{R}, \epsilon(t))$:

$$V(\mathbf{R}, \epsilon(t)) = E_0(\mathbf{R}, \epsilon(t)) \tag{33}$$

Equation (32) is the working equation that is used in the optimal control applications using the ENBO approximation. Further details of the theory are given in Ref. [41].

B. Illustrative Examples

In this section, we provide some examples of optimal control theory calculations using the ENBO approximation. The reader is referred to Ref. [42], from where all the examples are taken, for further details.

Figure 1 shows the results of an OCT calculation for the molecular vibrational excitation process $H_2(v = 0, j = 0) \rightarrow H_2(v' = 1, j' = 0)$. Panel (a)

Figure 1 (a) The optimized electric field as a function of time for the $H_2(v = 0, j = 0) \rightarrow H_2(v' = 1, j' = 0)$ excitation. (b) Absolute value of the Fourier transform of the optimized electric field. (c) The change in populations of the ground state and target excited state shown as a function of time. Taken from Ref. [24] with permission from Qinghua Ren, Gabriel G. Balint-Kurti, Frederick R. Manby, Maxim Artamonov, Tak-San Ho, and Herschel Rabitz, *J. Chem. Phys.* **124**, 014111 (2006). Copyright 2006, American Institute of Physics.

(a)

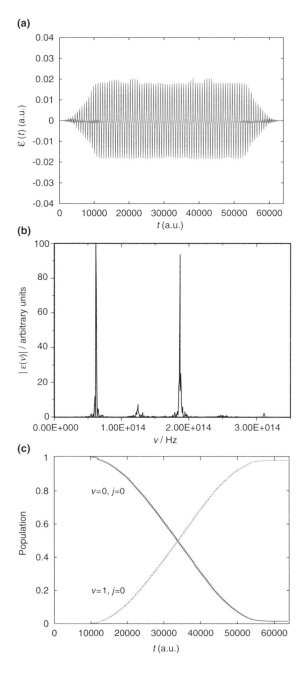

(b)

(c)

shows the electric field of the optimized laser pulse. The envelope of the initially guessed pulse had a $\sin^2$ shape with a single maximum. The optimization process, and in particular the effect of the penalty term, has resulted in an optimized pulse that shows a rapid build up of the magnitude of the electric field followed by a long plateau in this magnitude. By a suitable choice of the fluence penalty term [see Eqs. (1) and (3)] the magnitude of the electric field has been constrained always to be smaller than that which might lead to ionization or to electronic excitation. Panel (b) shows the frequency spectrum of the optimized laser pulse. It contains just two main frequencies as compared with the initially guessed pulse that contained only the lower of these two frequencies. The lower frequency corresponds to excitation via a two photon process, while the higher frequency corresponds to a Raman processes. Direct one photon vibrational excitation is not allowed in this case as the molecule is homonuclear and does not possess a dipole moment. Panel (c) shows the change in population of the $(v = 0, j = 0)$ and the $(v = 1, j = 0)$ quantum levels as a function of time during the laser pulse.

Figure 2 shows the OCT results for the rotational excitation process $H_2(v = 0, j = 0) \rightarrow H_2(v' = 0, j' = 2)$. The initial guessed laser field for this processes involved two frequencies and the beating of these two frequencies resulted in a pulsed structure for the electric field. The resulting optimized electric field is shown in the top panel of the figure. It retains the pulsed shape of the initial guessed field but, as in the case of the vibrational excitation process (Fig. 1), the maximum field strength quickly reaches a plateau rather than following the $\sin^2$ shape of the guessed field. The frequency spectrum of the optimized laser pulse, shown in the central panel, is now much more complicated than before and the population~time graph (bottom panel) has a stepped character, reflecting the pulsed nature of the electric field.

Figure 3 demonstrates the simplifications in the spectrum of an optimized laser pulse that can be achieved through the application of the sifting technique [see Eq. (7)]. The excitation efficiency of the pulse is only minimally reduced due to the additional restrictions imposed in the sifting procedure. The example used in this case is for a vibrational–rotational excitation process, $H_2(v = 0, j = 0) \rightarrow H_2(v' = 1, j' = 2)$.

Figure 2 (a) The optimized electric field as a function of time for the $H_2(v = 0, j = 0) \rightarrow$ $H_2(v' = 0, j' = 2)$ rotational excitation process. (b) Absolute value of the Fourier transform of the optimized electric field. (c) The change in populations of the ground-and target excited-state shown as a function of time. Taken from Ref. [24] with permission from Qinghua Ren, Gabriel G. Balint-Kurti, Frederick R. Manby, Maxim Artamonov, Tak-San Ho, and Herschel Rabitz, *J. Chem. Phys.* **124**, 014111 (2006). Copyright 2006, American Institute of Physics.

(a)

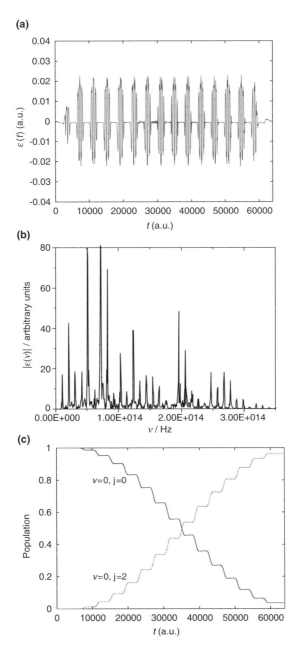

(b)

(c)

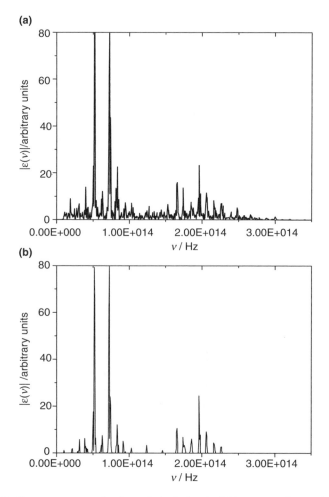

Figure 3 Frequency spectra for the optimized electric field corresponding to the excitation process $H_2(v = 0, j = 0) \rightarrow H_2(v' = 1, j' = 2)$. (a) without frequency sifting and (b) using frequency sifting [see Eq. (7)].

V. SOLUTION OF TIME-DEPENDENT SCHRÖDINGER EQUATION

The application of OCT requires the repeated solution of the time-dependent Schrödinger equation [see Eqs. (4.a, b)]. Our ability to solve this equation is therefore central to the application of the theory. If the Hamiltonian is independent of time, then the formal solution to the time-dependent Schrödinger equation can be written in the form:

$$\psi(t) = e^{-i\hat{H}t/\hbar}\,\psi(t=0) \tag{34}$$

Clearly, in OCT the Hamiltonian is intrinsically time dependent and this form of the solution is not valid for large t values. We can, however, use the formula to propagate the wave function forward by a small time interval δt.

$$\psi(t + \delta t) = e^{-i\hat{H}(t)\delta t/\hbar}\,\psi(t) \tag{35}$$

The short-time propagator $\hat{U}(\delta t) = e^{-i\hat{H}(t)\delta t/\hbar}$ becomes progressively more valid the smaller the time step δt.

A very large number of papers have been written on the solution of the time-dependent Schrödinger equation [125–134]. Of particular interest is a multi author collaborative paper (Ref. [130]) that compares several different methods. While this paper, in common with the majority of others, concentrates mainly on situations where the Hamiltonian is independent of time, it does consider time-dependent Hamiltonians as well. It concludes that for time-independent Hamiltonians the Chebyshev polynomial expansion method [128, 129] should be the method of choice and that for time-dependent Hamiltonians, as occur in OCT, the short iterative Lanczos propagation method [130] should be used. Despite this recommendation, the most commonly used methods for solving the time-dependent Schrödinger equation in OCT applications are the split-operator method of Feit and Fleck [125–127] or an expansion of the wave function, and indirectly therefore of the time-evolution operator $\hat{U}(t)$, in terms of the eigenfunctions of the isolated molecular Hamiltonian—see, however, Ref. [135] where Farnum and Mazziotti discuss a spectral difference Lanczos method for OCT problems. Kono et al. [134] compared various methods of solving the the time-dependent Schrödinger equation where the full electronic Hamiltonian is taken into account, including the various Coulomb interactions. Such Hamiltonians are needed for field strengths or frequencies higher than those on which we are concentrating in this chapter. One of the difficulties to be overcome is to have a suitable approach for treating optimal control problems in polyatomic (multidimensional) systems. To that end, both the time-dependent Hartree approach [136] and the multiconfigurational time-dependent Hartree approach [137] have been applied to OCT calculations for quantum dynamics.

The following two sections present a brief overview of the split-operator method, as used in several recent applications [41, 42, 61, 62], and of the basis set expansion approach.

A.   The Split-Operator Fourier Grid Method

This section outlines one possible numerical procedure for solving the time-dependent Schrödinger equation. An important aspect of the efficient implementation of the split-operator method is that it should go hand in hand with a grid representation of the wave function. Such grid representations were first popularized in this field by Light and co-workers [138]. In these methods, there is a one-to-one correspondence between the grid points and an orthonormal basis set. The representation of the wave function in terms of the grid points is termed the Discrete Variable Representation (DVR) and that in terms of the basis functions is called the Finite Basis Representation (FBR). There is a simple transformation of the wave function between the two bases using a unitary matrix. The key step in any quantum mechanical time-propagation procedure involves the action of Hamiltonian operator on the wave function. The utility of the dual DVR–FBR representation is that the FBR representation is chosen so that the kinetic energy operator should be local in this representation while the potential energy is always local in the DVR representation. Below we outline the application of these ideas to a one-dimensional (1D) system involving a radial coordinate [125–129]. Similar representations are available also for angular grids [138–141].

The transformations connecting the coordinate-space wave function, $\psi(R)$, to the momentum-space wave function, $\psi^k(k)$, are

$$\psi^k(k) = \frac{1}{\sqrt{2\pi}} \int_{R=0}^{R=L} \exp(-ikR)\psi(R)dR \tag{36}$$

and

$$\psi(R) = \frac{1}{\sqrt{2\pi}} \int_{k=k_{min}}^{k=k_{max}} \exp(ikR)\psi^k(k)dk \tag{37}$$

The coordinate grid is defined as [95]:

$$R_i = R_0 + i\,\delta R; \quad i = 0, \ldots, N-1 \tag{38}$$

where $\delta R = L/N$ and L is the length of the 1D grid. The grid representation of the wave function is written as:

$$\psi_i = \psi(R_i)\sqrt{\delta R}$$
$$= \left(\frac{L}{N}\right)^{\frac{1}{2}} \psi(R_i) \tag{39}$$

The normalization condition can therefore be written as $\sum_i |\psi_i|^2 = 1$. As we have used an evenly spaced grid, the corresponding finite basis representation is the set of linear momentum eigenfunctions $\exp(ikR)$. Because the spatial grid consists of N spatial grid points, the conjugate grid in momentum space is also composed of N discrete points. The largest possible wavelength on the grid is $\lambda_{\max} = L$. As the momentum is related to the wavelength by $k = (2\pi)/\lambda$; the increment on the discretized momentum grid is $\delta k = (2\pi)/\lambda_{\max} = (2\pi)/L$ (see Ref. [95]). The discrete grid in momentum space is therefore:

$$\begin{aligned} k_\ell &= k_{\min} + \ell\,\delta k \\ &= -\frac{N\pi}{L} + \ell\,\frac{2\pi}{L}; \quad \ell = 0, \cdots, N-1 \end{aligned} \tag{40}$$

The discretized momentum-space wave function corresponding to a momentum of $k_\ell \hbar$ is denoted by ψ_ℓ^k. As with the discretized spatial wave function [Eq. (37)], the discretized momentum wave functions are also normalized so that $\sum_\ell |\psi_\ell^k|^2 = 1$ (i.e., $\psi_\ell^k = \psi^k(k_\ell)\sqrt{\delta k}$).

The wave function in momentum space is given by the Fourier transform of the coordinate-space wave function

$$\begin{aligned} \psi^k(k_\ell) &= \frac{1}{\sqrt{2\pi}} \sum_{i=0}^{i=N-1} \exp(-ik_\ell R_i)\,\psi_i\sqrt{\delta R} \\ &= \frac{\psi_\ell^k}{\sqrt{\delta k}} = \left(\frac{L}{2\pi}\right)^{\frac{1}{2}} \psi_\ell^k \end{aligned} \tag{41}$$

The discretized momentum space wave function, ψ_ℓ^k, is therefore given by

$$\begin{aligned} \psi_\ell^k &= \left(\frac{1}{L}\right)^{\frac{1}{2}} \sum_{i=0}^{i=N-1} \exp(-ik_\ell R_i)\psi_i\sqrt{\delta R} \\ &= \left(\frac{1}{L}\right)^{\frac{1}{2}} \left(\frac{L}{N}\right)^{\frac{1}{2}} \sum_{i=0}^{i=N-1} \exp(-ik_\ell R_i)\,\psi_i \\ &= \frac{1}{\sqrt{N}} \sum_{i=0}^{i=N-1} \exp(-ik_\ell R_i)\,\psi_i \end{aligned} \tag{42}$$

The inverse Fourier transform is

$$\psi_i = \frac{1}{\sqrt{N}} \sum_{\ell=0}^{\ell=N-1} \exp(ik_\ell R_i)\,\psi_\ell^k \tag{43}$$

We now express the Hamiltonian operator as a sum of kinetic and potential energy operators, $\hat{H}(t) = \hat{T} + \hat{V}(t)$, and rewrite the short-time propagator [see Eq. (35) and also Ref. [92]] in the form:

$$
\begin{aligned}
\hat{U}(\delta t) &= e^{-i\hat{H}(t)\delta t/\hbar} \\
&= e^{-i\hat{T}\delta t/2\hbar} e^{-i\hat{V}(t+\delta t/2)\delta t/\hbar} e^{-i\hat{T}\delta t/2\hbar}
\end{aligned}
\tag{44}
$$

This expansion of the propagator is not completely valid as the kinetic energy, $\hat{T} = -\frac{\hbar^2}{2\mu}\frac{\partial^2}{\partial R^2}$, and potential energy, $\hat{V}(t)$, operators do not commute. The way in which we have split the operators up, with the potential energy term in the middle, is referred to as the potential energy referenced split operator method [132, 142]. The method has been shown to have an error, arising from the noncommutation of the kinetic energy and potential energy operators, depending on the third power of the time interval δt [127, 143].

The application of the kinetic energy part of the short-time propagator proceeds as follows:

$$
\phi_i = [e^{-i\hat{T}\delta t/2\hbar}\psi(R)]_{R=R_i}\sqrt{\delta R}
\tag{45}
$$

By substituting for $\psi(r)$ using Eq. (37), we obtain

$$
\begin{aligned}
\phi_i &= \left[e^{-i\hat{T}\delta t/2\hbar} \frac{1}{\sqrt{2\pi}} \int_{k=k_{min}}^{k=k_{max}} dk\, \exp(ikR)\, \psi^k(k) \right]_{R=R_i} \sqrt{\delta R} \\
&= \left[\frac{1}{\sqrt{2\pi}} \int_{k=k_{min}}^{k=k_{max}} dk\, e^{-i\hat{T}\delta t/2\hbar} \exp(ikR)\, \psi^k(k) \right]_{R=R_i} \sqrt{\delta R} \\
&= \left[\frac{1}{\sqrt{2\pi}} \int_{k=k_{min}}^{k=k_{max}} dk\, e^{-ik^2\hbar\delta t/4\mu} \exp(ikR)\, \psi^k(k) \right]_{R=R_i} \sqrt{\delta R}
\end{aligned}
\tag{46}
$$

By using Eq. (36) to substitute for $\psi^k(k)$ we obtain

$$
\begin{aligned}
\phi_i &= \left[\frac{1}{\sqrt{2\pi}} \int_{k=k_{min}}^{k=k_{max}} dk\, e^{-ik^2\hbar\delta t/4\mu} \exp(ikR) \frac{1}{\sqrt{2\pi}} \int_{R'=0}^{R'=L} dR'\, \exp(-ikR')\, \psi(R') \right]_{R=R_i} \sqrt{\delta R} \\
&= \frac{1}{\sqrt{2\pi}} \int_{k=k_{min}}^{k=k_{max}} dk\, e^{-ik^2\hbar\delta t/4\mu} \exp(ikR_i) \left[\frac{1}{\sqrt{2\pi}} \int_{R'=0}^{R'=L} dR'\, \exp(-ikR')\, \psi(R') \right] \sqrt{\delta R}
\end{aligned}
\tag{47}
$$

Now replacing the continuous integral (or Fourier transform) in the square brackets by discrete Fourier transforms as in Eqs. (41) we obtain

$$\phi_i = \frac{1}{\sqrt{2\pi}} \int_{k=k_{min}}^{k=k_{max}} dk \, e^{-ik^2\hbar\delta t/4\mu} \exp(ikR_i) \left[\frac{1}{\sqrt{2\pi}} \sum_{j=0}^{j=N-1} \exp(-ikR_j) \, \psi_j \sqrt{\delta R}\right] \sqrt{\delta R}$$

$$= \frac{1}{2\pi} \int_{k=k_{min}}^{k=k_{max}} dk \, e^{-ik^2\hbar\delta t/4\mu} \exp(ikR_i) \left[\sum_{j=0}^{j=N-1} \exp(-ikR_j) \, \psi_j\right] \delta R \tag{48}$$

Now discretizing the integral over k we obtain

$$\phi_i = \frac{1}{2\pi} \sum_{\ell=0}^{\ell=N-1} \delta k \, e^{-ik_\ell^2\hbar\delta t/4\mu} \exp(ik_\ell R_i) \left[\sum_{j=0}^{j=N-1} \exp(-ik_\ell R_j) \, \psi_j\right] \delta R$$

$$= \frac{1}{2\pi} \frac{2\pi}{L} \frac{L}{N} \sum_{\ell=0}^{\ell=N-1} e^{-ik_\ell^2\hbar\delta t/4\mu} \exp(ik_\ell R_i) \left[\sum_{j=0}^{j=N-1} \exp(-ik_\ell R_j) \, \psi_j\right] \tag{49}$$

$$= \frac{1}{\sqrt{N}} \sum_{\ell=0}^{\ell=N-1} e^{-ik_\ell^2\hbar\delta t/4\mu} \exp(ik_\ell R_i) \left[\frac{1}{\sqrt{N}} \sum_{j=0}^{j=N-1} \exp(-ik_\ell R_j) \, \psi_j\right]$$

The parameter ϕ_i is now the value of the wave function, in the DVR grid representation, at the ith grid point that results from acting with the operator $\exp(-i\hat{T}\delta t/2\hbar)$ on the wave function ψ.

The step that has just been outlined in detail is the most difficult step in the propagation of the wave function. The action with the operator $\exp(-i\hat{V}(R,t)\delta t/2\hbar)$ is straightforward as this operator is a local operator in the grid representation and we just multiply the grid representation of the wave function at grid point "i" by the value of the operator at the same grid point.

B. Expansion in a Basis

Perhaps the most straightforward method of solving the time-dependent Schrödinger equation and of propagating the wave function forward in time is to expand the wave function in the set of eigenfunctions of the unperturbed Hamiltonian [41], $\hat{H}_0$, which is the Hamiltonian in the absence of the interaction with the laser field.

Thus the complete Hamiltonian may be written in the form:

$$\hat{H} = \hat{H}_0 + \hat{V}(\epsilon(t)) \tag{50}$$

If $\{\psi_i^0\}$ form the complete set of eigenfunctions of $\hat{H}_0$ with eigenvalues E_i^0, that is,

$$\hat{H}_0 \psi_i^0 = E_i^0 \psi_i^0 \tag{51}$$

and if we start off with a wave function $\Psi_0(t = 0) = \sum_i c_i \psi_i^0$ at $t = 0$. The solution of the unperturbed time-dependent Schrödinger equation

$$i\hbar \frac{\partial}{\partial t} \Psi_0(t) = \hat{H}_0 \Psi_0(t) \tag{52}$$

is $\Psi_0(t = 0) = \sum_i c_i \exp(-iE_i^0 t/\hbar)\psi_i^0$.

One way in which we can solve the problem of propagating the wave function forward in time in the presence of the laser field is to utilize the above knowledge. In order to solve the time-dependent Schrödinger equation, we normally divide the time period into small time intervals. Within each of these intervals we assume that the electric field and the time-dependent interaction potential is constant. The matrix elements of the interaction potential in the basis of the zeroth-order eigenfunctions ψ_i^0; $V_{ij} = \langle \psi_i^0 | V(\epsilon(t)) | \psi_j^0 \rangle$ are then evaluated and we can use an eigenvector routine to compute the eigenvectors, $\chi_\lambda = \sum_j b_{\lambda j} \psi_j^0$ of the full Hamiltonian, $\hat{H}(\epsilon(t))$, in the zeroth-order basis.

$$\sum_j \hat{H}(\epsilon(t))_{ij} b_{j\lambda}^T = E_\lambda(\epsilon(t)) b_{i\lambda}^T \tag{53}$$

where both $E_\lambda(\epsilon(t))$ and $b_{i\lambda}^T$ depend on the instantaneous value of the electric field $\epsilon(t)$.

If the small time interval in question extends from t to $t + \delta t$ then the matrix elements of the propagator for this short time interval may be written as a matrix in the basis of the zeroth-order eigenfunctions:

$$
\begin{aligned}
U(\epsilon, \delta t)_{ij} &= \langle \psi_i^0 | \exp[-i\hat{H}(\epsilon(t))\delta t/\hbar] | \psi_j^0 \rangle \\
&= \langle \psi_i^0 | \sum_\gamma |\chi_\gamma\rangle\langle\chi_\gamma| \exp[-i\hat{H}(\epsilon(t))\delta t/\hbar] \sum_\lambda |\chi_\lambda\rangle\langle\chi_\lambda| |\psi_j^0\rangle \\
&= \sum_\gamma \sum_\lambda \langle \psi_i^0 | \chi_\gamma\rangle\langle\chi_\gamma| \exp[-iE_\lambda \delta t/\hbar] |\chi_\lambda\rangle\langle\chi_\lambda | \psi_j^0\rangle \\
&= \sum_\gamma \langle \psi_i^0 | \chi_\gamma\rangle \{\exp[-iE_\gamma \delta t/\hbar]\} \langle\chi_\gamma | \psi_j^0\rangle \\
&= \sum_\gamma b_{\gamma i} \{\exp[-iE_\gamma \delta t/\hbar]\} b_{\gamma j}
\end{aligned}
\tag{54}
$$

where we have used the fact that within the space of the set of eigenfunctions $\{\chi_i\}$ the identity operator can be written in the form $\hat{I} = \sum_\lambda |\chi_\lambda\rangle\langle\chi_\lambda|$.

The short time propagator, $U(\epsilon, \delta t)_{ij}$ depends on the electric field strength, $\epsilon(t)$. This field strength oscillates very many times during a laser pulse. A valuable technique that has been proposed to minimize the computational cost of performing OCT calculations was proposed by Yip et al. [144]. This technique involves calculating and storing the matrix representation of the short-time propagator (as discussed above) for a range of field strengths. During the calculation the same field strength occurs many times. Instead of repeatedly calculating the propagator at each time step and for each field strength, the propagator for the closest stored value is selected and used to propagate the wave function forward in time by the constant time increment δt. This toolkit concept has been utilized in optimal control calculations for the vibrational excitation of H_2 [41].

VI. ANALYTIC DESIGN OF OPTIMAL LASER PULSES

One of the difficulties with optimal control theory is in identifying the underlying physical mechanism, or mechanisms, leading to control. Methods [2, 7, 9, 14, 26–29], that utilize a small number of interfering pathways reveal the mechanism by construction. On the other hand, while there have been many successful experimental and theoretical demonstrations of control based on OCT, there has been little analytical work to reveal the mechanism behind the complicated optimal pulses. In addition to reducing the complexity of the pulses, the many methods for imposing explicit restrictions on the pulses, see Section II.B, can also be used to dictate the mechanisms that will be operative. However, in this section we discuss some of the analytic approaches that have been used to understand the mechanisms of optimal control or to analytically design optimal pulses. Note that we will not discuss numerical methods that have been used to analyze control mechanisms [145–150].

A powerful tool in the analytic design and understanding of optimal pulses is the well-known pulse area theorem [151]. The pulse area theorem states that, within the rotating wave approximation (RWA) and for an on-resonance frequency, the final transition probability from one initial state to a final state in a two-level model is solely determined by the area of the pulse, A. For a dipole transition $A = \int_0^T \mu\epsilon_0 s(t)dt$, where μ is the transition dipole between the two states, T is the pulse duration, ϵ_0 is the field strength, and $s(t)$ is the pulse envelope. For a pulse with an area of π (a so-called π-pulse), 100% of the population will be transferred. It has been demonstrated [152] that within the RWA, the π-pulse is the globally optimized field for a two-level system. In a recent paper [153], Cheng and Brown formulated the design of laser pulses for performing quantum gate operations in terms of the analytic pulse area theorem rather than in the context of numerical OCT. In this work, it was shown that the π-pulse could be used to readily determine the effects of both laser parameters and molecular properties on the control process. Also, an analytic result was obtained emphasizing the critical role the penalty parameter β [see Eq. (2)], plays in determining the final population transfer (quantum gate fidelity).

The general concept of the π-pulse has recently been exploited in the analytical control of molecular excitation involving polarization effects [62]. In this paper, an analytic laser pulse was designed based on an adiabatic two-state approximation within the Floquet picture [154–156] to control population transfer between two rovibrational states of H_2. In this work, a general π-pulse involving the polarizability was introduced, (see Eq. (7) of Ref. [62]): Recall that for the homonuclear diatomic H_2 there are no dipole transitions. By using the general π-pulse, the proper time-dependent change in the laser frequency to account for the dynamic Stark shifts could be determined analytically. These analytic results were then confirmed via exact numerical simulations. Floquet theory has been very extensively used as a tool for designing laser pulses to perform molecular excitations and transformation processes [152, 157–159].

The π-pulse has proved extremely useful in interpreting population transfer between two levels. However, the combination of analytic and OCT techniques has also proven useful in multilevel systems. One of the simplest multilevel systems is the three-level Λ or ladder system where an intermediate state is dipole coupled with both the initial and final states, but there is no direct coupling between the initial and final states. Stimulated Raman adiabatic passage (STIRAP) [160] has proven to be a simple and robust method for transferring population directly from the initial to final state without ever significantly populating the intermediate state. However, the connection between OCT and STIRAP was difficult to obtain. The primary difficulty is that adiabatic passage is an energetically expensive method involving pulse areas many times π. Therefore, with penalty parameters, such as Eq. (2), restricting the total energy, the location of STIRAP using OCT is not favored. However, using a version of "local" optimization and a specially tailored objective functional, Tannor and co-workers showed that STIRAP could be obtained for three-level models [161, 162]. Using OCT, they were also able to obtain the alternating-STIRAP (A-STIRAP) mechanism [163] for population transfer in odd number of level systems where all pulses corresponding to even transitions precede all odd transitions. More interesting, the entirely new paradigm of straddling-STIRAP (S-STIRAP) for adiabatic population transfer in both odd and even level systems was identified from the OCT calculations, and then subsequently verified with an analytic formulation [162]. In S-STIRAP, there are partially overlapping pulses for the Stokes and pump transitions, that is, the transitions between the second to last and final states and between the initial and second states, respectively, straddled by pulses on-resonance with all of the intermediate transitions.

From these few examples, it is clear that OCT can be used to guide, or even identify, new methods for analytic pulse design. Clearly, this is important as one would like to understand the general mechanisms for control in addition to simply achieving the control objective. However, analytic pulse design can also be utilized to provide starting guidelines for OCT determinations of control fields. These can provide initial guess fields for numerical optimization but analytic methods could also be used to simplify the numerical effort required in the optimization.

VII. SUMMARY

This chapter has provided a brief overview of the application of optimal control theory to the control of molecular processes. It has addressed only the theoretical aspects and approaches to the topic and has not covered the many successful experimental applications [33, 37, 164–183], arising especially from the closed-loop approach of Rabitz [32]. The basic formulae have been presented and carefully derived in Section II and Appendix A, respectively. The theory required for application to photodissociation and unimolecular dissociation processes is also discussed in Section II, while the new equations needed in this connection are derived in Appendix B. An exciting related area of coherent control which has not been treated in this review is that of the control of bimolecular chemical reactions, in which both initial and final states are continuum scattering states [7, 14, 27–29, 184–188].

Due to their intrinsic nature the OCT equations must be solved iteratively. Their solution results in a laser pulse, specified by means of a time-dependent electric field, designed to bring about a particular, desired molecular transformation. The various methods that can be used for solving the OCT equations and the methods available for imposing physically desirable restrictions on the theoretically designed laser pulse are presented in Section III. In particular, the conjugate gradient method and iterative schemes are discussed and important details of these methods are presented in Appendixes C–F. These appendixes include a discussion of the different conjugate gradient methods available, of the Krotov iterative method and a proof that the generalized iterative method [see Eq. (21)] is guaranteed to converge monotonically.

The ENBO method, which is a method for incorporating polarization and higher order electric field–molecule interaction terms into the theory, is discussed in Section IV. Nearly all OCT experiments actually use laser pulses that give rise to strong electric fields that are sufficiently strong to significantly distort the molecular electronic structure and often lead to ionization. The ENBO approximation must be used with care, as it is only valid for sufficiently low field strengths such that ionization is not significant and for sufficiently low frequencies such that the electronic wave function can adjust instantaneously to the changing electric field. As a rough guide, the region of validity of the ENBO approximation may be taken to be for frequencies in the infrared (IR) regions or below and for field strengths up to $2.06 \times 10^8 \, \mathrm{V \, cm^{-1}}$ ($5.6 \times 10^{13} \, \mathrm{W \, cm^{-2}}$ or 0.04 a.u.). This chapter does not discuss the very good theoretical work aimed at understanding and modeling situations where the laser intensity is greater than these limits and where ionization plays an important role [106, 114, 189–192]. Some examples of the application of the ENBO method to the controlled excitation of the molecular motion of H_2 are also described in this section.

An essential part of the theoretical solution of the OCT equations is the solutions of the time-dependent Schrödinger equation. This must be carried out very many times during the iterative solution of the OCT equations. Methods available for the solution of these equations are discussed in Section V. The necessity of solving the time-dependent Schrödinger equation arises in many areas of chemistry and physics, in most cases, however, the Hamiltonian is independent of time. This expands the choice of methods available for solving the problem. Because of the presence of the fast oscillating electric field of the laser light, the time-steps taken in the solution of the equation must be sufficiently small to follow the oscillations of the electric field. This section discusses the split-operator method and a basis set expansion method for solving the time-dependent Schrödinger equation.

The penultimate section of the chapter, Section VI, briefly describes the role of analytic solutions to the problem of designing laser pulses to control molecular processes. The validity of the analytic solutions generally depend on the applicability of various approximations. These may include, for example, a two-level approximation in which only two quantum levels are taken to play a significant role in the process under discussion. While the underlying approximations of these analytic approaches are rarely totally valid, the results of the theories hold valuable lessons for the more general case where brute force OCT methods need to be applied. They can be invaluable in understanding control mechanisms and in appreciating the role of different variables, such as laser intensity and pulse duration, on achieving desired control objectives.

In this chapter, we have presented many of the relevant working equations of OCT including details of their derivations and methods for their solution. As such, we hope that this chapter will serve as a guidepost to those interested in entering this exciting research area and as a useful reference to those already pursuing work in the field.

APPENDIX A: VARIATION OF THE OBJECTIVE FUNCTIONAL AND DERIVATION OF THE OPTIMAL CONTROL EQUATIONS

The objective functional is written as [see Eq. (1)]:

$$
J = |\langle \psi(T)|\Phi \rangle|^2 - \beta \int_0^T dt\, \epsilon(t)^4 - 2\Re \left\{ \int_0^T dt\, \langle \chi(t)| \left[\frac{\partial}{\partial t} + \frac{i}{\hbar} \hat{H}(\epsilon(t)) \right] |\psi(t)\rangle \right\}
$$

$$(A.1)$$

where the meanings of the various quantities have been defined in the main body of the text below Eq. (1).

Expanding out the first and the third terms in Eq. (A.1) we can rewrite it as:

$$
J = \langle \psi(T)|\Phi \rangle \langle \Phi|\psi(T)\rangle - \beta \int_0^T dt\, \epsilon(t)^4
$$

$$
- \int_0^T dt \left\{ \langle \chi(t)| \frac{\partial}{\partial t} |\psi(t)\rangle + \frac{i}{\hbar} \langle \chi(t)|\hat{H}(\epsilon(t))|\psi(t)\rangle \right\}
$$

$$
- \int_0^T dt \left\{ \langle \chi(t)| \frac{\partial}{\partial t} |\psi(t)\rangle^* - \frac{i}{\hbar} \langle \chi(t)|\hat{H}(\epsilon(t))|\psi(t)\rangle^* \right\} \qquad (A.2)
$$

Now, consider independently small changes in $\epsilon(t)$, ψ and χ. These variations are now allowed to be independent because of the use of the Lagrange multiplier, $\chi(t)$ to impose the constraint equation.

First, consider a small change in ψ. We get

$$
\frac{\partial J}{\partial \psi(t)} \delta\psi(t) = \langle \delta\psi(T)|\Phi \rangle \langle \Phi|\psi(T)\rangle + \langle \psi(T)|\Phi \rangle \langle \Phi|\delta\psi(T)\rangle
$$

$$
- \int_0^T dt \left\{ \langle \chi(t)| \frac{\partial}{\partial t} |\delta\psi(t)\rangle + \frac{i}{\hbar} \langle \chi(t)|\hat{H}(\epsilon(t))|\delta\psi(t)\rangle \right\}
$$

$$
- \int_0^T dt \left\{ \langle \chi(t)| \frac{\partial}{\partial t} |\delta\psi(t)\rangle^* - \frac{i}{\hbar} \langle \chi(t)|\hat{H}(\epsilon(t))|\delta\psi(t)\rangle^* \right\} \qquad (A.3)
$$

Note that

$$
\langle \chi(t)| \frac{\partial}{\partial t} |\delta\psi(t)\rangle = \frac{\partial}{\partial t} \langle \chi(t)|\delta\psi(t)\rangle - \left\langle \frac{\partial}{\partial t} \chi(t)|\delta\psi(t) \right\rangle \qquad (A.4)
$$

By using Eq. (A-4) in Eq. (A-3) we obtain

$$
\begin{aligned}
\frac{\partial J}{\partial \psi(t)} \delta\psi(t) =& \langle \delta\psi(T)|\Phi\rangle\langle\Phi|\psi(T)\rangle + \langle\psi(T)|\Phi\rangle\langle\Phi|\delta\psi(T)\rangle \\
& - \int_0^T dt \left\{ \frac{\partial}{\partial t}\langle\chi(t)|\delta\psi(t)\rangle + \frac{\partial}{\partial t}\langle\chi(t)|\delta\psi(t)\rangle^* \right\} \\
& - \int_0^T dt \left\{ -\left\langle \frac{\partial}{\partial t}\chi(t)\Big|\delta\psi(t)\right\rangle + \frac{i}{\hbar}\langle\chi(t)|\hat{H}(\epsilon(t))|\delta\psi(t)\rangle \right\} \\
& - \int_0^T dt \left\{ -\left\langle \frac{\partial}{\partial t}\chi(t)\Big|\delta\psi(t)\right\rangle^* - \frac{i}{\hbar}\langle\chi(t)|\hat{H}(\epsilon(t))|\delta\psi(t)\rangle^* \right\} \\
=& \langle\delta\psi(T)|\Phi\rangle\langle\Phi|\psi(T)\rangle + \langle\psi(T)|\Phi\rangle\langle\Phi|\delta\psi(T)\rangle \\
& - \{\langle\chi(T)|\delta\psi(T)\rangle - \langle\chi(0)|\delta\psi(0)\rangle + \langle\chi(T)|\delta\psi(T)\rangle^* - \langle\chi(0)|\delta\psi(0)\rangle^*\} \\
& - \int_0^T dt \left\{ -\left\langle \frac{\partial}{\partial t}\chi(t)\Big|\delta\psi(t)\right\rangle + \frac{i}{\hbar}\langle\chi(t)|\hat{H}(\epsilon(t))|\delta\psi(t)\rangle \right\} \\
& - \int_0^T dt \left\{ -\left\langle \frac{\partial}{\partial t}\chi(t)\Big|\delta\psi(t)\right\rangle^* - \frac{i}{\hbar}\langle\chi(t)|\hat{H}(\epsilon(t))|\delta\psi(t)\rangle^* \right\} \quad (A.5)
\end{aligned}
$$

We set $\delta\psi(0) = 0$ as $\psi(t = 0) = \phi_i$, this being the initial boundary condition of the problem, so no variation is allowed for $\psi(t)$ at $t = 0$.

$$
\begin{aligned}
\frac{\partial J}{\partial \psi(t)} \delta\psi(t) =& \langle \delta\psi(T)|\Phi\rangle\langle\Phi|\psi(T)\rangle + \langle\psi(T)|\Phi\rangle\langle\Phi|\delta\psi(T)\rangle \\
& - \{\langle\chi(T)|\delta\psi(T)\rangle + \langle\chi(T)|\delta\psi(T)\rangle^*\} \\
& - \int_0^T dt \left\{ -\left\langle \frac{\partial}{\partial t}\chi(t)\Big|\delta\psi(t)\right\rangle - \left\langle \frac{i}{\hbar}\hat{H}(\epsilon(t))\chi(t)\Big|\delta\psi(t)\right\rangle \right\} \\
& - \int_0^T dt \left\{ -\left\langle \frac{\partial}{\partial t}\chi(t)\Big|\delta\psi(t)\right\rangle^* - \langle\frac{i}{\hbar}\hat{H}(\epsilon(t))\chi(t)|\delta\psi(t)\rangle^* \right\} \\
=& \langle\delta\psi(T)|\Phi\rangle\langle\Phi|\psi(T)\rangle + \langle\psi(T)|\Phi\rangle\langle\Phi|\delta\psi(T)\rangle \\
& - \{\langle\chi(T)|\delta\psi(T)\rangle + \langle\delta\psi(T)|\chi(T)\rangle\} \\
& - \int_0^T dt \left\{ -\left\langle \left\{\frac{\partial}{\partial t} + \frac{i}{\hbar}\hat{H}(\epsilon(t))\right\}\chi(t)\Big|\delta\psi(t)\right\rangle \right\} \\
& - \int_0^T dt \left\{ -\left\langle \left\{\frac{\partial}{\partial t} + \frac{i}{\hbar}\hat{H}(\epsilon(t))\right\}\chi(t)\Big|\delta\psi(t)\right\rangle^* \right\} \\
=& \langle\delta\psi(T)|\{|\Phi\rangle\langle\Phi|\psi(T)\rangle - |\chi(T)\rangle\}\} + \{\langle\psi(T)|\Phi\rangle\langle\Phi - \langle\chi(T)|\}|\delta\psi(T)\rangle \\
& - \int_0^T dt \left\{ -\left\langle \left\{\frac{\partial}{\partial t} + \frac{i}{\hbar}\hat{H}(\epsilon(t))\right\}\chi(t)\Big|\delta\psi(t)\right\rangle \right\} \\
& - \int_0^T dt \left\{ -\left\langle \left\{\frac{\partial}{\partial t} + \frac{i}{\hbar}\hat{H}(\epsilon(t))\right\}\chi(t)\Big|\delta\psi(t)\right\rangle^* \right\} \quad (A.6)
\end{aligned}
$$

Therefore we conclude that the variation of J with respect to an arbitrary change in ψ is zero if:

$$\left\{\frac{\partial}{\partial t} + \frac{i}{\hbar}\hat{H}(\epsilon(t))\right\}\chi(t) = 0 \qquad (A.7)$$

that is, $\chi(t)$ obeys the time-dependent Schrödinger equation. And if

$$\{|\Phi\rangle\langle\Phi|\psi(T)\rangle - |\chi(T)\rangle\} = 0 \qquad (A.8)$$

Or

$$\chi(T) = \Phi\langle\Phi|\psi(T)\rangle \qquad (A.9)$$

We can now undertake the simpler task of varying the other variables, $\chi(t)$ and ϵ. Variation of $\chi(t)$ gives

$$\begin{aligned}
\frac{\partial J}{\partial \chi(t)}\delta\chi(t) &= -\int_0^T dt\,\langle\delta\chi(t)|\left\{\frac{\partial}{\partial t} + \frac{i}{\hbar}\hat{H}(\epsilon(t))\right\}|\psi(t)\rangle \\
&\quad - \int_0^T dt\,\langle\delta\chi(t)|\left\{\frac{\partial}{\partial t} + \frac{i}{\hbar}\hat{H}(\epsilon(t))\right\}|\psi(t)\rangle^* \\
&= 0
\end{aligned} \qquad (A.10)$$

The variation of the variable $\chi(t)$ therefore gives, as expected, the condition that $\psi(t)$ must obey the time-dependent Schrödinger equation with the boundary condition $\psi(t = 0) = \phi_i$ that is:

$$\left\{\frac{\partial}{\partial t} + \frac{i}{\hbar}\hat{H}(\epsilon(t))\right\}\psi(t) = 0 \qquad (A.11)$$

with $\psi(t = 0) = \phi_i$.

Variation of ϵ yields

$$\begin{aligned}
\frac{\partial J}{\partial \epsilon(t)}\delta\epsilon(t) &= -4\,\beta\int_0^T dt\,\epsilon(t)^3\delta\epsilon(t) \\
&\quad - \int_0^T dt\left\{\frac{i}{\hbar}\langle\chi(t)|\frac{\partial\hat{H}}{\partial\epsilon(t)}\delta\epsilon(t)|\psi(t)\rangle\right\} \\
&\quad + \int_0^T dt\left\{\frac{i}{\hbar}\langle\chi(t)|\frac{\partial\hat{H}}{\partial\epsilon(t)}\delta\epsilon(t)|\psi(t)\rangle^*\right\} \\
&= \int_0^T dt\left\{-4\,\beta\epsilon(t)^3 - \frac{i}{\hbar}\left\{\langle\chi(t)|\frac{\partial\hat{H}}{\partial\epsilon(t)}|\psi(t)\rangle\right\} + \frac{i}{\hbar}\left\{\langle\chi(t)|\frac{\partial\hat{H}}{\partial\epsilon(t)}|\psi(t)\rangle^*\right\}\right\}\delta\epsilon(t) \\
&= \int_0^T dt\left\{-4\,\beta\epsilon(t)^3 + \frac{2}{\hbar}\,\text{Im}\left\{\langle\chi(t)|\frac{\partial\hat{H}}{\partial\epsilon(t)}|\psi(t)\rangle\right\}\right\}\delta\epsilon(t) \\
&= 0
\end{aligned} \qquad (A.12)$$

The variation of $\epsilon(t)$ therefore leads to the condition:

$$4\,\beta\epsilon(t)^3 = +\frac{2}{\hbar}\,\mathrm{Im}\left\{\langle\chi(t)|\frac{\partial\hat{H}}{\partial\epsilon(t)}|\psi(t)\rangle\right\} \tag{A.13}$$

Equations (A.7), (A.11), and (A.13), together with boundary conditions for the first two of these equations, form the basic equations that must be satisfied in order for the objective functional J to be a maximum.

Let us now examine what happens when we assume that the Hamiltonian for the interaction of the system with the electric field of the laser takes on different forms. The most common analytic form for the Hamiltonian is

$$\hat{H} = \hat{H}_0 - \mu\epsilon(t) \tag{A.14}$$

in this case;

$$\frac{\partial\hat{H}}{\partial\epsilon(t)} = -\mu \tag{A.15}$$

Substituting this into Eq. (A.13) we obtain

$$\epsilon(t)^3 = -\frac{\mathrm{Im}\{\langle\chi(t)|\mu|\psi(t)\rangle\}}{2\,\hbar\,\beta} \tag{A.16}$$

If we had used the penalty term $-\beta\int_0^T dt\,\epsilon(t)^2$ we would have obtained a similar equation to Eq. (A.16), but without the factor of 2 in the denominator.

Let us suppose that the system of interest does not possess a dipole moment; as in the case of a homonuclear diatomic molecule. In this case, the leading term in the electric field-molecule interaction involves the polarizability, α, and the Hamiltonian is of the form:

$$\hat{H} = \hat{H}_0 - \frac{\alpha}{2}\epsilon(t)^2 \tag{A.17}$$

with a Hamiltonian of this form, the derivative of the Hamiltonian becomes:

$$\frac{\partial\hat{H}}{\partial\epsilon(t)} = -\alpha\epsilon(t) \tag{A.18}$$

and Eq. (A-13) becomes:

$$\epsilon(t)^2 = -\frac{\mathrm{Im}\{\langle\chi(t)|\alpha|\psi(t)\rangle\}}{2\,\hbar\,\beta} \tag{A.19}$$

Use of the normal penalty term for the laser fluence in Eq. (A.1) [i.e., $-\beta \int_0^T dt\, \epsilon(t)^2$] would lead to the equation:

$$1 = -\frac{\mathrm{Im}\{\langle \chi(t)|\alpha|\psi(t)\rangle\}}{\beta\,\hbar} \tag{A.20}$$

The parameter $\epsilon(t)$ has now canceled out on the two sides of Eq. (A.13) and the equation is no longer an equation for $\epsilon(t)$. Our conclusion is that it is not appropriate to use the standard form for the fluence penalty term if the laser molecule interaction contains a significant contribution from the molecular polarizabilty.

APPENDIX B: OPTIMAL CONTROL EQUATIONS FOR PHOTODISSOCIATION

The objective functional for the optimal control of photodissociation may be defined as:

$$J = \int_0^T dt\langle\psi(t)|\Lambda\hat{F}|\psi(t)\rangle - \beta \int_0^T dt|\epsilon(t)|^2$$
$$- 2\Re\left\{\int_0^T dt\langle\chi(t)|\frac{\partial}{\partial t} + \frac{i}{\hbar}\hat{H}|\psi(t)\rangle\right\} \tag{B.1}$$

where $\hat{H}$ is the total Hamiltonian of the molecule in the presence of external control field $\epsilon(t)$; $\chi(t)$ is an undetermined Lagrange multiplier, ensuring that $\psi(t)$ obeys the time-dependent Schrödinger equation; β is the penalty associated with the pulse fluence; and Λ is a diagonal matrix whose elements Λ_n specify the weights of product channels $|n\rangle$ in the optimization calculation. The flux, or current density operator, $\hat{F}$ is defined as [193].

$$\hat{F} = \frac{i}{\hbar}\{\hat{H}, \hat{I}\,h(R_d)\}$$
$$= \frac{1}{2}\left\{\frac{\hat{p}}{m}\delta(R - R_d) + \delta(R - R_d)\frac{\hat{p}}{m}\right\} \tag{B.2}$$

where $\hat{F}$, $\hat{H}$, and $\hat{I}$ are all matrices each of whose rows and columns are associated with a particular product channel $|n\rangle$. h is the Heaviside function

$$h(\xi) = \begin{cases} 1, & \xi > 0 \\ 0, & \xi < 0 \end{cases} \tag{B.3}$$

and R_d is the value of the scattering coordinate on the dividing surface that separates the products from reactants.

As before, we now consider the variation of J with the small independent changes in ψ, χ, and ϵ. These variations are allowed to be independent because of the use of the undetermined Lagrange multiplier. Starting with the variation in $\psi(t)$, we have

$$\frac{\partial J}{\partial \psi} \delta\psi = \int_0^T dt\{\langle \delta\psi(t)|\Lambda\hat{F}|\psi(t)\rangle + \langle \psi(t)|\Lambda\hat{F}|\delta\psi(t)\rangle\}$$
$$- \int_0^T dt\left\{ \langle \chi(t)|\frac{\partial}{\partial t} + \frac{i}{\hbar}\hat{H}|\delta\psi(t)\rangle + \text{c.c.}\right\} \tag{B.4}$$

where c.c. indicates the complex conjugate. Noting that

$$\langle \chi(t)|\frac{\partial}{\partial t}|\delta\psi(t)\rangle = \frac{\partial}{\partial t}\langle \chi(t)|\delta\psi(t)\rangle - \langle \frac{\partial}{\partial t}\chi(t)|\delta\psi(t)\rangle \tag{B.5}$$

we have

$$\frac{\partial J}{\partial \psi}\delta\psi = \int_0^T dt\{\langle \delta\psi(t)|\Lambda\hat{F}|\psi(t)\rangle + \langle \psi(t)|\Lambda\hat{F}|\delta\psi(t)\rangle\}$$
$$- \langle \chi(T)|\delta\psi(T)\rangle + \langle \chi(0)|\delta\psi(0)\rangle$$
$$+ \int_0^T dt\left\langle \frac{\partial}{\partial t}\chi(t)|\delta\psi(t)\right\rangle - \int_0^T dt\langle \chi(t)|\frac{i}{\hbar}\hat{H}|\delta\psi(t)\rangle$$
$$- \langle \chi(T)|\delta\psi(T)\rangle^* + \langle \chi(0)|\delta\psi(0)\rangle^*$$
$$+ \int_0^T dt\left\langle \frac{\partial}{\partial t}\chi(t)|\delta\psi(t)\right\rangle^* - \int_0^T dt\langle \chi(t)|\frac{i}{\hbar}\hat{H}|\delta\psi(t)\rangle^*$$
$$= -\langle \chi(T)|\delta\psi(T)\rangle - \langle \chi(T)|\delta\psi(T)\rangle^*$$
$$+ \int_0^T dt\{\langle \delta\psi(t)|\Lambda\hat{F}|\psi(t)\rangle + \langle \psi(t)|\Lambda\hat{F}|\delta\psi(t)\rangle\}$$
$$+ \int_0^T dt\left\langle \frac{\partial}{\partial t}\chi(t)|\delta\psi(t)\right\rangle - \int_0^T dt\langle \chi(t)|\frac{i}{\hbar}\hat{H}|\delta\psi(t)\rangle$$
$$+ \int_0^T dt\left\langle \frac{\partial}{\partial t}\chi(t)|\delta\psi(t)\right\rangle^* - \int_0^T dt\langle \chi(t)|\frac{i}{\hbar}\hat{H}|\delta\psi(t)\rangle^* \tag{B.6}$$

In Eq. (B.6), we have set that $\delta\psi(0) = 0$ and dropped the terms $\langle \chi(0)|\delta\psi(0)\rangle$ and $\langle \chi(0)|\delta\psi(0)\rangle^*$. This is because $\psi(0) = \phi_i$ being the initial condition of the problem and no variation is allowed at $t = 0$ in ψ. If the Hamiltonian $\hat{H}$ is a hermitian operator, the flux operator $\hat{F}$ will also be hermitian. The above equation can then be rewritten as

$$
\frac{\partial J}{\partial \psi} \delta\psi = -\langle \chi(T)|\delta\psi(T)\rangle - \langle \chi(T)|\delta\psi(T)\rangle^*
$$

$$
+ \int_0^T dt \{\langle \delta\psi(t)|\Lambda\hat{F}\psi(t)\rangle + \langle \Lambda\hat{F}\psi(t)|\delta\psi(t)\rangle\}
$$

$$
+ \int_0^T dt \left\langle \left[\frac{\partial}{\partial t} + \frac{i}{\hbar}\hat{H}\right]\chi(t)|\delta\psi(t)\right\rangle
$$

$$
+ \int_0^T dt \left\langle \left[\frac{\partial}{\partial t} + \frac{i}{\hbar}\hat{H}\right]\chi(t)|\delta\psi(t)\right\rangle^* \tag{B.7}
$$

The maximal condition for J requires that $\delta J = 0$ for an arbitrary $\delta\psi$. This in turn requires that the following equations be obeyed

$$
i\hbar\frac{\partial}{\partial t}\chi(t) = \hat{H}\chi(t) - i\hbar\Lambda\hat{F}\psi(t) \tag{B.8}
$$

with the boundary condition

$$
\chi(t = T) = 0. \tag{B.9}
$$

Variation of χ yields

$$
\frac{\partial J}{\partial \chi}\delta\chi = -2\Re\left\{\int_0^T dt\langle\delta\chi(t)|\frac{\partial}{\partial t} + \frac{i}{\hbar}\hat{H}|\psi(t)\rangle\right\}
$$

$$
= 0 \tag{B.10}
$$

Therefore the variation gives, as expected, the condition that $\psi(t)$ must obey the time-dependent Schrödinger equation

$$
i\hbar\frac{\partial}{\partial t}\psi(t) = \hat{H}\psi(t) \tag{B.11}
$$

with the boundary condition

$$
\psi(t = 0) = \phi_i. \tag{B.12}
$$

The variation of $\epsilon(t)$ gives

$$
\frac{\partial J}{\partial \epsilon}\delta\epsilon = -2\beta\int_0^T dt\,\epsilon(t)\cdot\delta\epsilon(t)
$$

$$
- 2\Re\left\{\int_0^T dt\,\langle\chi(t)|\frac{i}{\hbar}\frac{\partial\hat{H}}{\partial\epsilon}\cdot\delta\epsilon(t)|\psi(t)\rangle\right\}
$$

$$
= 0 \tag{B.13}
$$

Then the optimal field $\epsilon(t)$ is written as

$$\epsilon(t) = \frac{1}{\hbar\beta}\mathrm{Im}\langle\chi(t)|\frac{\partial\hat{H}}{\partial\epsilon(t)}|\psi(t)\rangle. \tag{B.14}$$

and the gradient of J with respect to ϵ becomes:

$$\frac{\partial J}{\partial\epsilon(t)} = -2\beta\epsilon(t) + \frac{2}{\hbar}\mathrm{Im}\langle\chi(t)|\frac{\partial\hat{H}}{\partial\epsilon(t)}|\psi(t)\rangle. \tag{B.15}$$

We may express the total Hamiltonian $\hat{H}$ as

$$\hat{H} = \hat{H}_0 - \boldsymbol{\mu}\cdot\epsilon(t) \tag{B.16}$$

where $\hat{H}_0$ is the field-free Hamiltonian of molecule, $\boldsymbol{\mu}$ is the dipole moment function that includes both the permanent and transition dipole moments. The optimal field may be rewritten as

$$\epsilon(t) = -\frac{1}{\hbar\beta}\mathrm{Im}\langle\chi(t)|\boldsymbol{\mu}|\psi(t)\rangle \tag{B.17}$$

The *new* pulse design equations for the optimal control of photodissociation may be summarized as

$$i\hbar\frac{\partial}{\partial t}\psi(t) = \hat{H}\psi(t), \quad \psi(0) = \phi_i \tag{B.18a}$$

$$i\hbar\frac{\partial}{\partial t}\chi(t) = \hat{H}\chi(t) - i\hbar\Lambda\hat{F}\psi(t), \quad \chi(T) = 0 \tag{B.18b}$$

$$\epsilon(t) = -\frac{1}{\hbar\beta}\mathrm{Im}\langle\chi(t)|\boldsymbol{\mu}|\psi(t)\rangle \tag{B.18c}$$

APPENDIX C: DERIVATIVE OF THE OBJECTIVE FUNCTIONAL

To carry out the optimization of the objective functional, J, we need to calculate the derivative of the functional with respect to the field at a specified time. From Eq. (A.12), this is clearly given by

$$\frac{\partial J}{\partial\epsilon(t)} = \int_0^T dt\left\{-4\,\beta\epsilon(t)^3 + \frac{2}{\hbar}\,\mathrm{Im}\left\{\langle\chi(t)\left|\frac{\partial\hat{H}}{\partial\epsilon(t)}\right|\psi(t)\rangle\right\}\right\} \tag{C.1}$$

If the time is divided into small finite increments of δt, and the field is considered to possess a fixed value during each of these finite increments of time, then we can vary the field strength during each of these increments separately and obtain the equation:

$$\left[\frac{\partial J}{\partial \epsilon(t)}\right]_{t=t_i} = \left\{ -4\,\beta\epsilon(t_i)^3 + \frac{2}{\hbar}\,\mathrm{Im}\left\{ \langle\chi(t_i)| \left[\frac{\partial \hat{H}}{\partial \epsilon(t)}\right]_{t=t_i} |\Psi(t_i)\rangle \right\} \right\} \delta t \qquad \text{(C.2)}$$

APPENDIX D: VARIOUS CONJUGATE GRADIENT METHODS

Conjugate gradient (CG) methods comprise a class of unconstrained optimization algorithms that are characterized by low memory requirements and strong local and global convergence properties. A nonlinear conjugate gradient method generates a sequence of fields ϵ_k, $k \geq 1$, starting from an initial guess ϵ_0, using the formula

$$\epsilon_{k+1} = \epsilon_k + \lambda_k \mathbf{d}_k \qquad \text{(D.1)}$$

where the positive step size λ_k is obtained by a line search, and the directions $\mathbf{d}_k$ are generated by the rule

$$\mathbf{d}_{k+1} = \mathbf{g}_{k+1} + \zeta_k \mathbf{d}_k, \qquad \mathbf{d}_0 = \mathbf{g}_0 \qquad \text{(D.2)}$$

Here ζ_k is the CG update parameter. In the above equations, $\epsilon_k = \{\epsilon^k(t_j)\}_{0 \leq j \leq N}$ is the vector notation for the discretized electric field strength, $\mathbf{g}_k = \{g^k(t_j)\}_{0 \leq j \leq N}$ for the gradient of objective functional J with respect to the field strength (evaluated at a field strength of ϵ_k) and $\mathbf{d}_k = \{d^k(t_j)\}_{0 \leq j \leq N}$ for the search direction at the kth iteration. The time has been discretized into N time steps, such as that $t_j = j \times \delta t$, where $j = 0, 1, 2, \cdots, N$. Different CG methods correspond to different choices for the scalar ζ_k.

Let $\| \cdot \|$ denote the Euclidean norm and define $\mathbf{y}_k = \mathbf{g}_{k+1} - \mathbf{g}_k$. Table I provides a chronological list of some choices for the CG update parameter. If the objective function is a strongly convex quadratic, then in theory, with an exact line search, all seven choices for the update parameter in Table I are equivalent. For a nonquadratic objective functional J (the ordinary situation in optimal control calculations), each choice for the update parameter leads to a different performance. A detailed discussion of the various CG methods is beyond the scope of this chapter. The reader is referred to Ref. [194] for a survey of CG methods. Here we only mention briefly that despite the strong convergence theory that has been developed for the Fletcher–Reeves, [195],

TABLE I

Various Choices for the CG Update Parameter

Update Parameter	Year	Origin
$\zeta_k^{HS} = \dfrac{g_{k+1}^T y_k}{d_k^T y_k}$	1952	From the original (linear) CG paper of Hestenes and Stiefel (Ref. [198])
$\zeta_k^{FR} = \dfrac{\\| g_{k+1} \\|^2}{\\| g_k \\|^2}$	1964	First nonlinear CG method, proposed by Fletcher and Reeves (Ref. [195])
$\zeta_k^{PRP} = \dfrac{g_{k+1}^T y_k}{\\| g_k \\|^2}$	1969	Proposed by Polak and Ribière and by Polak (Ref. [100])
$\zeta_k^{CD} = \dfrac{\\| g_{k+1} \\|^2}{-d_k^T g_k}$	1987	Proposed by Fletcher (Ref. [196]), CD stands for "Conjugate Descent"
$\zeta_k^{LS} = \dfrac{g_{k+1}^T y_k}{-d_k^T g_k}$	1991	Proposed by Liu and Storey (Ref. [199])
$\zeta_k^{DY} = \dfrac{\\| g_{k+1} \\|^2}{d_k^T y_k}$	1999	Proposed by Dai and Yuan (Ref. [197])
$\zeta_k^{HZ} = \left(y_k - 2d_k \dfrac{\\| y_k \\|^2}{d_k^T y_k} \right) \dfrac{g_{k+1}}{d_k^T y_k}$	2005	Proposed by Hager and Zhang (Ref. [194])

(FR) Conjugate Descent [196] (CD) and Dai–Yuan [197] (DY) methods, these methods are all susceptible to jamming (i.e., they begin to take small steps without making significant progress to the maximum). The Polak–Ribière–Polyak [100] (PRP), Hestenes–Stiefel [198] (HS), and Liu–Storey [199] (LS) methods possess a built-in restart feature that addresses the jamming problem: When the factor $y_k = (g_{k+1} - g_k)$ in the numerator of ζ^k [see Eq. (12)] is small and ζ^k consequently is also small; the new search direction d_{k+1} becomes essentially the gradient direction g_{k+1}. In general, the performance of these methods are better than that of the FR, CD, and DY methods. On the other hand, Powell [200] showed, using a three-dimensional example, that with an exact line search, the PRP method could sometimes get into an infinite cycle without converging to a stationary point. He suggested that the following modification be made to the update parameter for the PRP method:

$$\zeta_k^{PRP+} = \max \left(\zeta_k^{PRP}, 0 \right) \tag{D.3}$$

ensuring a positive value for the parameter.

Note that with an exact line search, $\zeta_k^{HS} = \zeta_k^{LS} = \zeta_k^{PRP}$. Hence, the convergence properties of the HS and LS methods should be similar with the PRP method.

APPENDIX E: DETAILED DESCRIPTION
OF KROTOV ITERATIVE METHOD FOR SOLVING
THE OPTIMAL CONTROL EQUATIONS

Adopting the time discretization described in Section III.A (i.e., $t_j = j \times \delta t$, where $j = 0, 1, 2, \ldots, N$ and $N\delta t = T$) and the first-order split-operator method for the wavepacket propagation, the algorithm for the Krotov iteration method [81,93] may be written as:

Step 1: Guess an initial electric field $\bar{\epsilon}(t)$.

Step 2: Set $k = 0$, $\epsilon^k = \bar{\epsilon}(t)$ and $\psi^k(t_0) = \phi_i$.

Step 3: Propagate $\psi^k(t)$ forward in time up to T according Eq. (18.a) with the field $\epsilon^k(t)$ to obtain $\psi^k(t_N)$.

Step 4: Evaluate the objective functional J^k according to Eq. (1). Note that the last term in the equation is zero, as $\psi^k(t)$ is a solution of the time-dependent Schrödinger equation.

Step 5: If $k \geq 1$, compute $\delta J^k = J^k - J^{k-1}$ and compare δJ^k with the convergence threshold, ϵ. If $\delta(J^k) \leq \epsilon$, then stop the iteration and declare that the optimal pulse has been obtained.

Step 6: Set $\chi^k(t_N) = \langle \Phi | \psi^k(t_N) \rangle \Phi$.

Step 7: Propagate $\chi^k(t)$ backward in time according to Eq. (18.c) with the field $\epsilon^k(t)$ to obtain $\chi^k(t_0)$.

Step 8.1: Set $j = 0$ and $\psi^{k+1}(t_0) = \phi_i$.

Step 8.2: Compute the "new" electric field $\epsilon^{k+1}(t_j)$ from $\chi^k(t_j)$ and $\psi^{k+1}(t_j)$ according to Eq. (18.b)

$$\epsilon^{k+1}(t_j) = -\frac{1}{\beta\hbar} \text{Im}\langle \chi^k(t_j) | \mu | \psi^{k+1}(t_j) \rangle \tag{E.1}$$

Step 8.3: Calculate $\psi^{k+1}(t_{j+1})$ with the "new" field $\epsilon^{k+1}(t_j)$ as

$$\psi^{k+1}(t_{i+1}) = e^{-i\hat{H}_0\delta t/\hbar} e^{+i[\mu\epsilon^{k+1}(t_i)]\delta t/\hbar} \psi^{k+1}(t_i). \tag{E.2}$$

Step 8.4: To avoid prohibitively large use of computer memory in storing $\chi^k(t_j)_{0 \leq j \leq N}$ at every time step during the backward propagation in step 7, $\chi^k(t)$ is propagated simultaneously as well with the "old" electric field $\epsilon^k(t)$. $\chi^k(t_{j+1})$ is calculated as

$$\chi^k(t_{j+1}) = e^{-i\hat{H}_0\delta t/\hbar} e^{+i[\mu\epsilon^k(t_j)]\delta t/\hbar} \chi^k(t_j). \tag{E.3}$$

Step 8.5: If $j \geq N$, set $k = k + 1$ and go back to Step 4; else, set $j = j + 1$ and go back to Step 8.2.

It can be seen from the algorithm model stated above that in the Krotov method the electric field obtained in the kth iteration is used immediately to propagate $\psi(t)$, which has a direct contribution to the new electric field in the next time step. In one iteration, the Krotov method involves three wave packet propagations, that is, the forward propagations of $\psi^{k+1}(t)$ and $\chi^k(t)$ in Steps 8.3 and 8.4 and the backward propagation of $\chi^k(t)$ in Step 7. If the second-order split-operator method is employed for the wave packet propagation, a more efficient discrete implementation of the Krotov method may be generated. The second-order scheme for the Krotov method coincides with the first-order scheme described above except for Eqs. (E.1–E.3) in Steps 8.2, 8.3, and 8.4. The new equations for these steps now become:

Step 8.2: Calculate the "new" electric field $\epsilon^{k+1}(t + \delta t/2)$

$$\epsilon^k(t_j + \delta/2) = \frac{1}{\beta\hbar}\text{Im}\langle\chi^k(t_j)|\mu + \frac{i\delta t}{2\hbar}[\hat{H}_0,\mu]|\psi^{k+1}(t_j)\rangle \qquad \text{(E.4)}$$

Step 8.3: Propagate $\psi^{k+1}(t_j)$ forward one step in time with the "new" field $\epsilon^{k+1}(t_j + \delta t/2)$,

$$\psi^{k+1}(t_{j+1}) = e^{-i\hat{H}_0\delta t/2\hbar}e^{+i[\mu\epsilon^{k+1}(t_j+\delta t/2)]\delta t/\hbar}e^{-i\hat{H}_0\delta t/2\hbar}\psi^{k+1}(t_j) \qquad \text{(E.5)}$$

Step 8.4: $\chi^k(t_{j+1})$ is calculated as

$$\chi^k(t_{j+1}) = e^{-i\hat{H}_0\delta t/2\hbar}e^{+i[\mu\epsilon^k(t_j+\delta t/2)]\delta t/\hbar}e^{-i\hat{H}_0\delta t/2\hbar}\chi^k(t_j) \qquad \text{(E.6)}$$

Equation (E.4) may be obtained by considering the Taylor expansion of the laser field up to first order

$$\epsilon(t + \delta t) = \epsilon(t) + \frac{\partial\epsilon(t)}{\partial t}\delta t \qquad \text{(E.7)}$$

and evaluating the time derivative $\partial\epsilon(t)/\partial t$ according to Eq. (18.b).

Roughly, in order to reach the same accuracy, a time step in the second-order scheme can be 10 times larger than that in the first-order scheme, while each iteration step in the second-order scheme will cost about three times as much as that in the first-order scheme [i.e., the cost of second-order propagator is about twice as much as that of first-order propagator and the cost for computing the "new" field according to Eq. (E.4) is about the same as that of a single propagation with the first-order propagator]. Thus we anticipate that the efficiency of the second-order scheme will about three times higher than that of the first-order scheme. Similarly, discrete iteration schemes may be established for the methods suggested by Rabitz et al. [44, 92] and Maday and Turinici [94].

APPENDIX F: CONVERGENCE OF THE ITERATIVE SOLUTION
OF THE OPTIMAL CONTROL THEORY EQUATIONS

In this appendix we follow the treatment of Maday and Turinici [94], and show that the iterative scheme laid out in Eqs. (21.a–d) is guaranteed to converge. The convergence of the algorithm can be proved by evaluating the difference between the values of the objective functional between two successive iterations. Suppose that $\delta \neq 0$ and $\eta \neq 0$, then,

$$
\begin{aligned}
J(\epsilon^{k+1}) - J(\epsilon^k) &= |\langle \psi^{k+1}(T)|\Phi\rangle|^2 - \beta \int_0^T \epsilon^{k+1}(t)^2 dt - |\langle \psi^k(T)|\Phi\rangle|^2 + \beta \int_0^T \epsilon^k(t)^2 dt \\
&= |\langle(\psi^{k+1}(T) - \psi^k(T))|\Phi\rangle|^2 + 2\mathrm{Re}\{\langle(\psi^{k+1}(T) - \psi^k(T))|\Phi\rangle\langle\Phi|\psi^k(T)\rangle\} \\
&\quad - \beta \int_0^T \epsilon^{k+1}(t)^2 + \beta \int_0^T \epsilon^k(t)^2
\end{aligned}
\tag{F.1}
$$

By using Eqs. (21.a–d) we can show that the second term in Eq. (F.1) above can be rewritten in the form:

$$
\begin{aligned}
2\mathrm{Re}&\{\langle(\psi^{k+1}(T) - \psi^k(T))|\Phi\rangle\langle\Phi|\psi^k(T)\rangle\} = 2\mathrm{Re}\{\langle(\psi^{k+1}(T) - \psi^k(T))|\chi^k(T)\rangle\} \\
&= 2\mathrm{Re}\left\{\int_0^T \left[\left\langle \frac{\partial(\psi^{k+1}(t) - \psi^k(t))}{\partial t}\Big|\chi^k(t)\right\rangle + \left\langle (\psi^{k+1}(t) - \psi^k(t))\Big|\frac{\partial\chi^k(t)}{\partial t}\right\rangle\right] dt\right\} \\
&= 2\mathrm{Re}\left\{\int_0^T \left[\left\langle \left(\frac{\hat{H}_0 - \mu\epsilon^{k+1}(t)}{i\hbar}\psi^{k+1}(t) - \frac{\hat{H}_0 - \mu\epsilon^k(t)}{i\hbar}\psi^k(t)\right)\Big|\chi^k(t)\right\rangle \right.\right. \\
&\quad \left.\left. + \left\langle (\psi^{k+1}(t) - \psi^k(t))\Big|\frac{\hat{H}_0 - \mu\tilde{\epsilon}^k(t)}{i\hbar}\chi^k(t)\right\rangle\right] dt\right\} \\
&= 2\mathrm{Re}\left\{\int_0^T \left[\epsilon^{k+1}(t)\left\langle \frac{-\mu}{i\hbar}\psi^{k+1}(t)\Big|\chi^k(t)\right\rangle - \epsilon^k(t)\left\langle \frac{-\mu}{i\hbar}\psi^k(t)\Big|\chi^k(t)\right\rangle \right.\right. \\
&\quad \left.\left. + \tilde{\epsilon}^k(t)\left\langle (\psi^{k+1}(t) - \psi^k(t))\Big|\frac{-\mu}{i\hbar}\chi^k(t)\right\rangle\right] dt\right\} \\
&= \frac{2}{\hbar}\int_0^T [\epsilon^{k+1}(t)\mathrm{Im}\{\langle\psi^{k+1}(t)|\mu|\chi^k(t)\rangle\} - \epsilon^k(t)\mathrm{Im}\{\langle\psi^k(t)|\mu|\chi^k(t)\rangle\} \\
&\quad - \tilde{\epsilon}^k(t)\mathrm{Im}\{\langle\psi^{k+1}(t)|\mu|\chi^k(t)\rangle\} + \tilde{\epsilon}^k(t)\mathrm{Im}\{\langle\psi^k(t)|\mu|\chi^k(t)\rangle\}] \, dt \\
&= \frac{2}{\hbar}\int_0^T [-\epsilon^{k+1}(t)\mathrm{Im}\{\langle\chi^k(t)|\mu|\psi^{k+1}(t)\rangle\} + \epsilon^k(t)\mathrm{Im}\{\langle\chi^k(t)|\mu|\psi^k(t)\rangle\} \\
&\quad + \tilde{\epsilon}^k(t)\mathrm{Im}\{\langle\chi^k(t)|\mu|\psi^{k+1}(t)\rangle\} - \tilde{\epsilon}^k(t)\mathrm{Im}\{\langle\chi^k(t)|\mu|\psi^k(t)\rangle\}] \, dt \\
&= 2\beta\left\{\int_0^T \left[\epsilon^{k+1}(t)\frac{(\epsilon^{k+1} - (1-\delta)\tilde{\epsilon}^k)}{\delta} - \epsilon^k(t)\frac{(\tilde{\epsilon}^k - (1-\eta)\epsilon^k)}{\eta} \right.\right. \\
&\quad \left.\left. - \tilde{\epsilon}^k(t)\frac{(\epsilon^{k+1} - (1-\delta)\tilde{\epsilon}^k)}{\delta} + \tilde{\epsilon}^k(t)\frac{(\tilde{\epsilon}^k - (1-\eta)\epsilon^k)}{\eta}\right] dt\right\}
\end{aligned}
\tag{F.2}
$$

By substituting this back into Eq. (F.1) we obtain

$$J(\epsilon^{k+1}) - J(\epsilon^k) = |\langle \psi^{k+1}(T) - \psi^k(T)|\Phi\rangle|^2 + \beta \int_0^T dt \left(\frac{2}{\delta} - 1\right)(\epsilon^{k+1} - \tilde{\epsilon}^k)^2$$

$$+ \left(\frac{2}{\eta} - 1\right)(\tilde{\epsilon}^k - \epsilon^k)^2 \qquad (F.3)$$

which is positive for any $\eta, \delta \in [0, 2]$. Note that the case $\delta = 0$ yields $\epsilon^{k+1} = \tilde{\epsilon}^k$ and $\eta = 0$ yields $\tilde{\epsilon}^k = \epsilon^k$. In both these cases, the conclusion that the right-hand side of Eq. (F.3) is positive remains valid. Each step of this algorithm will therefore result in an increase of the value of the objective functional for any $\eta, \delta \in [0, 2]$. Then we have that for any $\eta, \delta \in [0, 2]$ the algorithm given in Eqs. (21.a–d) converges monotonically in the sense that $J(\epsilon^{k+1}) > J(\epsilon^k)$.

Acknowledgments

The authors GGBK and SZ are grateful to the EPSRC for financial support. GGBK would like to thank H. Rabitz, Tak-San Ho, and M. Artamonov who were instrumental in introducing him to this field of research. He is also grateful to F.R. Manby and Qinghua Ren for many valuable discussions and ideas and for their collaboration in much of his initial work in this field. GGBK and SZ also thank D.J. Tannor for useful discussions. AB thanks the Natural Sciences and Engineering Research Council of Canada and the Canadian Foundation for Innovation for financial support. He thanks Dr. Taiwang Cheng for many useful discussions and insights on optimal control theory.

References

1. S. A. Rice and M. Zhao, *Optical Control of Molecular Dynamics*, Wiley-Interscience, New York, 2000.

2. M. Shapiro and P. Brumer, *Principles of the Quantum Control of Molecular Processes*, John Wiley & Sons, Canada, Ltd., 2003.

3. A. Bandrauk, *Molecules in Laser Fields*, Nato ASI series C *Mathematical and Physical Sciences*, Kluwer Academic Publishers, 1995.

4. N. P. Moore, G. M. Menkir, A. N. Markevitch, and P. G. R. J. Levis, in *Laser Control and Manipulation of Molecules*, R. J. Gordon (ed.), American Chemical Society, Washington, DC, 2001, ACS Symposium Series in Chemistry.

5. *Coherent Control of Photochemical and Photobiological Systems, J. Photochem. Photobiol. A*, **180**(3), 225–334 (2006).

6. *Coherent Control with Femtosecond Laser Pulses, Eur. Phys. J. Sci. D*, **14**(2), (2001).

7. D. J. Tannor and S. A. Rice, *Adv. Chem. Phys.* **70**, 441 (1988).

8. H. Rabitz and S. Shi, in *Advances in Molecular Vibrations and Collision Dynamics*, J. Bowman (ed.), JAI Press 1991, vol. 1, p. 187.

9. P. Brumer and M. Shapiro, *Ann. Rev. Phys. Chem.* **43**, 257 (1992).

10. R. Gordon and S. A. Rice, *Ann. Rev. Phys. Chem.* **48**, 601 (1997).

11. H. Rabitz and W. Zhu, *Acc. Chem. Res.* **33**, 572 (2000).

12. S. A. Rice and S. Shah, *Phys. Chem. Chem. Phys.* **4**, 1683 (2002).

13. A. Auger, A. Yedder, E. Cances, C. le Bris, C. Dion, A. Keller, and O. Atabek, *Math. Models Methods Appl. Sci.* **12**, 1281 (2002).

14. M. Shapiro and P. Brumer, *Phys. Rep.* **425**, 195 (2006).

15. M. Shapiro and P. Brumer, *Rep. Prog. Phys.* **66**, 859 (2003).

16. M. Shapiro and P. Brumer, *Adv. At. Mol. Opt. Phys.* **42**, 287 (2000).

17. H. Rabitz, *Theor. Chem. Acc.* **109**, 64 (2003).

18. H. Rabitz, R. de Vivie-Riedle, M. Motzkus, and K. Kompa, *Science*, **288**, 824 (2000).

19. B. Kohler, J. L. Krause, F. Raksi, K. R. Wilson, V. V. Yakovlev, R. M. Whitnell, and Y. Yan, *Acc. Chem Res.* **28**, 133 (1995).

20. M. Shapiro and P. Brumer, *Int. Rev. Phys. Chem.* **31**, 187 (1994).

21. M. Dantus and V. V. Lozovoy, *Chem. Rev.* **104**, 1813 (2004).

22. T. Brixner and G. Gerber, *Chem. Phys. Chem.* **4**, 418 (2003).

23. T. Brixner, N. H. Damrauer, and G. Gerber, *Adv. Atom. Mol. Opt. Phys.* **46**, 1 (2001).

24. A. M. Weiner, *Rev. Sci. Instr.* **71**, 1929 (2000).

25. H. Kawashima, M. M. Wefers, and K. A. Nelson, *Annu. Rev. Phys. Chem.* **46**, 625 (1995).

26. P. Brumer and M. Shapiro, *Chem. Phys. Lett.* **126**, 541 (1986).

27. D. J. Tannor and S. A. Rice, *J. Chem. Phys.* **83**, 5013 (1985).

28. D. J. Tannor, R. Kosloff, and S. A. *Rice, J. Chem. Phys.* **85**, 5805 (1986).

29. R. Kosloff, S. A. Rice, P. Gaspard, S. Tersigni, and D. J. Tannor, *Chem. Phys.* **139**, 201 (1989).

30. A. P. Peirce, M. A. Dahleh, and H. Rabitz, *Phys. Rev., A* **37**, 4950 (1988).

31. A. P. Peirce, M. A. Dahleh, and H. Rabitz, *Phys. Rev., A* **42**, 1065 (1990).

32. R. S. Judson and H. Rabitz, *Phys. Rev. Lett.* **68**, 1500 (1992).

33. A. Assion, T. Baumert, M. Bergt, T. Brixner, B. Kiefer, V. Seyfried, M. Strehle, and G. Gerber, *Science* **282**, 919 (1998).

34. T. C. Weinacht, J. L. White, and P. H. Bucksbaum, *J. Phys. Chem. A* **103**, 10166 (1999).

35. R. Bartels, S. Backus, E. Zeek, L. Misoguti, G. Vdovin, I. P. Christov, M. M. Murnane, and H. C. Kapteyn, *Nature* (London) **406**, 164 (2000).

36. T. Hornung, R. Meier, and M. Motzkus, *Chem. Phys. Lett.* **326**, 445 (2000).

37. R. J. Levis, G. M. Menkir, and H. Rabitz, *Science*, **292**, 709 (2001).

38. Š. Vajda, A. Bartelt, E. Kaposta, T. Leisner, C. Lupulescu, S. Minemoto, P. Rosendo-Francisco, and L. Wöste, *Chem. Phys.* **267**, 231 (2001).

39. C. J. Bardeen, V. V. Yakovlev, K. R. Wilson, S. D. Carpenter, P. M. Weber, and W. S. Warren, *Chem. Phys. Lett.* **280**, 151 (1997).

40. T. Baumert, T. Brixner, V. Seyfried, M. Strehle, and G. Gerber, *Appl. Phys. B* **65**, 779 (1997).

41. G. G. Balint-Kurti, F. Manby, Q. Ren, M. Artamonov, T. Ho, H. Rabitz, *J. Chem. Phys.* **122**, 084110 (2005). Note that a minus sign should be inserted on the right-hand side of Eqs. (15) and (16) of this paper to make them consistent with later equations.

42. Q. Ren, G. G. Balint-Kurti, F. Manby, M. Artamonov, T. Ho, and H. Rabitz, *J. Chem. Phys.* **124**, 014111 (2006).

43. S. Shi and H. Rabitz, *J. Chem. Phys.* **92**, 364 (1990).

44. W. Zhu and H. Rabitz, *J. Chem. Phys.* **109**, 385 (1998).

45. W. Zhu and H. Rabitz, *Phys. Rev. A* **58**, 4741 (1998).

90   GABRIEL G. BALINT-KURTI, SHIYANG ZOU AND ALEX BROWN

46. Y. Ohtsuki, K. Nakagami, Y. Fujimura, W. S. Zhu, and H. Rabitz, *J. Chem. Phys.* **114**, 8867 (2001).
47. C. Tesch and R. de Vivie-Riedle, *Phys. Rev. Lett.* **89**, 157901 (2002).
48. J. Palao and R. Kosloff, *Phys. Rev. Lett.* **89**, 188301 (2002).
49. C. Tesch, L. Kurtz, and R. de Vivie-Riedle, *Chem. Phys. Lett.* **343**, 633 (2001).
50. U. Troppmann, C. Tesch, and R. de Vivie-Riedle, *Chem. Phys. Lett.* **378**, 273 (2003).
51. U. Troppmann and R. de Vivie-Riedle, *J. Chem. Phys.* **122**, 154105 (2005).
52. S. Suzuki, K. Mishima, and K. Yamashita, *Chem. Phys. Lett.* **410**, 358 (2005).
53. Y. Ohtsuki, *Chem. Phys. Lett.* **404**, 126 (2005).
54. D. Babikov, *J. Chem. Phys.* **121**, 7577 (2004).
55. R. Xu, Y. Yan, Y. Ohtsuki, Y. Fujimura, and H. Rabitz, *J. Chem. Phys.* **120**, 6600 (2004).
56. Y. Ohtsuki, G Turinici, and H. Rabitz, *J. Chem. Phys.* **120**, 5509 (2004).
57. A. Kaiser and V. May, *J. Chem. Phys.* **121**, 2528 (2004).
58. I. Serban, J. Werschnik, and E. K. U. Gross, *Phys. Rev. A* **71**, 053810 (2005).
59. A. Kaiser and V. May, *Chem. Phys.* **320**, 95 (2006).
60. K. Sundermann and R. de Vivie-Riedle, *J. Chem. Phys.* **110**, 1896 (1999).
61. Q. Ren, G. G. Balint-Kurti, F. Manby, M. Artamonov, T. Ho, and H. Rabitz, *J. Chem. Phys.* **125**, 021104 (2006).
62. S. Zou, Q. Ren, G. G. Balint-Kurti, and F. Manby, *Phys. Rev. Lett.* **96**, 243003 (2006).
63. P. Gross, D. Neuhauser, and H. Rabitz, *J. Chem. Phys.* **96**, 2834 (1992).
64. L. R. Rabiner and C. M. Rader, (eds.), *Digital Signal Processing*, IEEE Press selected reprint series, IREEE Press: New York, 1972.
65. T. Hornung, M. Motzkus and R. de Vivie-Riedle, *J. Chem. Phys.* **115**, 3105 (2001).
66. A. F. Barlelt, T. Feurer, and L. Wöste, *Chem. Phys.* **318**, 207 (2005).
67. A. Lindinger, S. m. Weber, C. Lupulescu, F. Vetler, M. Plewicki, A. Merli, L. Wöste, A. F. Bartelt, and H. Rabitz, *Phys. Rev. A* **71**, 013419 (2005).
68. L. Shen and H. Rabitz, *J. Phys. Chem.* **100**, 4811 (1994).
69. J. Farnum and D. Mazziotti, *Chem. Phys. Lett.* **416**, 142 (2005).
70. T. Cheng and A. Brown, *J. Chem. Phys.* **124**, 144109 (2006).
71. J. Werschnik and E. Gross, *J. Opt. B* **7**, S300 (2005).
72. S. Shi and H. Rabitz, *J. Phys. Chem.* **92**, 2927 (1990).
73. B. Amstrup, J. D. Doll, R. A. Sauerbey, G. Szabó, and A. Lorincz, *Phys. Rev. A* **48**, 3830 (1993).
74. B. A. G. J. Tóth, G. Szabó, H. Rabitz, and A. Lorincz, *J. Phys. Chem.* **99**, 5206 (1993).
75. Y. S. Kim and H. Rabitz, *J. Chem. Phys.* **117**, 1024 (2002).
76. J. M. Geremia, W. Zhu, and H. Rabitz, *J. Chem. Phys.* **113**, 10841 (2000).
77. D. Zeidler, S. Frey, K. L. Kompa, and M. Motzkus, *Phys. Rev. A* **64**, 023420 (2001).
78. T. Mancal and V. May, *Chem. Phys. Lett.* **362**, 407 (2002).
79. M. Shapiro and P. Brumer, *J. Chem. Phys.* **84**, 4103 (1986).
80. A. Amstrup, R. Carlson, A. Matro, and S. Rice, *J. Phys. Chem.* **95**, 8019 (1991).
81. J. Somoloi, V. Kazakov, and D. Tannor, *Chem. Phys.* **172**, 85 (1993).
82. M. Kaluza, J. Muckerman, P. Gross, and H. Rabitz, *J. Chem. Phys.* **100**, 4211 (1994).
83. P. Gross, D. Bairagi, M. Mishra, and H. Rabitz, *Chem. Phys. Lett.* **223**, 263 (1994).

84. I. Andrianov and G. Paramonov, *Phys. Rev. A* **59**, 2134 (1999).
85. K. Nakagami, Y. Ohtsuki, and Y. Fujimura, *J. Chem. Phys.* **117**, 6429 (2002).
86. B. Hosseini, H. Sadeghpour, and N. Balakrishnan, *Phys. Rev. A* **71**, 023402 (2005).
87. M. Abe, Y. Ohtsuki, Y. Fujimura, Z. Lan, and W. Domcke, *J. Chem. Phys.* **124**, 224316 (2006).
88. A. Vibòk and G. Balint-Kurti, *J. Chem. Phys.* **96**, 7615 (1992).
89. A. Vibòk and G. Balint-Kurti, *J. Phys. Chem.* **96**, 8712 (1992).
90. D. Manolopoulos, *J. Chem. Phys.* **117**, 9552 (2002).
91. T. Gonzalez-Lezana, E. Rackham, and D. Manolopoulos, *J. Chem. Phys.* **120**, 2247 (2004).
92. W. Zhu, J. Botina, and H. Rabitz, *J. Chem. Phys.* **108**, 1953 (1998).
93. D. Tannor, V. Kazakov, and V. Orlov, in *Time-Dependent Quantum Molecular Dynamics*, J. Broeckhove and L. Lathouwers (eds.), Plenum, New York, 1992, pp. 347–360.
94. Y. Maday and G Turinici, *J. Chem. Phys.* **118**, 8191 (2003).
95. W. H. Press, B. P. Flannery, S. A. Teukolsky, and W. T. Vetterling, in *Numerical Recipes*, Cambridge U. P., Cambridge, MA, 1986.
96. S. Shi, A. Woody, and H. Rabitz, *J. Chem. Phys.* **88**, 6870 (1988).
97. S. Shi and H. Rabitz, *Comp. Phys. Commun.* **63**, 71 (1991).
98. W. Jakubetz, B. Just, J. Manz, and H. Schreier, *J. Phys. Chem.* **94**, 294 (1990).
99. J. Combariza, B. Just, J. Manz, and G. Paramonov, *J. Phys. Chem.* **95**, 10351 (1991).
100. E. Polak, *Computational Methods in Optimization*, vol. 77 of *Mathematics in Science and engineering*, Academic Press: New York, 1971.
101. E. G. Birgin, J. M. Martínez, and M. Raydan, *SIAM J. Optim.* **10**, 1196 (2000).
102. V. Krotov, *Automat. Remote Control* **34**, 1863 (1973).
103. Y. Maday, J. Salomon, and G. Turinici, *Numer. Math.* **103**, 323 (2006).
104. J. Salomon, C. M. Dion, and G. Turinici, *J. Chem. Phys.* **123**, 144310 (2005).
105. S. P. Shah and S. A. Rice, *J. Chem. Phys.* **113**, 6536 (2000).
106. Y. Sato, H. Kono, S. Koseki, and Y. Fujimura, *J. Am. Chem. Soc.* **125**, 8019 (2003).
107. R. Baer, D. J. Kouri, M. Baer, and D. K. Hoffman, *J. Chem. Phys.* **119**, 6998 (2003).
108. S. M. Smith, X. Li, A. N. Markevitch, D. A. Romanov, R. J. Levis, and H. B. Schlegel, *J. Phys. Chem. A* **109**, 10527 (2005).
109. S. M. Smith, X. Li, A. N. Markevitch, D. A. Romanov, R. J. Levis, and H. B. Schlegel, *J. Phys. Chem. A* **109**, 5176 (2005).
110. X. Li, S. M. Smith, A. N. Markevitch, D. A. Romanov, R. J. Levis, and H. B. Schlegel, *Phys. Chem. Chem. Phys.* **7**, 233 (2005).
111. A. N. Markevitch, D. A. Romanov, S. M. Smith, H. B. Schlegel, M. Y. Ivanov, and R. J. Levis, *Phys. Rev. A* **69**, 013401 (2004).
112. A. N. Markevitch, S. M. Smith, D. A. Romanov, H. B. Schlegel, M. Y. Ivanov, and R. J. Levis, *Phys. Rev. A* **68**, 011402(R) (2003).
113. X. Chu, and S.-I. Chu, *Phys. Rev. A* **64**, 063404 (2001).
114. H. Kono, S. Koseki, M. Shiota, and Y. Fujimura, *J. Phys. Chem. A* **105**, 5627 (2001).
115. H. Kono, Y. Sato, N. Tanaka, T. Kato, K. Nakai, S. Koseki, and Y. Fujimura, *Chem. Phys.* **304**, 203 (2004).
116. M. Born and J. R. Oppenheimer, *Ann. Phys.* **84**, 457 (1927).

117. H.-J. Werner et al., *Molpro, version 2002.1, a package of ab initio programs*, 2002, see "http://www.molpro.net".

118. M. J. Frisch et al., *Gaussian 03, Revision C.02*, Gaussian, Inc., Wallingford, CT, 2004.

119. M. W. Schmidt et al., *Gamess: General atomic and molecular electronic structure system* (2006), see:http://www.msg.ameslab.gov/GAMESS/GAMESS.html and *J. Comput. Chem.* **14**, 1347 (1993).

120. M. Guest, J. van Lenthe, J. Kendrick, K. Schffel, P. Sherwood, and R. Harrison, *Gamess-uk is a package of ab initio programs* (2006), with contributions from R. D. Amos, R. J. Buenker, M. Dupuis, N. C. Handy, I. H. Hillier, P. J. Knowles, V. Bonacic-Koutecky, W. von Niessen, V. R. Saunders, and A. Stone. The package is derived from the original GAMESS code due to M. Dupuis, D. Spangler and J. Wendoloski, NRCC Software Catalog, Vol. 1, Program No. QG01 (GAMESS), 1980. http://www.cse.clrc.ac.uk/qcg/gamess-uk/.

121. H. Lischka et al., *Columbus, an ab initio electronic structure program, release 5.9.1*, see http://www.univie.ac.at/columbus/.

122. J. D. Farnum, G. Gidofalvi, and D. A Mazziotti, *J. Chem. Phys.* **124**, 234103 (2006).

123. V. Usachenko and S.-I. Chu, *Phys. Rev. A* **71**, 063410 (2005).

124. M. Baer, *Chem. Phys.* **259**, 123 (2000).

125. J. A. Fleck Jr., J. Morris, and M. Feit, *Appl. Phys.* **10**, 129 (1976).

126. M. Feit, J. A. Fleck Jr., and A. Steiger, *J. Comp. Phys.* **47**, 412 (1982).

127. M. Feit and J. Fleck Jr., *J. Chem. Phys.* **78**, 301 (1983).

128. H. Tal-Ezer and R. Kosloff, *J. Chem. Phys.* **81**, 3967 (1984).

129. R. Kosloff, *J. Phys. Chem.* **92**, 2087 (1988).

130. C. Leforestier et al., *J. Comp. Phys.* **94**, 59 (1991).

131. A. Bandrauk and H. Shen, *Can. J. Chem.* **70**, 555 (1992).

132. T. Truong, J. Tanner, P. Bala, J. McCammon, D. Kouri, B. Lesyng, and D. Hoffman, *J. Chem. Phys.* **96**, 2077 (1992).

133. A. Ritchie and M. Riley, in *Sandia Report (SAND97.UC-401)*, (Sandia National Laboratories, Albuquerque, NM, 1997), ACS Symposium Series in Chemistry.

134. H. Kono, A. Kita, Y. Ohtsuki, and Y. Fujimura, *J. Comp. Phys.* **130**, 148 (1997).

135. J. D. Farnum and D. A. Mazziotti, *J. Chem. Phys.* **120**, 5962 (2004).

136. M. Messina, K. R. Wilson, and J. L. Krause, *J. Chem. Phys.* **104**, 173 (1996).

137. L. X. Wang, H. D. Meyer, and V. May, *J. Chem. Phys.* **125**, 014102 (2006).

138. J. C. Light, I. P. Hamilton, and V. J. Lill, *J. Chem. Phys.* **82**, 1400 (1985).

139. A. R. Offer and G. G. Balint-Kurti, *J. Chem. Phys.* **101**, 10416 (1994).

140. C. Leforestier, *J. Chem. Phys.* **94**, 6388 (1991).

141. G. Corey and D. Lemoine, *J. Chem. Phys.* **97**, 4115 (1992).

142. O. Sharafeddin, R. Judson, and D. Kouri, *J. Chem. Phys.* **93**, 5580 (1990).

143. M. D. Feit and J. A. Fleck, Jr., *J. Chem. Phys.* **80**, 2578 (1984).

144. F. Yip, D. Mazziotti, and H. Rabitz, *J. Chem. Phys.* **118**, 8168 (2003).

145. R. W. Sharp and H. Rabitz, *J. Chem. Phys.* **121**, 4516 (2004).

146. A. Mitra and H. Rabitz, *J. Phys. Chem. A* **108**, 4778 (2004).

147. A. Mitra, I. R. Sola, and H. Rabitz, *Phys. Rev. A* **67**, 043409 (2003).

148. A. Mitra and H. Rabitz, *Phys. Rev. A* **67**, 033407 (2003).

149. J. L. White, B. J. Pearson, and P. H. Bucksbaum, *J. Phys. B* **37**, L399 (2004).

150. E. Dennis and H. Rabitz, *Phys. Rev. A* **67**, 033401 (2003).

151. N. Rosen and C. Zener, *Phys. Rev.* **40**, 502 (1932).

152. M. E. Garcia and I. Grigorenko, *J. Phys. B* **37**, 2569 (2004).

153. T. Cheng and A. Brown, *J. Chem. Phys.* **124**, 034111 (2006).

154. J. Shirley, *Phys. Rev.* **138**, B979 (1965).

155. M. Grifoni and P. Hanggi, *Phys. Rep.* **304**, 229 (1998).

156. P. Kuchment, *Floquet Theory for Partial Differential Equations*, Birhä, Boston, 1993.

157. A. Bandrauk, J.-M. Gauthier, and J. McCann, *Chem. Phys. Lett.* **200**, 399 (1992).

158. I. Grigorenko and M. E. Garcia, *Phys. Rev. A* **74** 013404 (2006).

159. S. Zou, A. Kondorskly, G. Milnokov, and H. Nakamura, *J. Chem. Phys.* **122**, 084112 (2005).

160. N. V. Vitanov, T. Halfmann, B. W. Shore, and K. Bergmann, *Annu. Rev. Phys. Chem.* **52**, 763 (2001).

161. V. S. Malinovsky and D. J. Tannor, *Phys. Rev. A* **56**, 4929 (1997).

162. I. R. Solá, V. S. Malinovsky, and D. J. Tannor, *Phys. Rev. A* **60**, 3081 (1999).

163. B. W. Shore, K. Bergmann, J. Oreg, and S. Rosenwaks, *Phys. Rev. A* **44**, 7442 (1990).

164. K. Yokoyama et al., *J. Chem. Phys.* **120**, 9446 (2004).

165. T. Hornung, R. Meier, D. Zeidler, K. Kompa, D. Proch, and M. Motzkus, *Appl. Phys. B* **71**, 277 (2000).

166. D. Zeidler, S. Frey, W. Wohlleben, M. Motzkus, F. Busch, T. Chen, W. Kiefer, and A. Materny, *J. Chem. Phys.* **116**, 5231 (2002).

167. R. J. Levis and H. A. Rabitz, *J. Phys. Chem. A* **106**, 6427 (2002).

168. P. Graham, G. Menkir, and R. J. Levis, *Spectrochim. Acta P. B* **58**, 1097 (2003).

169. G. Vogt, G. Krampert, P. Niklaus, P. Nuernberger, and G. Gerber, *Phys. Rev. Lett.* **94**, 068305 (2005).

170. R. A. Bartels, T. C. Weinacht, S. R. Leone, H. C. Kapteyn, and M. M. Murnane, *Phys. Rev. Lett.* **88**, 033001 (2002).

171. T. Brixner, F. J. G. de Abajo, J. Schneider, C. Spindler, and W. Pfeiffer, *Phys. Rev. B* **73**, 125437 (2006).

172. T. Brixner, G. Krampert, T. Pfeifer, R. Selle, G. Gerber, M. Wollenhaupt, O. Graefe, C. Horn, D. Liese, and T. Baumert, *Phys. Rev. Lett.* **92**, 208301 (2004).

173. C. Daniel, J. Full, L. González, C. Lupulescu, J. Manz, A. Merli, L. Vajda, and Tefon Wöste, *Science*, **299**, 536 (2003).

174. N. Dudovich, D. Oron, and Y. Silberberg, *Phys. Rev. Lett.* **92**, 103003 (pages 4) (2004).

175. V. Kalosha, M. Spanner, J. Herrmann, and M. Ivanov, *Phys. Rev. Lett.* **88**, 103901 (2002).

176. A. Lindinger, C. Lupulescu, M. Plewicki, F. Vetter, A. Merli, S. M. Weber, and L. Woste, *Phys. Rev. Lett.* **93**, 033001 (2004).

177. C. P. J. Martiny and L. B. Madsen, *Phys. Rev. Lett.* **97**, 093001 (2006).

178. N. V. Morrow, S. K. Dutta, and G. Raithel, *Phys. Rev. Lett.* **88**, 093003 (2002).

179. H. Ohmura, T. Nakanaga, and M. Tachiya, *Phys. Rev. Lett.* **92**, 113002 (2004).

180. D. Oron, N. Dudovich, D. Yelin, and Y. Silberberg, *Phys. Rev. Lett.* **88**, 063004 (2002).

181. T. Schmidt, C. Figl, A. Grimpe, J. Grosser, O. Hoffmann, and F. Rebentrost, *Phys. Rev. Lett.* **92**, 033201 (2004).

182. B. J. Sussman, D. Townsend, M. Y. Ivanov, and A. Stolow, *Science*, **314**, 278 (2006).

183. T. Suzuki, S. Minemoto, T. Kanai, and H. Sakai, *Phys. Rev. Lett.* **92**, 133005 (2004).

184. T. Seideman and M. Shapiro, *J. Chem. Phys.* **94**, 7910 (1991).

185. M. Shapiro and P. Brumer, *Phys. Rev. Lett.* **77**, 2574 (1996).

186. P. Brumer, A. Abrashkevich, and M. Shapiro, *Faraday Dissc.* **113**, 291 (1999).

187. A. Abrashkevich, M. Shapiro, and P. Brumer, *Chem. Phys.* **267**, 81 (2001).

188. V. Zeman, M. Shapiro, and P. Brumer, *Phys. Rev. Lett.* **92**, 133204 (2004).

189. S. Chelkowski, C. Foisy, and A. Bandrauk, *Phys. Rev. A* **57**, 1176 (1998).

190. I. Kawata, H. Kono, and Y. Fujimura, *J. Chem. Phys.* **110**, 11152 (1999).

191. K. Harumiya, H. Kono, and Y. Fujimura, *Phys. Rev. A* **66**, 043403 (2002).

192. B. Scháfer-Bung, R. Mitric", V. Bonacic"-Boutecky", A. Bartelt, C. Lupulescu, A. Lindinger, S. Vajda, S. M. Weber, and L. Wöste, *J. Phys. Chem. A* **108**, 4175 (2004).

193. A. Messiah, *Quantum Mechanics*, North-Holland Publishing Co.; Amsterdam, 1961.

194. W. W. Hager and H. Zhang, *SIAM. J. Optim.* **16**, 170 (2005).

195. R. Fletcher and C. Reeves, *Comput. J.* **7**, 149 (1964).

196. R. Fletcher, *Practical Methods of Optimization: Unconstrained Optimization*, John Wiley & Sons, Inc., New York, 1987, vol. 1.

197. Y. Dai and Y. Yuan, *SIAM. J. Optim.* **10**, 177 (1999).

198. M. Hestenes and E. Stiefel, *J. Research Nat. Bur. Standards*, **49**, 409 (1952).

199. Y. Liu and C. Storey, *J. Opt. Theory Appl.* **69**, 129 (1991).

200. M. Powell, in *Lecture Notes in Mathematics*, Springer-Verlag, Berlin, 1984, vol. 1066, p. 122.

NONADIABATIC CHEMICAL DYNAMICS: COMPREHENSION AND CONTROL OF DYNAMICS, AND MANIFESTATION OF MOLECULAR FUNCTIONS

HIROKI NAKAMURA

Institute for Molecular Science National Institutes of Natural Sciences, Myodaiji, Okazaki 444-8585, Japan

CONTENTS

Advances in Chemical Physics, Volume 138, edited by Stuart A. Rice
Copyright © 2008 John Wiley & Sons, Inc.

I. INTRODUCTION

"Nonadiabatic transition" presents a vey general concept of state and phase changes in nature as well as in society, being an origin of mutability of this world [1]. Without exception nonadiabatic dynamics play crucial roles in physics, chemistry and biology, even if this fact may not be explicitly well recognized in some occasions [1–13]. A nonadiabatic transition makes a very basic mechanism not only to comprehend various chemical dynamic processes occurring in Nature, but also to manifest new molecular functions in nanospace and to control dynamic processes by applying an external field. Theory of nonadiabatic transition can play important roles to accomplish these purposes. A complete set of analytical formulas for the curve crossing problem derived by Zhu and Nakamura [Zhu–Nakamura (ZN) theory] [1, 2, 9, 10–14] is actually useful for these studies.

For example, the ZN theory, which overcomes all the defects of the Landau–Zener–Stueckelberg theory, can be incorporated into various simulation methods in order to clarify the mechanisms of dynamics in realistic molecular systems. Since the nonadiabatic coupling is a vector and thus we can always determine the relevant one-dimensional (1D) direction of the transition in multidimensional space, the 1 D ZN theory can be usefully utilized. Furthermore, the comprehension of reaction mechanisms can be deepened, since the formulas are given in simple analytical expressions. Since it is not feasible to treat realistic large systems fully quantum mechanically, it would be appropriate to incorporate the ZN theory into some kind of semiclassical methods. The promising semiclassical methods are (1) the initial value representation method [15] and (2) the frozen Gaussian propagation method [16–18]. These methods have been developed for the adiabatic processes, namely, for the dynamics on a single adiabatic potential energy surface. Incorporation of the ZN theoy into them can extend their applicability to electronically nonadiabatic dynamics beyond the perturbative treatment [17,19]. A much simpler simulation method is the trajectory surface hopping (TSH) [20,21]. This is a classical trajectory method with electronic transitions between the two potential energy surfaces treated by the Landau–Zener (LZ) formula or by solving the time-dependent coupled equations. After these original works many modifications and improvements have been introduced [22,23] and the widely spread applications have been made because of the simplicity [24–26]. However, some crucial problems have been left unsolved, such as how to deal with the classically forbidden nonadiabatic transitions and how to define classical trajectories uniquely. These defects can be removed by using the ZN theory in the *adiabatic* state representation.

The ZN formulas can also be utilized to formulate a theory for the direct evaluation of thermal rate constant of electronically nonadiabatic chemical reactions based on the idea of transition state theory [27]. This formulation can be further utilized to formulate a theory of electron transfer and an improvement of the celebrated Marcus formula can be done [28].

Needless to say, multidimensional tunneling is another important quantum mechanical effect that should also be incorporated into the simulations, such as those mentioned above. In order to do that, the so-called caustics, which are nothing but turning points in the case of ordinary 1D system and from which tunneling trajectories emanate, should be properly detected along classical trajectories. An efficient method to do this has recently been devised by Oloyede et.al [29]. This method can be carried out by solving the first-order Riccati-type nonlinear time-dependent differential equations along trajectories and the method can be incorporated into the above mentioned simulations. Furthermore, quantum mechanical tunneling naturally plays crucial roles in energy splitting in a symmetric double-well potential and also in predissociation of a metastable state. Without doubt, these processes play important roles in molecular spectroscopy. Recently, we have developed the powerful semiclassical theories applicable to energy splitting and predissociation of realistic polyatomic molecules [30–32].

Once the mechanisms of dynamic processes are understood, it becomes possible to think about controlling them so that we can make desirable processes to occur more efficiently. Especially when we use a laser field, nonadiabatic transitions are induced among the so-called dressed states and we can control the transitions among them by appropriately designing the laser parameters [33–41]. The dressed states mean molecular potential energy curves shifted up or down by the amount of photon energy. Even the ordinary type of photoexcitation can be enhanced very much by appropriately designing the quadratically chirped pulses [36,37,42,43]. Nonadiabatic transitions at naturally existing conical intersections can also be controlled by preparing a wave packet with an appropriate magnitude and direction of momentum. This can be realized again by laser [38,41]. For the creation of new molecular functions, on the other hand, we can think of various methods. For example, the intriguing phenomenon of complete reflection found in the nonadiabatic tunneling (NT) type of transition in which the two diadiabatic potential curves cross with opposite signs of slopes may be actively utilized [44–47]. The laser control methods mentioned above can also be employed to enhance molecular functions, such as photochromism [48,49]. Namely, we can think of controlling nanoscale molecular machines by manipulating nonadiabatic transitions. In these studies of control of chemical dynamics and manifestation of molecular functions, the ZN formulas again play significant roles. The appropriate laser parameters can be designed by using them.

This chapter describes a summary of the recent activities done in the author's research group concerning the subjects mentioned above and is organized as follows. Incorporation of the ZN formulas into the TSH method and the semiclassical frozen Gaussian wavepacket propagation method will be described in Section II. The theory of nonadiabatic thermal rate constant is also presented in this section. Some numerical examples will be provided. Semiclasscal tunneling theory is explained in Section III with practical applications to realistic polyatomic molecules. The application of the thermal rate constant theory to electron transfer will be discussed in Section IV. The famous Marcus theoy can be improved so as to be applicable to the intermediate to strong electronic coupling regimes, which cannot be done by the presently available theories. Section V is devoted to controlling chemical dynamics by lasers. The quadratic chirping of laser frequency we have proposed is shown to be very effective to enhance electronic transitions and pump–dump processes. The newly formulated semiclassical "guided" optimal control theory is also presented. Manifestation and control of molecular functions are discussed in Section VI. Photochromism and hydrogen transmission through a five-membered carbon ring are taken as examples. Section VII concludes the paper. The whole set of Zhu–Nakamura formulas is presented in appendix.

II. NONADIABATIC CHEMICAL REACTION DYNAMICS

A. Generalized Trajectory Surface Hopping Method

It is getting more and more important to treat realistic large chemical and even biological systems theoretically by taking into account the quantum mechanical effects, such as nonadiabatic transition, tunneling, and intereference. The simplest method to treat nonadiabatic dynamics is the TSH method introduced by Bjerre and Nikitin [20] and by Tully and Preston [21], in which classical trajectories hop at the potential energy surface crossing according to the nonadiabatic transition probability evaluated by the LZ formula or the solution of time-dependent coupled equations. A random number is generated whenever the trajectory reaches the potential enery surface crossing region and the nonadiabatic surface hopping is made, if the calculated probability is larger than the random number. This procedure is called the anteater procedure. Because of its simplicity, the method is applicable to large systems and actually has enjoyed wide-spread applications [24–26]. Various modifications from the original version have been made especially by Tully and by Truhlar and co-workers [22,23]. There still remain, however, some crucial problems related to (1) definition of classical trajectory, (2) localizability of the transition, (3) energy and angular momentum conservation, and (4) treatment of classically forbidden transition in which the energy is lower than the energy at surface crossing point. In the *adiabatic* state representation, a classical trajectory runs on a single adiabatic potential energy surface until it reaches the surface crossing region and the nonadiabatic transition can be assumed to occur locally there. It is not conveneint, however, to solve the time-dependent coupled equations in the adiabatic representation to estimate the transition probability. The time-dependent coupled equations convenient to solve are given in the *diabatic* state representation. In this representation, however, the localizability of the transition cannot hold well and the unique definition of classical trajectory becoms questionable. If one uses the LZ formula in the *adiabatic* state representation, these problems are not very serious; but the LZ formula does not work well when the energy is close to the surface crossing energy. The most serious problem is that the classically forbidden transitions cannot be treated by any one of these methods. It is well known that the LZ formula cannot treat those transitions; but even the solutions of the time-dependent coupled equations and the widely used fewest switches method of Tully [22] cannot properly take into accout those classically forbidden transitions.

The ZN formulas in the *adiabatic* state representation enable us to solve all the problems mentioned above. Since the localizability holds well, classical trajectories can run on a single adaiabatic potential energy surface and thus the effects of relaxation can be taken into account easily. A whole chemical process can be divided into the following two processes: propagation on a single adiabatic potential energy surface and a localized nonadiabatic transition at the minimum energy separation position. The transition can be classified as hopping, reflection, and passing (see Fig. 1).

First, we have applied the ZN formulas to the DH_2^+ system to confirm that the method works well in comparison with the exact quantum mechanical numerical solutions [50]. Importance of the classically forbidden transitions has been clearly demonstrated. The LZ formula gives a bit too small results

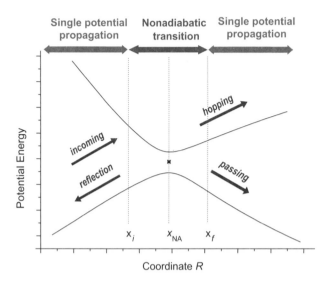

Figure 1. Two basic elements of dynamics:(1) propagation on a single adiabatic potential and (2) nonadiabatic transition. In the classically allowed case, the transition occurs at X_{NA}. In the classically forbidden case, on the other hand, the transition region spans the interval (x_i, x_f), where x_i and x_f are the turning points. Taken from Ref. [9].

even at high energies. It was found that in multidimensional systems classically forbidden transitions play relatively more important roles than in the case of 1D because of energy transfer among many degrees of freedom. Some of the results are shown in Figs. 2 and 3. The quantum mechanical exact results show violent oscillations that are resonances due to the potential well of the ground state. In the TSH calculations, all the long-lived trajectories are killed, since we are not interested in resonances here. This reaction system is, however, relatively simple, since the surface crossing seam is located a bit away from the reaction zone, only the LZ type of crossing in which the two diabatic potential curves have the same sign of slopes appears, and the geometry of the seam surface can be well analyzed in advance. Since these conditions are not generally satisfied, it is definitely required to develop a general TSH method that incorporates the ZN formulas and is directly applicable to large systems. Recently, we have developed such a method [51].

The method is composed of the following algorithms: (1) transition position is detected along each classical trajectory, (2) direction of transition is determined there and the 1D cut of the potential energy surfaces is made along that direction, (3) judgment is made whether the transition is LZ type or nonadiabatic tunneling type, and (4) the transition probability is calculated by the appropriate ZN formula. The transition position can be simply found by

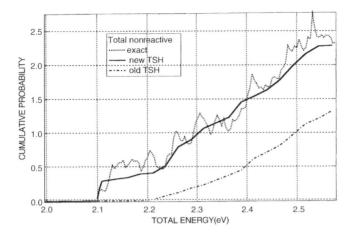

Figure 2. Total cumulative charge-transfer probabilities for $H_2 + D^+ \rightarrow H_2^+ + D$. Dashed line: exact quantum mechanical numerical solution. Solid line: TSH results with use of the Zhu-Nakamura formulas. Dash–dot line: TSH results with use of the LZ formula. Taken from Ref. [50].

detecting the minimum energy separation between the two adiabatic potential energies. The determination of transition direction has the following options: (1) direction perpendicular to the crossing seam surface, if the potential energy surface topography is well known and the seam surface can be well defined in advance; (2) direction of the nonadiabatic coupling vector, if it is available; (3) direction estimated from the Hessian, the second derivatives of adiabatic

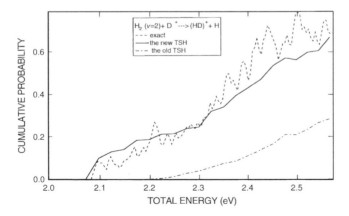

Figure 3. Initial vibrational state specified cummulative reaction probabilities for $v = 2$. Dashed line: exact quantum mechanical numerical solution. Solid line: TSH results with use of the Zhu–Nakamura formulas. Dash–dot line: TSH results with use of the Landau–Zener formula. Taken from Ref. [50].

potentials. If the seam surface is known well, method (1) is the best, since the nonadiabatic transition in the direction parallel to the seam surface is the so called noncrossing Rosen–Zener type and can usually be neglected. In general, however, the geometry of the seam surface cannot be known in advance. The second best in that case is the direction of the nonadiabatic coupling vector. However, the vector is not necessarily available in general, unfortunately, because it is not easy to compute.

In such a case the last choice is to take the direction of the eigenvector of the only one nonzero eigenvalue of the rank one Hessian matrix of the difference between the two adiabatic potential energies [51]. In the vicinity of conical intersection, the topology of the potential energy surface can be described by the diadiabatic Hamiltonian in the form

$$
\begin{vmatrix} A_i X_i & B_i X_i \\ B_i X_i & -A_i X_i \end{vmatrix} \tag{1}
$$

where the summation with respect to i is implied.

Here, A_i and B_i are some constants, and $X_i = R_i - R_i^0$, where R_i^0 represents the point of intersection. From this equation the direction of the nonadiabatic coupling vector is readily found as:

$$
e_i \sim \sum_{k=1}^{3} (A_i B_k - B_i A_k) X_k \tag{2}
$$

On the other hand, one can use Eq. (1) to calculate the Hessiam matrix of the difference between the two adiabatic potential energy surfaces, $\Delta V(R)$. Up to an irrelevant scalar factor, the result reads

$$
\frac{\partial^2 \Delta V}{\partial X_i \partial X_j} \sim e_i e_j \tag{3}
$$

This equation determines a rank-1 matrix, and the eigenvector of its only one nonzero eigenvalue gives the direction dictated by the nonadiabatic coupling vector. In the general case, the Hamiltonian differs from Eq.(1), and the Hessian matrix has the form

$$
\frac{\partial^2 \Delta V}{\partial X_i \partial X_j} \sim e_i e_j (1 + \epsilon_{ij}) \tag{4}
$$

The closer the trajectory approaches the conical intersection, the smaller ϵ_{ij} becomes. Since the nonadiabatic transitions are expected to take place in the close vicinity of the conical intersection, the nonadiabatic transition direction can be approximated by the eigenvector of the Hessian $\partial^2 \Delta V / \partial R_i \partial R_j$ corresponding to its maximum eigenvalue. Similar arguments hold for nonadiabatic transitions near the crossing seam surface, in which case the nondiagonal elements of the diabatic Hamiltonian of Eq. (1) should be taken as nonzero constant.

When the transitions are classically allowed, then the hops occur vertically. However, in the case of classically forbidden transitions, the transitions are not vertical anymore and the nonadiabatic electronic transition and the nuclear tunneling are coupled, namely, they cannot be treated separately. The ZN formulas properly describe these nonvertical transitions and provide the overall transition probability with both electronic transition and nuclear tunneling included. The position right after the transition is not the same anymore as the position before the transition. The total angular momentum conservation apparently violated by the nonvertical transition can be recovered by appropriately rotating the system. The detailed recipe is not described here, but can be found in Ref. [51]. The energy conservation is not a problem at all, since the classically forbidden transitions are treated properly and the total energy after the transition is conserved. As is shown in Fig. 4, the transition can be hopping onto the upper adiabatic potential energy surface or passing on the lower adiabatic potential energy surface when the transition is classically allowed. If it is classically forbidden, then reflection on the lower surface or hopping reflection onto the upper surface occurs in the LZ case and reflection or passing on the lower surface occurs in the NT case.

This generalized TSH (ZN–TSH) has been applied to a model triatomic system mimicking CH_2, which has a conical intersection in the reaction zone and both LZ and nonadiabatic tunneling types of transitions appear [51]. The ground and excited potential energy surfaces are constructed by using the Diatomics In Molecule (DIM) method (see Fig. 5). The numerical results in comparison with the exact quantum mechanical numerical solutions are shown in Fig. 6. The oscillations in the exact quantum mechanical results are again resonances due to the attractive well in the ground state. Apart from these oscillations the ZN–TSH works acceptably well. In these figures, the two ZN–TSH results are compared with respect to the choice of the transition direction:(1) the direct use of nonadiabatic coupling vector and (2) the Hesssian approximation. It is clearly seen that the Hessian approximation works well. These successful demonstrations confirm that the present method can be applied to any large systems of general potential energy surface topology. By using the present method, the very popular classical mechanical molecular dynamics (MD) simulation method could be improved easily and extended so

HIROKI NAKAMURA

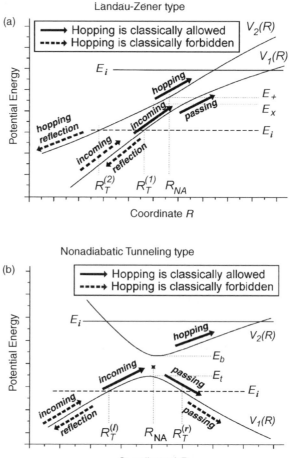

Figure 4. Interpretation of the transition modes: reflection, passing, and hopping along the adiabatic potential. Taken from Ref. [19].

as to take into account the nonadiabatic transitions properly. Finally, it would be worthwhile to mention that another important quantum mechanical effect, namely, quantum mechanical tunneling, can also be taken into account in the present methodology. In order to do that, it is crucial to detect caustics (turning points in 1D case) along trajectories. This can be easily done and will be explained later in Section III.

The advantages of this generalized TSH method can be summarized as follows: (1) both types of transitions in the potential curve crossing problems,

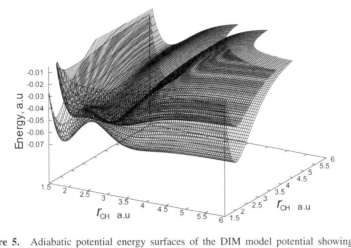

Figure 5. Adiabatic potential energy surfaces of the DIM model potential showing conical intersection at the C_{2v} symmetry. Taken from Ref. [51].

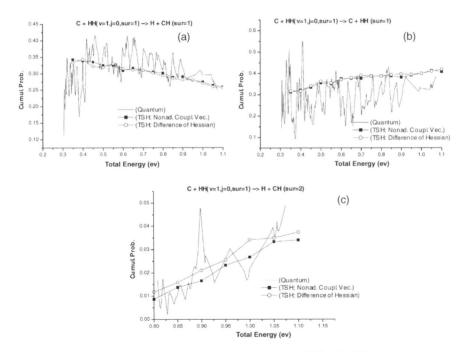

Figure 6. Initial rovibrational state specified reaction probabilities. Solid line:exact quantum mechanical numerical solution. Solid line with solid square: generalized TSH with use of the nonadiabatic coupling vector. Solid line with open circle: generalized TSH with use of Hessian. Sur = 1(2) means the ground (excited) potential energy surface. Taken from Ref. [51].

that is, LZ and NT types, can be treated by the full use of the ZN theory; (2) the Cartesian coordinate systems can be used; (3) a simple algorithm to conserve the total angular momentum for any large system is available, if necessary; (4) the transition direction can be estimated accurately from the Hessian matrix of the adiabatic potential energy difference; and (5) quantum mechanical tunneling effects can be incorporated on-the-fly. In order to save the cpu time, it is better to have some knowledge in advance about the potential energy surface topology such as in what configurations nonadiabatic transitions and tunneling are important. However, if necessary, all the dynamics calculations can be done on-the-fly together with the quantum chemical computations of potential energies.

B. Semiclassical Herman–Kluk-Type Frozen Gaussian Wavepacket Propagation Method

If it is necessary to take into account the third quantum mechanical effect, that is, interference, we have to rely on the semiclassical wavepacket propagation method, such as the Initial Value Representation (IVR) method and the Herman–Kluk-type frozen Gaussian wavepacket propagation method [15–18]. The ZN formulas can also be incorporated into these methods to treat nonadiabatic dynamics. As mentioned before, in the adiabatic state representation a whole chemical process can be divided into the following two steps (see Fig. 1): (1) motion on a single adiabatic potential energy surface up to the region of nonadiabatic transition, and (2) electronically nonadiabatic transition at the potential surface crossing. The appropriate phases corresponding to these steps should be taken into account: the classical action along classical trajectory and the so-called dynamical phases induced by nonadiabatic transition. There are two types of semiclassical path integral methods for the propagation on a single adiabatic potential energy surface. One is the IVR theory with direct use of classical trajectories devised by Miller and others [15]. In this case, the nonadiabatic transition *amplitude* of the ZN theory can be directly incorporated into the framework, since classical trajectories are directly treated. The other is the frozen Gaussian wavepacket propagation method [16–18]. The Herman–Kluk propagator combined with the cellularization procedure works well, giving good agreement with full quantum calculations [17,18]. We have formulated the semiclassical approach by making use of the advantages of the Herman–Kluk theory for single surface propagation [17] and the ZN theory for nonadiabatic transition. The nonadiabatic transition amplitude of the ZN theory properly provides the dynamical phases induced by the nonadiabatic transition and can be incorporated into each frozen Gaussian wavepacket.

The outline of the formulation is described below so that the reader can grasp the essential ideas of the method. The details of the formulation can be found in

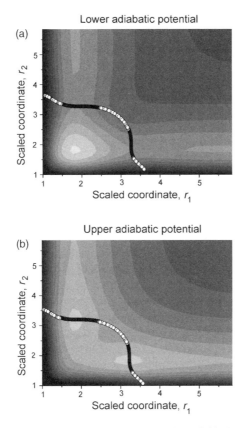

Figure 7. Two-dimensional (2D) H_2O system in the laser field. (a)excited and (b) ground adiabatic potentials. Filled circles nonadiabatic tunneling-type region. Open circles: LZ-type region. Taken from Ref. [19].

the original papers [19,38]. The total wave function at time t in the Herman–Kluk approach is expressed as:

$$\psi(\mathbf{r}, t) = \int_{\text{traj}} \frac{d\mathbf{q}_0 d\mathbf{p}_0}{(2\pi)^N} g(\mathbf{r}; \mathbf{q}_t, \mathbf{p}_t) C_{\mathbf{q}_0, \mathbf{p}_0, t} \exp[iS_{\mathbf{q}_0, \mathbf{p}_0, t}]$$
$$\times \int d\mathbf{r}_0 g^*(\mathbf{r}_0; \mathbf{q}_0, \mathbf{p}_0) \psi(\mathbf{r}_0, t = 0) \tag{5}$$

where $\psi(\mathbf{r}_0, t = 0)$ and $\psi(\mathbf{r}, t)$ are the wave functions at time zero and t, respectively, N is the dimensionality of configuration space, $S_{\mathbf{q}_0, \mathbf{p}_0, t}$ is the classical action along the trajectory from $(\mathbf{q}_0, \mathbf{p}_0, t = 0)$ to $(\mathbf{q}_t, \mathbf{p}_t, t)$, and $C_{\mathbf{q}_0, \mathbf{p}_0, t}$

is the Herman–Kluk preexponential factor along the trajectory [17]. The frozen
Gaussian wave packets are defined as:

$$g(\mathbf{r}; \mathbf{q}, \mathbf{p}) = \left(\frac{2\gamma}{\pi}\right)^{\frac{N}{4}} \exp[-\gamma(\mathbf{r} - \mathbf{q})^2 + i\mathbf{p} \cdot (\mathbf{r} - \mathbf{q})] \tag{6}$$

where γ is a constant parameter common for all wave packets. The
preexponential factor $C_{\mathbf{q},\mathbf{p},t}$ is given by

$$C_{\mathbf{q},\mathbf{p},t} = \pm \left| \frac{\partial \mathbf{p}_t}{\partial \mathbf{p}_0} + \frac{\partial \mathbf{q}_t}{\partial \mathbf{q}_0} - 2i\gamma \frac{\partial \mathbf{q}_t}{\partial \mathbf{p}_0} + \frac{i}{2\gamma} \frac{\partial \mathbf{p}_t}{\partial \mathbf{q}_0} \right|^{1/2} \tag{7}$$

where the sign is chosen to keep C differentiable at any time.

The above expression for $\psi(\mathbf{r}, t)$ is explained as follows: the initial wave
function is expanded in terms of the frozen Gaussian wave packets and each
packet is propagated by classical mechanics with its shape kept fixed. The final
wave function is expressed as a sum of thus propagated frozen wave packets
multiplied by the factor $C_{\mathbf{q}_0,\mathbf{p}_0,t} \exp[iS_{\mathbf{q}_0,\mathbf{p}_0,t}]$. The initial parameters $(\mathbf{q}_0, \mathbf{p}_0)$ of
trajectories are selected by the well-established Monte Carlo procedure. This
propagation is made on a single adiabatic potential energy surface and is carried
out up to the region of potential energy surface crossing. It is assumed that the
nonadiabatic transition occurs locally and instantaneously at the position $\mathbf{q}_I$
where the adiabatic potential energy difference becomes minimum along the
trajectory. Once this transition position is found, the nonadiabatic transition is
taken into account as follows. The local separability in the vicinity of $\mathbf{q}_I$
is assumed and the 1D direction of transition is determined. This can be done, as
explained in Section II.A from the three options. By using one of these methods,
we can reduce the problem to the 1D method and apply ZN theory.

First, the frozen Gaussian wavepackets just before the transition on the
initial adiabatic surface i are expanded as:

$$g_I(\mathbf{r}; \mathbf{q}_I, \mathbf{p}_I, t) = \int dE \alpha^i(E) \phi^i(E, \mathbf{r}) \tag{8}$$

where $\{\phi^i(E, \mathbf{r})\}$ are the energy normalized eigenfunctions in the electronic state
i at $\mathbf{q} \sim \mathbf{q}_I$. Right after the transition the coefficient $\alpha^i(E)$ changes to

$$\alpha^f_m(E) = T^m_{fi} \alpha^i(E) \tag{9}$$

where f and m specify the final electronic state and the mode on that state,
respectively. The mode specifies one of the following three:reflection, passing, or
hopping (see Fig. 4). Namely, the coefficient T^m_{fi} represents the transition

amplitude for the reflection or passing on the same adiabatic potential energy surface as the initial one $f = i$, or hopping or hopping reflection to the other potential energy surface $f \neq i$ after the transition, and is directly given by the ZN formulas including the dynamical phases.

The final wave function $\varphi^f(\mathbf{r})$ right after the transition is thus given by

$$\varphi^f(\mathbf{r}) = \sum_m \int dE \alpha^f_m(E) \phi^f(E, \mathbf{r}) \tag{10}$$

where $\{\phi^f(E, \mathbf{r})\}$ are the energy eigenfunctions in the final electronic state f. The function $\varphi^f(\mathbf{r})$ is expanded in terms of the frozen Gaussian wave packets $g_f(\mathbf{r}; \mathbf{q}_F, \mathbf{p}_F)$ of the same shape as before and the latter Gaussians are propagated on the new surface f. Then the final wave function at time t after the transition is expressed as

$$
\begin{aligned}
\psi_f(\mathbf{r}, t) = &\int_{\text{traj}} \frac{d\mathbf{q}_0\mathbf{p}_0}{(2\pi)^N} \int_{\text{traj}} \frac{d\mathbf{q}_F d\mathbf{p}_F}{(2\pi)^N} g_F(\mathbf{r}; \mathbf{q}_t, \mathbf{p}_t) C_{\mathbf{q}_F, \mathbf{p}_F, t} \\
&\times \exp[iS_{\mathbf{q}_F, \mathbf{p}_F, t}] F_{fi}(\mathbf{q}_F, \mathbf{p}_F, \mathbf{q}_I, \mathbf{p}_I) C_{\mathbf{q}_0, \mathbf{p}_0, t_{NA}} \exp[iS_{\mathbf{q}_0, \mathbf{p}_0, t_{NA}}] \\
&\times \int d\mathbf{r}_0 g_I^*(\mathbf{r}_0; \mathbf{q}_0, \mathbf{p}_0) \psi_i(\mathbf{r}_0, t = 0)
\end{aligned}
\tag{11}
$$

where t_{NA} represents the time of nonadiabatic transition. The expansion coefficients F_{fi} are given by

$$F_{fi}(\mathbf{q}_F, \mathbf{p}_F, \mathbf{q}_I, \mathbf{p}_I) = \int d\mathbf{r} g_F^*(\mathbf{r}; \mathbf{q}_F, \mathbf{p}_F) \varphi^f(\mathbf{r}, \mathbf{q}_F, \mathbf{p}_F : \mathbf{q}_I, \mathbf{p}_I) \tag{12}$$

where $\mathbf{q}_F$ and $\mathbf{p}_F$ are the position and momentum right after the transition. The position $\mathbf{q}_F$ is not necessarily the same as $\mathbf{q}_I$, since in the case of classically forbidden transition when the energy E is lower than the crossing point the positions $\mathbf{q}_I$ and $\mathbf{q}_F$ are turning points on the respective potential energy curve and are different from each other (see Fig. 4). The parameter $R_T^{(1,2)}$ and $R_T^{(l,r)}$ are used in this figure to denote these turning points. The new trajectories starting from $(\mathbf{q}_F, \mathbf{p}_F, t = t_{NA})$ reach $(\mathbf{q}_t, \mathbf{p}_t, t)$. In the actual computations, the integrals with respect to $\mathbf{q}_F$ and $\mathbf{p}_F$ are replaced by the sum of main components of the wave packets right after the transition [19,38]. The analysis of the function $|F_{fi}(\mathbf{q}_F, \mathbf{p}_F, \mathbf{q}_I, \mathbf{p}_I)|$ indicates that the main components can be found from the general principle of nonadiabatic transition in the 1D system.

Numerical examples are shown in Figs. 7–9. The model system used is a 2D model of H_2O in a continuous wave (CW) laser field of wavelength 515nm and intensity 10^{13}W/cm^{-2}. The ground electronic state $\tilde{X}$ and the first excited state $\tilde{A}$ are considered. The bending and rotational motions are neglected for

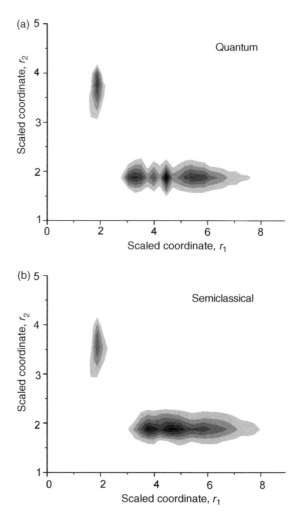

Figure 8. Final wave packets on the ground electronic stste. (a) exact quantum and (b) semiclassical results. Propagation duration is 20 fs. Taken from Ref. [19].

simplicity with the bending angle fixed at the equilibrium structure of the ground electronic state, $\theta = 104.52°$. The potentials and dipole moments used are the same as those in Ref.[42]. The kinetic energy operator,

$$T(r_1, r_2) = -\frac{1}{2m_H}\frac{\partial^2}{\partial r_1^2} - \frac{1}{2m_H}\frac{\partial^2}{\partial r_2^2} - \frac{\cos\theta}{m_O}\frac{\partial^2}{\partial r_1 \partial r_2} \qquad (13)$$

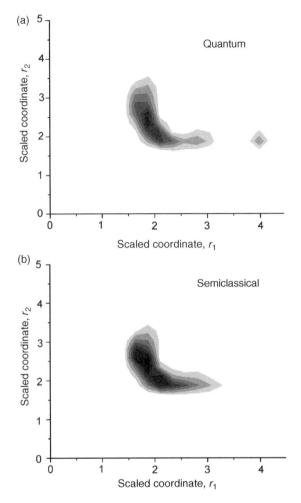

Figure 9. Final wave packets on the excited state. (a) exact quantum and (b) semiclassical results. Propagation time is 20 fs. Taken from Ref. [19].

is diagonalized by using the transformation

$$(r_1, r_2) = D(\tilde{r}_1, \tilde{r}_2) = \begin{pmatrix} -\dfrac{1}{2m_H}, & -\dfrac{\cos\theta}{2m_o} \\[2ex] -\dfrac{\cos\theta}{2m_O}, & -\dfrac{1}{2m_H} \end{pmatrix} (\tilde{r}_1, \tilde{r}_2) \tag{14}$$

as

$$D^{-1}TD = -\frac{1}{2m_H}\left(\frac{\partial^2}{\partial r_1^2} + \frac{\partial^2}{\partial r_2^2}\right) \tag{15}$$

where $\tilde{r}_{1,2}$ stand for the two H—O bond lengths. Dressed adiabatic potential energy surfaces are shown in Fig. 7. The nonadiabatic transition regions are marked as filled circles for the NT type and open circles for the LZ type. The initial wave packet is set to be a symmetric Gaussian of the full width at half-maximum (FWHM) $= 0.5$ a.u. centered at $r_1 = 5$ a.u. and $r_2 = 2$ a.u. on the upper (dressed) adiabatic potential energy surface. Figures 8 and 9 present a comaprison of our semiclassical results with the exact quantum mechanical ones. The final population on the excited electronic state is 34% in the exact quantum mechanical calculations and 40% in the semiclassical approximation. The number of trajectories used for propagation is 7000.

The similar method as that mentioned above can be applied to laser control of chemical dynamics. This will be discussed in Section V.

C. Semiclassical Theory of Thermal Rate Constant

As discussed by Miller and co-workers [52,53], it is worthwhile to develop theories that enable us to evaluate thermal reaction rate constants directly and not to rely on the calculations of the most detailed scattering matrix or the state-to-state reaction probability. Here, our formulation of the nonadiabatic transition state theory is briefly described for the simplest case in which the transition state is created by potential surface crossing [27].

We start from the quantum mechanically exact flux–flux correlation function expression [53]

$$kZ_r^q = \lim_{t \to \infty} Tr(\exp[-\beta H]F \exp[iHt/\hbar]h\exp[-iHt/\hbar]) \tag{16}$$

where Z_r^q is the quantum mechanical partition function of reactants, H is the Hamiltonian of the system, h is the Heaviside step function operator, $F = (i/\hbar)[H, h]$ is the operator of flux through the dividing surface $S(\mathbf{Q}) = 0$ between reactants and products, and $\beta = 1/\kappa T$. Replacing the Heaviside step function operator and the quantum mechanical trace by the ZN nonadiabatic transition probability and the phase integral, respectively, and carrying out the integration with respect to the momenta except for the component normal to the seam surface, one can finally obtain the semiclassical version of the thermal rate constant for nonadiabatic process as

$$k = \frac{Z_q^\dagger}{Z_r^q} \sqrt{\frac{1}{2\pi\beta}} \frac{\int d\mathbf{Q}P(\beta, \mathbf{Q})|\nabla S(\mathbf{Q})|\delta[S(\mathbf{Q})]\exp[-\beta V(\mathbf{Q})]}{\int d\mathbf{Q}\delta[S(\mathbf{Q})]\exp[-\beta V(\mathbf{Q})]} \tag{17}$$

where $Z_q^\dagger$ is the quantum mechanical partition function of activated complex. The effective coordinate-dependent nonadiabatic transmission probability

$P(\beta, \mathbf{Q})$ as a function of temperature T is defined by

$$P(\beta, \mathbf{Q}) = \beta \int_0^\infty dE_s \exp(-\beta[E_s - V(\mathbf{Q})])P_{ZN}(E_s, \mathbf{Q}) \qquad (18)$$

where E_s is the translational energy component perpendicular to the seam surface and $P_{ZN}(E_s, \mathbf{Q})$ is the nonadiabatic transmission probability at position $\mathbf{Q}$ on the seam surface. This probability is given by the ZN formula.

Examples of numerical applications are shown in Fig. 10, where the collinear (2D) model potentials in Ref. [54] are employed. The diabatic potentials actually used are explicitly given by

$$V_1(r, R) = D(1 - \exp[-\beta(r - r_e)])^2$$
$$+ \frac{1}{2}D(1 + \exp[-\beta(R + r/2 - r_e)])^2 - D/2 \qquad (19)$$

$$V_2(r, R) = D(1 - \exp[-\beta(R - r/2 - r_e)])^2$$
$$+ \frac{1}{2}D(1 + \exp[-\beta(R + r/2 - r_e)])^2 - D/2 \qquad (20)$$

$$V_{12}(r, R) = A \exp[-\gamma((r - r_c)^2 + (R - R_c)^2)] \qquad (21)$$

where r and R are Jacobi coordinates for A + BC arrangement and the parameters are chosen to mimic the H + H_2 system. The actual values of the parameters

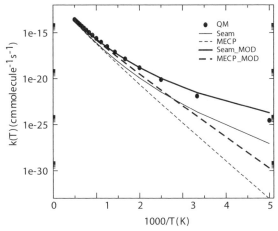

Figure 10. Arrhenius plot of the thermal rate constants for the 2D model system. Circles-full quantum results. Thick solid (dashed) curve: present nonadiabatic transition state theory by using the seam surface [the minimum energy crossing point (MECP)] approximation. Thin solid and dashed curves are the same as the thick ones except that the classical partition functions are used. Taken from Ref. [27].

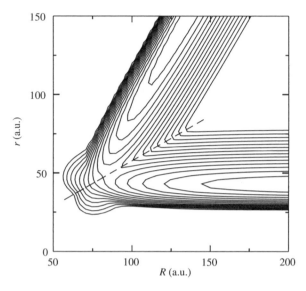

Figure 11. Contour plot of the adiabatic ground potential energy surface of the 2D model. The dashed line shows the seam surface. Taken from Ref. [27].

are $D = 4.9\,\text{eV}$, $\beta = 1.877\text{Å}$, $r_e = 0.7417\text{Å}$, $r_c = 1.5707 r_e$, $R_c = 1.5 r_c$, $\gamma = 0.01\text{Å}^{-1}$, and $A = 0.1\,\text{eV}$. The adiabatic ground potential energy surface is shown in Fig. 11. The present results (solid line) are in good agreement with the quantum mechanical ones (solid circles). The minimum energy crossing point (MECP) is conventionally used as the transition state and the transition probability is represented by the value at this point. This is called the MECP approximation and does not work well, as seen in Fig. 10. This means that the coordinate dependence of the nonadiabatic transmission probability on the seam surface is important and should be taken into account as is done explicitly in Eq. (18).

It is definitely necessary to extend this kind of theory to a general case in which the ordinary transition state and the potential surface crossing position are separated from each other.

III. MULTIDIMENSIONAL TUNNELING
IN CHEMICAL DYNAMICS

Needless to say, tunneling is one of the most famous quantum mechanical effects. Theory of multidimensional tunneling, however, has not yet been completed. As is well known, in chemical dynamics there are the following three kinds of problems: (1) energy splitting due to tunneling in symmetric double-well potential, (2) predissociation of metastable state through tunneling, and (3) tunneling in chemical reaction. It is strongly desired that the semiclassical theories, such as those explained in the previous section, can incorporate the tunneling effects so that chemical dynamics can be correctly described and comprehended. Recently, we have successfully formulated the powerful semiclassical theories for (1) and (2) mentioned above that can be applied with high accuracy to real polyatomic systems of vertually any dimension [30–32]. On the other hand, concerning the problem (3) mentioned above, we have successfully developed a method to efficiently detect caustics (turning points in the ordinary 1D system) while running classical trajectories in multidimensional space [29]. This fundamental problem takes into account the tunneling effects in the reactions. In this section, these theories and their applications to realistic multidimensional systems will be explained.

A. Tunneling Splitting in Polyatomic Molecules

1. General Formulation

There are many polyatomic molecules that show energy splittings due to proton transfer in symmetric double-well potentials. The most well treated molecule in this subject is a nine-atomic malonaldehyde $(C_3O_2H_4)$ in which proton transfer occurs between two oxygen atoms through the hydrogen bond. There have been proposed various semiclassical theories to treat this problem [55–67], but all of them are applicable only to 1D–2D systems unless some simplification tricks are employed [68, 69]. In the actual applications to real molecules, some kinds of effective 1D theories are usually used. Note that there is an intresting multidimensional effect that the tunneling splitting, or the tunneling probability, oscillates or decreases with the energy given to the system. In other words, the tunneling splitting does not incease exponentially as in the 1D case with vibrational excitation. This cannot be explained at all by the 1D theory [62, 70].

The WKB theory we developed and briefly described below is formally theoretically equivalent to the instanton theory [56–58, 71–76], but is more straightforward and practical, and probably easier to understand. Let us start with the 1D case in order to comprehend the basic ideas. The semiclassical wave function is given as usual by

$$\Psi = \exp[-\frac{1}{\hbar} W_0 - W_1] \tag{22}$$

The key observation here is that the ground-state energy is given by

$$E_0 = \frac{\hbar \omega_0}{2} + O(\hbar^2) \tag{23}$$

and is treated as the first-order term with respect to $\hbar$. Thus the energy E_0 does not come into the Hamilton–Jacobi (HJ) equation, but comes into the transport equation. Namely, we have

$$\frac{1}{2}\left(\frac{\partial W_0}{\partial x}\right)^2 - mV(x) = 0 \tag{24}$$

and

$$\frac{\partial W_1}{\partial x}\frac{\partial W_0}{\partial x} - \frac{\partial^2 W_0}{2\partial x^2} + \frac{\omega}{2} = 0 \tag{25}$$

These equations can be easily solved as

$$W_0(x) = \int_{-x_m}^{x} p_0(z)dz \tag{26}$$

and

$$W_1(x) = \frac{1}{2}\ln p_0(x) - \frac{1}{2}\int^{x}\frac{\omega}{p_0(z)}dz \tag{27}$$

where x_m is the potential minimum and $p_0(x) = \sqrt{2mV(x)}$. By removing the divergence in $\int_{-x_m}^{x} dz/p_0(z)$, we have

$$\Psi(x) = \sqrt{\frac{x + x_m}{p_0(x)}}\exp\left(-\int_{x_m}^{x} p_0(z)dz\right)\exp\left(\frac{1}{2}\int_{x_m}^{x}\left[\frac{\omega}{p_0(z)} - \frac{1}{z + x_m}\right]dz\right) \tag{28}$$

which gives the ground-state harmonic oscillator at x close to $-x_m$

$$\Psi \sim \frac{1}{\sqrt{\omega}}\exp\left(-\frac{\omega(x + x_m)^2}{2}\right) \tag{29}$$

By inserting Eq. (28) into the Herring formula,

$$\Delta_0 = \frac{2\Psi(x)\frac{d\Psi(x)}{dx}|_{x=0}}{\int dx\Psi^2(x)} \tag{30}$$

and evaluating the normalization factor using Eq. (29), we obtain the energy splitting as

$$\Delta_0 = \sqrt{\frac{4\omega\hbar}{\pi m}}p_0(0)\exp\left(-\int_{-\infty}^{0} d\tau\left[\omega - \frac{\partial p_0(z)}{m\partial z}\right]\exp(-2W_0(0)/\hbar)\right) \tag{31}$$

In the same way as described above, we can formulate the multidimensional theory without relying on the complex-valued Lagrange manifold that constitutes one of the main obstacles of the conventional multidimensional WKB theory [62,63,77,78]. Another crucial point is that the theory should not depend on any local coordinates, which gives a cumbersome problem in practical applications. Below, a general formulation is described, which is free from these difficuluties and applicable to vertually any multidimensional systems [30].

We start from a very general Hamiltonian,

$$H = -\frac{1}{2\sqrt{g}}\frac{\partial}{\partial q^i}\left(\sqrt{g}g^{ij}\frac{\partial}{\partial q^j}\right) + V(\mathbf{q}) \tag{32}$$

where g^{ij} is the metric tensor and $g = \det g_{ij}$ with g_{ij} being the inverse of g^{ij} and $\mathbf{q} = (q_1, q_2, \ldots, q_N)$ is the N-dimensional coordinate vector. Hereafter, the energy is measued from the bottom of the potential minima $\mathbf{q}_m$, that is, $V(\mathbf{q}_m) = 0$. The energy splitting in this general case is calculated from

$$\Delta_0 = \frac{2\int\sqrt{g}d\delta(f)\Psi g^{ij}\partial_i\Psi\partial_j f}{\int\sqrt{g}d\mathbf{q}\Psi^2} \tag{33}$$

where $\partial_i \equiv \partial/\partial q_i$ and $f(\mathbf{q}) = 0$ defines the dividing surface Σ that is nothing but the $N - 1$ dimensional potential barrier separating the two potential minima. The HJ and transport equations are given by

$$H\left(\mathbf{q}, \frac{\partial W_0}{\partial \mathbf{q}}\right) = 0 \tag{34}$$

and

$$g^{ij}\partial_i W_0\partial_j W_1 - \frac{1}{2}g^{ij}\partial_{ij}^2 W_0 + \frac{\omega}{2} - \frac{1}{2}\Lambda^i\partial_i W_0 = 0 \tag{35}$$

where

$$H(\mathbf{q}, \mathbf{p}) = \frac{1}{2m}g^{ij}p_ip_j - V(\mathbf{q}) \tag{36}$$

is the classical Hamiltonian with the upside down potential and

$$\Lambda^i(\mathbf{q}) = \frac{1}{\sqrt{g}}\partial_j(\sqrt{g}g^{ij}) \tag{37}$$

In order to solve Eq. (34), we use the method of characteristics and consider a family of classical trajectories on the inverted potential $\{\mathbf{q}(\beta, \tau), \mathbf{p}(\beta, \tau)\}$, where β is an $(N - 1)$-dimensional parameter to characterize the trajectory and τ is the time running for the infinite interval along the trajectory, where $\tau = -\infty$ corresponds to the minimum of the potential $\mathbf{q}(\beta, -\infty) = \mathbf{q}_m, \mathbf{p}(\beta, -\infty) = 0$. The solution we want is the trajectory that connects the two potential minima and along which the action becomes minimum. This is called the instanton trajectory and belongs to the above mentioned family $\mathbf{q}_0(\tau) = \mathbf{q}(\beta_0, \tau)$. At $\mathbf{q}$ close to the potential minimum $\mathbf{q}_m$, the momentum $\mathbf{p}(\mathbf{q})$ is linear with respect to the deviation $(\mathbf{q} - \mathbf{q}_m)$ and $W_0(\mathbf{q})$ is quadratic,

$$W_0 = \frac{1}{2}(\mathbf{q} - \mathbf{q}_m)^T \mathbf{A}_m (\mathbf{q} - \mathbf{q}_m) + o((\mathbf{q} - \mathbf{q}_m)^2) \tag{38}$$

Inserting this equation into the HJ equation, we obtain

$$\mathbf{A}_m \mathbf{g}_m \mathbf{A}_m = V_{\mathbf{qq}}(\mathbf{q}_m) \tag{39}$$

where $V_{\mathbf{qq}}$ is the matrix of the second derivatives of the potential, $\mathbf{g}_m$ is the compact notation for $g^{ij}(\mathbf{q}_m)$, and $\mathbf{A}_m = \mathbf{A}(\mathbf{q}_m)$. Once the instanton trajectory $\mathbf{q}_0(\tau)$ is found, then $2W_0$ can be calculated as the action integral along that. The quantity W_1 along the instanton trajectory is found from the transport equation [Eq. (35)]. Since $g^{ij}\partial_i W_0 \partial_j$ is nothing but the time derivative along the instanton trajectory $d/d\tau$, the direct integration using the Gaussian approximation in the direction perpendicular to the instanton gives

$$2W_{1\Sigma} = \int_{-\infty}^{0} d\tau [\mathrm{Tr}(\tilde{\mathbf{A}}(\tau) - \mathbf{A}_m + \mathbf{p}^T \mathbf{\Lambda}] \tag{40}$$

where $\mathrm{Tr}(\tilde{\mathbf{A}}) \equiv \tilde{A}_i^i = g^{ij}\tilde{A}_{ij}$, and we took into account the following relation

$$\omega = \frac{1}{2}\mathrm{Tr}(\mathbf{A}_m) + O(\hbar) \tag{41}$$

The matrix $\tilde{\mathbf{A}}$ is defined as

$$\tilde{A}_{ij}(\tau) = \frac{\partial^2 W_0}{\partial q^i \partial q^j} = \frac{\partial p_i}{\partial q^j} \tag{42}$$

and this matrix satisfies the first-order differential equation,

$$\frac{d\tilde{\mathbf{A}}}{dt} = -H_{\mathbf{qq}} - H_{qp}\tilde{\mathbf{A}} - \tilde{\mathbf{A}}H_{pq} - \tilde{\mathbf{A}}H_{pp}\tilde{\mathbf{A}} \tag{43}$$

where H_{qq}, and so on, are the matrices of the second derivatives of the classical Hamiltonian, $\partial^2 H / \partial q^i q^j , \ldots$. The initial condition is

$$\tilde{\mathbf{A}}(-\infty) = \mathbf{A}_m \tag{44}$$

Finally, the splitting can be expressed as

$$\Delta_0 = \sqrt{\frac{4 g_\Sigma \det \mathbf{A}_m}{\pi g_m \det \mathbf{A}_\Sigma}} \frac{(\mathbf{p}^T \mathbf{g} \mathbf{p})_\Sigma}{\sqrt{(\mathbf{p}^T \mathbf{A}^{-1} \mathbf{p})_\Sigma}} \exp[-2 W_{0\Sigma} - 2 W_{1\Sigma}] \tag{45}$$

where

$$\mathbf{A}_\Sigma \equiv \tilde{\mathbf{A}}_\Sigma + \delta \tilde{\mathbf{A}}_\Sigma \tag{46}$$

with

$$(\delta \tilde{\mathbf{A}}_\Sigma)_{ij} = -p_k(\mathbf{q}_\Sigma) \Gamma_{ij}^k(\mathbf{q}_\Sigma) \tag{47}$$

where

$$\Gamma_{ij}^k = \frac{1}{2} g^{ks} \left(\frac{\partial g_{is}}{\partial q_j} + \frac{\partial g_{js}}{\partial q_i} - \frac{\partial g_{ij}}{\partial q_s} \right) \tag{48}$$

are the Christoffel symbols.

Now, the general formulation of the problem is finished and ready to be applied to real systems without relying on any local coordinates. The next problems to be solved for practical applications are (1) how to find the instanton trajectory $\mathbf{q}_0(\tau)$ efficiently in multidimensional space and (2) how to incorporate high level of accurate *ab initio* quantum chemical calculations that are very time consuming. These problems are discussed in the following Section III. A. 2.

2. *How to Find Instanton Trajectory and How to Incorporate*
Accurate ab initio Quantum Chemical Calculations

The instanton trajectory $\mathbf{q}_0(\tau)$ to be found satisfies the boundary conditions,

$$\mathbf{q}_0(-\infty) = \mathbf{q}_m \qquad \mathbf{q}_0(\infty) = \tilde{\mathbf{q}}_m \tag{49}$$

and minimizes the classical action

$$S[\mathbf{q}(\tau)] = \int_{-\infty}^{\infty} L(\dot{\mathbf{q}}(\tau), \mathbf{q}(\tau)) d\tau \tag{50}$$

where $\mathbf{q}_m$ and $\tilde{\mathbf{q}}_m$ are the positions of the two symmetric potential minima, $V(\mathbf{q}_m) = V(\tilde{\mathbf{q}}_m) = 0$, and $L(\dot{\mathbf{q}}, \mathbf{q})$ is the classical Lagrangian for the inverted potential,

$$L(\dot{\mathbf{q}}, \mathbf{q}) = \frac{m}{2} g_{ij}(\mathbf{q}) \dot{q}_i \dot{q}_j + V(\mathbf{Q}) \tag{51}$$

One might think that it would be easy to find the instanton trajectory by running classical trajectories even in a multidimensional space. This is actually not true at all. Instead of doing that, we introduce a new parameter z, which spans the interval $[-1, 1]$ instead of using the time τ and employ the variational principle using some basis functions to express the tarjectory. The 1:1 correspondence between τ and z can be found from the energy conservation and the time variation of z is expressed as

$$\dot{z}(z) = \sqrt{\frac{2V(\mathbf{q}_0(z))}{g_{ij}(\mathbf{q}_0(z)) \frac{dq_0^i(z)}{dz} \frac{dq_0^j}{dz}}} \tag{52}$$

where the overdot means the derivative with respect to time. The proposed variational method is to minimize the classical action in the space of the paths expanded in terms of the basis functions as

$$q_0^i(z, \{C\}) = \frac{1}{2}[(\tilde{q}_m^i + q_m^i) + (\tilde{q}_m^i - q_m^i)z] + \sum_{n=1}^{N_b} C^{in} \phi_n(z) \tag{53}$$

where $\{\phi_n(z)\}$ is a certain set of smooth basis functions under the boundary condition $\phi_n(\pm 1) = 0$. The first two terms in the above equation represent a straight line connecting the two minima. The expansion coefficients $\{C\}$ are determined variationally. In the actual calculations, we have used the basis $\phi_n(z) = (1 - z^2)P_n(z)$, where $P_n(z)$ are the Legendre functions. Starting from a certain initial guess of the instanton path, for example, the straight line connecting the two minima, the path is improved iteratively by minimizing the action $S(\{C\})$ with respect to the coefficients $\{C\}$. In this procedure, the explicit time dependence of $z(\tau)$ does not have to be determined. Since $\dot{z}(z)$ is given analytically as a function of z, all the calculations can be carried out in terms of the parameter z. For example, the action S is given by

$$S = \int_{-1}^{1} dz \frac{1}{2} \dot{z}(z) g_{kl}(\mathbf{q}_0(z)) q_0'^k(z) q_0'^l(z) + \int_{-1}^{1} \frac{dz}{\dot{z}(z)} V(\mathbf{q}_0(z)) \tag{54}$$

where the "$\prime$" stands for the derivative with respect to the parameter z. In order to minimize the action S, we assume the variation of the path by shifting the coefficients, $C \rightarrow C + \delta C$. The action up to the second order with respect to δC is given by

$$S(\{\delta C\}) = S(\{C\}) + \delta C^T \frac{\partial S}{\partial \delta C} + \frac{1}{2} \delta C^T \frac{\partial^2 S}{\partial \delta C \partial \delta C} \delta C \qquad (55)$$

Minimization of this quantity gives a set of new coefficents and the improved instanton trajecotry. The second and third terms in the above equation require the gradient and Hessian of the potential function $V(\mathbf{q})$. For a given approximate instanton path, we choose N_r values of the parameter $\{z_n\}_{n=1,2,...N_r}$ and determine the corresponding set of N_r reference configurations $\{\mathbf{q}_0(z_n)\}$. The values of the potential, first and second derivatives of the potential at any intermediate z, can be obtained easily by piecewise smooth cubic interpolation procedure.

Although accurate determination of electronic states of large molecules is now possible with the use of the high level of *ab initio* quantum chemical methods, it is still a formidable task to obtain a *global* potential energy surface of such accuracy. Since the tunneling splitting dynamics is predominantly determined by the potential energies along the instanton trajectory and the global topology of the trajectory does not depend strongly on the quality of the *ab initio* method, fortunately, it is recommended to find the instanton path first by using a nontime-consuming low level *ab initio* method. In this first step, even the simple straight-line path can be assumed as an initial guess. As seen in the actual examples shown later, however, the high level *ab initio* method is definitely required eventually to obtain the reliable accurate values of tunneling splitting. Since the number of iterations strongly depend on the quality of the initial guess, it is recommended to improve the accuracy of the *ab initio* method steadily so that the computational efforts at the final stage can be reduced. The typical example of the low level *ab initio* methods is the MP2, and the coupled-cluster singles and doubles including a perturbational estimate of triple excitations [CCSD(T)] with the basis set of the Dunning's cc-pVDZ can be used as an intermediate level.

3. Numerical Applications

a. Malonaldehyde. Tunneling splitting in malonaldehyde (see Fig. 12) was discussed well both experimentally [79–82] and theoretically [83–88]. Malonaldehyde, as depicted in Fig. 12, is a nine-atom molecule and has a hydrogen bond of O–H. . .O where the tunneling of a hydrogen atom between two oxygen atoms occurs to cause the splitting in the vibrational energy levels. Tunneling splittings have been determined experimentally by the microwave spectroscopic studies by Wilson's group [79, 82] and by other groups [80, 81]. In such heavy–light–heavy systems, the reaction path is sharply curved near the transition state

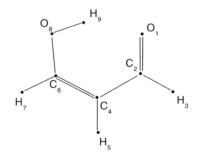

Figure 12.   Malonaldehyde molecule. Taken from Ref. [94].

and one has to take into account the multidimensional effects on the tunneling splitting. Carrington and Miller [84] first estimated the ground-state tunneling splitting based on the reaction surface Hamiltonian formalism that utilized two large amplitude coordinates and the extensions to three-dimensional (3D) models were done by Shida et al. [85]. Full-dimensional calculations were carried out by the semiclassical trajectory method [86] and by the instanton theory [83,87]. A full dimensional potential energy surface for malonaldehyde has been recently generated in [89] using the modified-Shepard interpolation (MSI) technique [90–92] at the level of the second-order Moller–Plesset perturbation theory (MP2) [93] with 6-31G(d,p) basis set, which was then used within the semiclassical trajectory method of Makri and Miller [68] to estimate the tunneling splitting. In this study, we discovered that the accuracy of the MSI (MP2/6-31G) potential energy function is not good enough and a much higher level *ab initio* method is required. Besides, the various dynamics methods used so far to estimate the tunneling splitting are also not accurate enough.

Applications of the theory described in Section III.A.2 to malonaldehyde with use of the high level *ab initio* quantum chemical methods are reported below [94,95]. The first necessary step is to define 21 internal coordinates of this nine-atom molecule. The nine atoms are numerated as shown in Fig. 12 and the Cartesian coordinates x_{in} in the body-fixed frame of reference (BF) $\mathbf{r}_n = \sum_{i=1}^{3} x_{in}\mathbf{e}_i$ where $n = 1, 2, \ldots 9$ numerates the atoms are introduced. This BF frame is defined by the two conditions. First, the origin is put at the center of mass of the molecule,

$$\sum_{n=1}^{9} M_n x_{in} = 0 \quad \text{for} \quad (i = 1, 2, 3, \cdots) \tag{56}$$

where M_n is the mass of the nth atom. The second condition fixes the orientation of the BF axis $\{\mathbf{e}_i\}$. We require that (1) the "tunneling" hydrogen atom (H_9) lies in the $(\mathbf{e}_1, \mathbf{e}_2)$ plane and (2) $\mathbf{e}_1$ is directed along the line connecting the two oxygen atoms O_1 and O_8. In terms of x_{in}, these two conditions read

$$x_{39} = 0, x_{21} = x_{28}, x_{31} = x_{38} \tag{57}$$

Equations (56) and (57) give six constrains and define the BF-system uniquely. The internal coordinates $q_k (k = 1, 2, \cdots, 21)$ are introduced so that the functions satisfy these equations at any q_k. In the present calculations, 6 Cartesian coordinates $(x_{19}, x_{29}, x_{18}, x_{11}, x_{21}, x_{31})$ from the triangle $O_8 - H_9 - O_1$ and 15 Cartesian coordinates of 5 atoms C_2, C_4, C_6, H_3, H_7 are taken. These 21 coordinates are denoted as q_k. Their explicit numeration is immaterial. Equations (56) and (57) enable us to express the rest of the Cartesian coordinates $(x_{39}, x_{28}, x_{38}, \mathbf{r}_5)$ in terms of $\{q_k\}$. With this definition, $x_{in}(q_1, q_2, \ldots, q_{21})$ are just linear functions of $\{q_k\}$, which is convenient for constructing the metric tensor. Note also that the symmetry of the potential is easily established in terms of these internal coordinates. This naturally reduces the numerical effort to one-half. Construction of the Hamiltonian for zero total angular momentum $(J = 0)$ is now straightforward. First, let us consider the metric,

$$ds^2 = \sum_n M_n d\mathbf{r}_n^2 \tag{58}$$

and perform the transformation $\{d\mathbf{r}_n\} \rightarrow \{dq_k, d\mathbf{\Omega}, d\mathbf{R}_{\text{c.m.}}\}$ where $d\mathbf{\Omega} = (d\omega_1, d\omega_2, \omega_3)$ represent the projections of the infinitesimal rotation onto the BF axes and $\mathbf{R}_{\text{c.m.}}$ is the center of mass position. The latter separates from the equation of motion and does not need to be considered. By inserting $d\mathbf{r}_n = (\partial \mathbf{r}_n / \partial q_k) dq_k + [d\mathbf{\Omega} \times \mathbf{r}_n]$ into Eq. (58), we obtain

$$ds^2 = \sum M_n [dx_{in} dx_{in} + 2e_{ijk} x_{kn} dx_{in} d\omega_j + (\mathbf{r}_n^2 \delta_{ij} - x_{in} x_{jn}) d\omega_i d\omega_j] \tag{59}$$

$$= \sum M_n \left(\frac{\partial x_{in}}{\partial q_k} \frac{\partial x_{in}}{\partial q_l} dq_k dq_l + 2e_{ijs} x_{sn} \frac{\partial x_{in}}{\partial q_l} dq_l d\omega_j + (\mathbf{r}_n^2 \delta_{ij} - x_{in} x_{jn}) d\omega_i d\omega_j \right) \tag{60}$$

where e_{ijk} is a fully antisymmetric tensor with respect to its three indices. Equation (59) explicitly determines the covariant metric tensor $g_{ij}(\mathbf{q})$ in terms of the 24 generalized coordinates $d\mathbf{Q} = \{d\mathbf{q}, d\mathbf{\Omega}\}$:

$$ds^2 = (d\mathbf{q}^T, d\mathbf{\Omega}^T) \begin{pmatrix} G_{qq}, G_{q\omega} \\ G_{\omega q}, G_{\omega\omega} \end{pmatrix} \begin{pmatrix} d\mathbf{q} \\ d\mathbf{\Omega} \end{pmatrix} \tag{61}$$

With the above choice of coordinates, the internal (G_{qq}), Coriolis $(G_{q\omega})$, and rotational $(G_{\omega\omega})$, parts of the metric are constant, linear, and quadratic functions of $\mathbf{q}$, respectively. The expression of the kinetic energy operator now reads

$$T = -\frac{\hbar^2}{2\sqrt{G}} \sum_{i,j=1}^{24} \frac{\partial}{\partial Q_i} \left(\sqrt{G} G^{ij} \frac{\partial}{\partial Q_j} \right) \qquad (62)$$

where $\partial/\partial \mathbf{Q} = (\partial/\partial \mathbf{q}, \partial/\partial \mathbf{\Omega})$, and G^{ij} are the elements of the inverse of the covariant metric tensor

$$\sum_{k=1}^{24} G_{ik} G^{kj} = \delta_{ij} \qquad (63)$$

and $G = |g_{ij}|$ is the determinant of the metric. By setting $\partial/\partial \mathbf{\Omega} \equiv \mathbf{J} = 0$ in Eq. (62), we obtain the kinetic energy operator for zero angular momentum $(J = 0)$ in the form of Eq. (32), where $g = G$ and g^{kl} is a 21×21 block of G^{ij}. Now, all the necessary computations can be carried out according to the formulas presented before.

The actual computations are performed as follows. The employed *ab initio* methods are the MP2, the quadratic configuration interaction method, including single and double substitutions (QCISD) [96] and the coupled-cluster singles and doubles including a perturbational estimate of triple excitations (CCSD(T)) [97] with the basis sets of the Dunning's cc-pVDZ and aug-cc-pVDZ [98, 99]. We also attempted the hybrid basis set of aug-cc-pVDZ (for two oxygen atoms and the transferring hydrogen atom H_9 in Fig. 12) and cc-pVDZ (for other atoms), denoted as (aug-)cc-pVDZ. *Ab initio* calculations were carried out by using GAUSSIAN 98 [100] and MOLPRO [101]. Table I gives a summary of barrier heights obtained by each electronic structure method, as well as that by the MSI(MP2/6-31G) potential energy function [88]. Comparisons of the results obtained by cc-pVDZ with those by aug-cc-pVDZ show that the calculations with augmented basis sets yield larger barriers by 0.38 and 0.61 kcal mol^{-1} at the MP2 and QCISD levels, respectively. The results with aug-cc-pVDZ and (aug-)cc-pVDZ are almost the same at both MP2 and QCISD levels, which indicates that the augmented basis sets need to be added only on two oxygen atoms and the moving hydrogen atom. Then, the coupled-cluster singles and doubles including a perturbational estimate of triple excitations [CCSD(T)] method with the (aug-)cc-pVDZ basis sets and with the hybrid basis set of aug-cc-pVTZ and cc-pVTZ, denoted as (aug-)cc-pVTZ, is applied to locate the stationary points. Table I shows the resulting BH at the CCSD(T)/(aug-)cc-pVDZ level as 4.53 kcal mol^{-1} which is between those of MP2 and QCISD obtained as 3.31 and 5.44 kcal mol^{-1}, respectively. The best estimation for the barrier height was obtained as 3.81 kcal mol^{-1} at CCSD(T)/(aug-)cc-pVTZ level of theory; the effect of triple-zeta level of basis sets decreases the BH.

TABLE I
Comparison of Barrier Heights (BH) in kcal mol^{-1} for the
Proton Transfer in Malonaldehyde Computed by
Different Quantum Chemical Methods

Barrier Height	
Quantum Chemical Method	BH
MP2/6-31G(d,p)[a]	3.62
/cc-pVDZ	2.91
/aug-cc-pVDZ	3.29
/(aug-)cc-pVDZ	3.31
QCSID/(aug-)cc-pVDZ	5.44
/cc-pVDZ	4.80
/aug-cc-pVDZ	5.41
CCSD(T)/(aug-)cc-pVDZ	4.53
/(aug-)cc-pVTZ	3.81

[a]Ref. [88].

Computations at the level of CCSD(T)/(aug-)cc-pVTZ are, however, very much time consuming and in the following instanton calculations the best level of theory employed is CCSD(T)/(aug-)cc-pVDZ.

We computed the tunneling splitting for MP2/cc-pVDZ, QCISD/(aug-)cc-pVDZ and CCSD(T)/(aug-)cc-pVDZ methods. The *ab initio* calculations were performed on Pentium 4 1.8 GHz Linux PC computer. To evaluate the energy, gradient, and Hessian matrix at one point it took 14 m, 1 d 3 h 43 m, 10 d 16 h 28 m cpu time for MP2/cc-pVDZ, QCISD/(aug-)cc-pVDZ and CCSD(T)/ (aug-)cc-pVDZ, respectively. We used the instanton trajectory of the MSI potential energy function as an initial guess for MP2/cc-pVDZ calculations, the new convergent $q_0(z)$ as the first approximation for the QCISD/(aug-)cc-pVDZ level, and so on. In all these cases, four extra iterations were good enough to achieve convergence at each higher *ab initio* level. Figure 13 depicts the instanton trajectory. For illustration purpose, the motion of four different atoms in the molecular plane is shown. In this figure, we also compare the instantons for the present accurate *ab initio* potential, the global MP2-level [88], and the semiempirical analytical potential energy surface [86]. The shape of the instanton in the semiempirical case differs drastically from the correct one. This finding is also clearly seen for the large amplitude motion of the "tunneling" H_9 atom. The reasonable value of the tunneling splitting obtained in Ref. [30] is due to mutual cancellation of the short "tunneling path" and the overestimated barrier height. Our final results for the tunneling splitting are presented in Table II. The values of $B, S_0 = 2W_0$ and $S_1 = 2W_{1\Sigma}$ are also shown, when the splitting is expressed as $\Delta_0 = B \exp(-S_0 - S_1)$. The MP2/cc-pVDZ and QCISD/(aug-)cc-pVDZ methods are not accurate enough. Both methods give

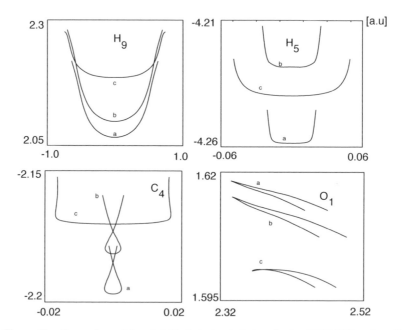

Figure 13. Comparison of the *ab initio* instanton trajectory by the CCSD(T)/(aug-)cc-pVDZ method with those obtained by (b) the global MP2 method and (c) analytic potential energy function. Taken from Ref. [95].

about four times of discrepancy compared with the experimental value $\Delta_0 = 21.6\,\text{cm}^{-1}$. This deviation is mainly due to the incorrect barrier height, which in MP2/cc-pVDZ and QCISD/(aug-)cc-pVDZ methods constitute 2.91 and $5.41\,\text{kcal}\,\text{mol}^{-1}$, respectively. The correct value lies in between these two (see Table I). The error also correlates with the behavior of the potential along the instanton path. However, the deviation of the results for the different *ab initio* methods comes mostly from the principal exponential factor S_0, which strongly depends on the barrier height. The preexponential factors are not that sensitive, indicating that the different *ab initio* methods produce the potential functions with similar topology. As a result, the shapes of the instanton trajectories are close to each other. The main differences among the three trajectories in Fig. 13 are actually due to their starting positions (potential minima). This explains the fact that at the higher *ab initio* computational level the convergence is easily achieved with a few extra step of the iteration. Thus, although the MP2/cc-pVDZ and QCISD/(aug-)cc-pVDZ calculations cannot produce a reliable value of the tunneling splitting, they enable us to reduce the numerical efforts on the higher *ab initio* levels. In this case, the instanton trajectory for CCSD(T)/ (aug-)cc-pVDZ was obtained by four extra steps of

TABLE II
Comparison of the Tunneling Splittings (Δ_0) in Malonaldehyde

(A) Case of Hydrogen			
$\Delta_0 = B\exp(-S_1 - S_0)$			
Method	$\Delta_0(\text{cm}^{-1})$	$B(\text{cm}^{-1})$	S_0	S_1
MP2/6-31G(p,d)[a]	30.7	1855	5.56	−1.46
/cc-pVDZ	77	1642	4.53	−1.48
QCISD/(aug-)cc-pVDZ	4.5	2200	7.38	−1.18
CCSD(T)/(aug-)cc-pVDZ	16.4	1922	6.14	−1.37
/(aug-)cc-pVTZ[b]	21.2	1642	5.83	−1.48
/(aug-)cc-pVTZ[c]	22.2	1922	5.83	−1.37
Tautermann(CCSD(T))[d]	24.7			
Empirical[e]	57.7			
Experiment[f]	21.6			

[a]Ref. [88].
[b]The parameter Δ_0 is estimated by taking B,S_1 from MP2/cc-pVDZ(4,6).
[c]The parameter Δ_0 is estimated by taking B,S_1 from CCSD(T)/(aug-)cc-pVDZ(5).
[d]Ref. [87].
[e]Ref. [83].
[e]Ref. [80].

(B) Case of Deuterium			
$\Delta_0 = B\exp(-S_1 - S_0)$			
Method	$\Delta_0(\text{cm}^{-1})$	$B(\text{cm}^{-1})$	S_0	S_1
MP2/6-31G(p,d)[a]	4.58	1408	7.09	−1.31
/cc-pVDZ	14.9	1269	5.74	−1.30
CCSD(T)/(aug-)cc-pVTZ[b]	3.0	1269	7.35	−1.30
Empirical[c]	8.63			
Experiment[d]	2.9			

[a]Ref. [88].
[b]The parameter Δ_0 is estimated by taking B,S_1 from MP2/cc-pVDZ(4,6).
[c]Ref. [83].
[d]Ref. [82].

iteration using only 33 *ab initio* points in total. We have obtained the tunneling splitting $16.4\,\text{cm}^{-1}$ in a relatively good agreement with the experimental value $21.6\,\text{cm}^{-1}$ [80]. As shown in Table I, the potential barrier is overestimated in the CCSD(T)/(aug-)cc-pVDZ method, which is likely to be the reason for conservative calculated value.

To study the accuracy of the above result we estimated the splitting for the higher CCSD(T)/(aug) cc-pVTZ level of electronic structure theory. The full implementation of the present method with this *ab initio* level is still too time consuming. The calculations of Hessian matrix would require ~ 140 days for one reference point or > 5 years for two steps of iteration. However, as noted above, the instanton paths for different *ab initio* methods show fairly similar topology. Thus, we can estimate Δ_0 by using the CCSD(T)/(aug) cc-pVTZ energy along the previously obtained instanton path. This gives a classical action S_0 at higher *ab initio* level while the other two factors B and S_1 can be taken from the lower level. In this way, we have obtained $\Delta_0 = 21.2(22.2)\,\mathrm{cm}^{-1}$ by taking the preexponential factors from the MP2 (CCSD) calculations. The isotope effect was estimated in the same way. For deuterium isotope of malonaldehyde, we first found the accurate instanton trajectory and calculated the splitting by the MP2/cc-pVDZ ab initio method. The action S_0 was recalculated for this path by using CCSD(T)/(aug-)cc-pVTZ *ab initio* potential points while two other factors B and S_1 remained unchanged. This gives $\Delta_0(D) = 3.0\,\mathrm{cm}^{-1}$ in very good agreement with the experimental data $2.9\,\mathrm{cm}^{-1}$ [82].

As mentioned above, it is unfortunate that the fully converged *ab initio* potential energies cannot be obtained because of too much cpu time consumption. From our experience in the *ab initio* calculations, however, we can roughly guess that the error of the effective barrier height would be $0.1-0.2\,\mathrm{kcal\,mol}^{-1}$, which leads to a possible error of $\sim 2-3\,\mathrm{cm}^{-1}$ in $\Delta_0(H)$. Of course, the accuracy of the present method is also affected by the instanton approach itself. For 1D models the latter typically gives 5% level of error or even better for large values of S_0. In Ref. [30], we have calculated the splitting in the H_2O complex and found the same accuracy of the semi-classical result. In the same reference, we have estimated both $\Delta_0(H)$ and $\Delta_0(D)$ of malonaldehyde using the empirical potential in Ref. [86] as $\Delta_0(H, D) = 57.7, 8.63\,\mathrm{cm}^{-1}$ in comparison with the values, $\Delta_0(H, D) = 21.8, 5.2\,\mathrm{cm}^{-1}$, reported in Ref. [86]. The absolute values are very different from each other, but the isotope effect is actually better reproduced by our calculations. We believe that the present theory is quite accurate and a practically useful theoy, being a kind of foolproof theory, and that the quality of the *ab initio* data is the most decisive factor for obtaining reliable values of tunneling splitting. By using the present framework of the theory, we can investigate the importance of various degrees of freedom. We actually investigated the effect of the out-of-plane vibration on the splitting [102]. The calculations can be carried out in the same way as before except that the metric g^{kl} is restricted to the 15×15 block of the G^{ij} in order to restrict the motion onto the molecular plane. Although the computations were carried out not at the highest *ab initio* level, but at the level of CCSD(T)/(aug-)cc-pVDZ, the result obtained is $\Delta_0(H) = 160\,\mathrm{cm}^{-1}$ in comparison with $16.4\,\mathrm{cm}^{-1}$ at the same *ab initio* level. This clearly indicates that the in plane and out-of-plane modes of the hydrogen atom are strongly coupled.

Before concluding, it is probably worthwhile to make the following comment concerning the computational strategy. In the present work, we first resorted to a hybrid scheme of the calculations by using the globally analytically fitted potential energy surface. Although this kind of preliminary step is impossible in general case (due to the absence of such kind of potential energy surface), it is actually not needed. According to the present methodology to find the instanton trajectory, the latter can be found after 10–12 steps of iteration by taking even the straight line as the initial guess. This requires 100–120 reference points in total and any non-time consuming *ab initio* level of quantum chemical computations can be readily used to accomplish this step. The obtained result can now be taken as the initial approximation for the highest possible *ab initio* level. This is now the second stage that requires only three to four extra steps of iteration to get the final result.

b. Vinyl Radical C$_2$H$_3$. The vinyl radical is well known to be an important intermediate in combustion chemistry (see Fig. 14). Very recently, Tanaka et al. [103] investigated this radical by millimeter-wave spectroscopy and reported a set of precise molecular constants together with the tunneling splitting. This five-atom molecule is a good example so that we can accomplish the full-scale high level quantum chemical calculations and confirm the accuracy of our semiclassical theory of tunneling splitting. The calculations were carried out in the same way as in malonaldehyde. The iterative finding of the instanton path is shown in Fig. 15, where the straight line is assumed as the initial guess as before at the MP2 level and the final one was found at the level of CCSD(T)/aug-cc-pVTZ. Only two more steps are required at this level to achieve the convergence. The final results are $\Delta_0(H) = 0.14\,\mathrm{cm}^{-1}$ and $\Delta_0(H) = 0.53\,\mathrm{cm}^{-1}$ by the MP2/6-31G(d,p) and the CCSD(T)/aug-cc-pVTZ method, respectively. The converged result, $\Delta_0(H) = 0.53\,\mathrm{cm}^{-1}$, is in good agreement with the experimental value, $\Delta_0(H) = 0.54\,\mathrm{cm}^{-1}$ [103]. This guarantees the accuracy of the present semiclassical theory at least for the ground vibrational state.

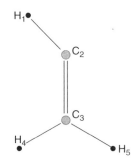

Figure 14. Vinyl radical. Taken from Ref. [104].

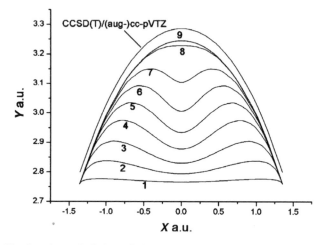

Figure 15. Iterative calculation of the instanton path. The labels 1–9 show gradual improvement of the instanton trajectory shape using the MP2/cc-pVDZ *ab initio* data. After switching to the CCSD(T)/(aug-)cc-pVDZ *ab initio* method, only two more steps needed to achieve convergence and obtain the final results. Taken from Ref. [104].

4. Tunneling Splitting in Low Vibrationally Excited States

The semiclassical theory introduced above can be extended to low vibrationally excited states [32]. The multidimensionality effects are more crucial in this case. As was found before [62, 70], the energy splitting may oscillate or even decrease against vibrational excitation. This cannot be explained at all by the effective 1D theory.

Let us first start with the simple 1D problem of the symmetric potential $V(-x) = V(x)$ with 2 equivalent potential minima $\pm x_m$. In our treatment, the principal exponential factor $W_0(x)$ does not depend on the energy E and thus the Hamilton–Jacobi equation does not change. Putting

$$E_{n=1} = E_{n=0} + \hbar\omega \tag{64}$$

and

$$W_1^{(n=1)} = W_1^{(n=0)} + w \tag{65}$$

then the additional factor w satisfies the following equation:

$$\frac{p_0(x)}{m}\frac{dw(x)}{dx} + \omega = 0 \tag{66}$$

This can be solved easily to give

$$
\begin{aligned}
w(x) &= -\int^x \frac{m\omega}{p_0(x)}\,dx = -\int^\tau \omega\,d\tau \\
&= \int_{-\infty}^\tau \left(\frac{1}{m}\frac{dp_0}{dx} - \omega\right) d\tau - \ln p_0(x)
\end{aligned}
\tag{67}
$$

At $x \to -x_m$, $\exp[-w(x)] \sim p_0(x)$ and the semiclassical wave function becomes

$$
\Psi_{(n=1)} = p_0(x)\exp[-m\omega(x+x_m)^2/2\hbar] \sim \omega(x+x_m)\exp[-n\omega(x+x_m)^2/2\hbar]
\tag{68}
$$

which coincides with the first excited state of harmonic oscillator. Taking into account the difference in the normalization factor $N_{(n=1)} = (\hbar m\omega/2)N_{(n=0)}$ from the ground state, we can finally obtain the splitting for the excited state as

$$
\begin{aligned}
\Delta_{(n=1)} &= 2\Delta_{(n=0)}\frac{p_0^2(0)}{\hbar m\omega}\exp\left[2\int_{-\infty}^0 d\tau\left(\omega - \frac{1}{m}\frac{dp_0}{dx}\right)\right] \\
&= \Delta_{(n=0)}\frac{V(0)}{\omega}\exp\left[2\int_{-\infty}^0 d\tau\left(\omega - \frac{1}{2V}\frac{dV}{d\tau}\right)\right]
\end{aligned}
\tag{69}
$$

Since $dp_0(x)/dx$ is a monotonically decreasing function of time and the barrier height $V(0)$ is always exceeds the excitation energy, we can see from Eq. (69)

$$
\Delta_{(n=1)} > \Delta_{(n=0)}
\tag{70}
$$

In the case of longitudinal excitation Eqs. (69) and (70) hold true in a general multidimensional system, as explained later.

For the purpose of illustration, let us next consider the simple 2D case. Introducing the local variables (s, ξ) where $s \in [-s_m, s_m]$ runs along the instanton path $q_0(s)$ and ξ is the coordinate perpendicular to the instanton, we obtain

$$
W_0(s, \chi) = \int_{-s_m}^s p_0(s)\,ds + \frac{m\theta(s)}{2}\chi^2
\tag{71}
$$

$$
W_1^{(n=1)}(s) = \int_{-s_m}^s ds\,\frac{m}{2p_0(s)}\left(\frac{1}{m}\frac{dp_0}{ds} + \theta(s) - \omega_\| - \omega_\perp\right)
\tag{72}
$$

where $\theta(s)$ plays a role of the effective frequency in the direction of the transversal local coordinate ξ, as given before. The additional term $w(s, \xi)$ in

Eq. (65) satisfies the equation,

$$\frac{p_0(s)}{m}\frac{\partial w}{\partial s} + \theta(s)\xi\frac{\partial w}{\partial \xi} + \frac{\Delta E}{\hbar} = 0 \tag{73}$$

where ΔE is the excitation energy.

In the case of longitudinal excitation ($n_\parallel = 1, n_\perp = 0$), $\Delta E = \hbar\omega_\parallel$ and the solution is the same as Eq. (69). In the transversal excitation case, $\Delta E = \hbar\omega_\perp$ and we obtain the solution of $w(s,\xi)$ as

$$w(s,\xi) = \int_{-s_m}^{s} ds\, \frac{m}{p_0(s)}[\theta(s) - \omega_\perp] - \ln\xi \tag{74}$$

The semiclassical wave function at $(s \to -s_m, \xi \to 0)$ becomes

$$\Psi = \xi \exp\left(-\frac{m\omega_\parallel(s + s_m)^2}{2\hbar} - \frac{m\omega_\perp\xi^2}{2\hbar}\right) \tag{75}$$

which coincides with the first excited state ($n_\parallel = 0, n_\perp = 1$) of the 2D harmonic oscillator. Considering the normalization factor $N_{n_\perp} = \hbar N_0/2m\omega_\perp$, we obtain finally

$$\Delta_{n_\perp=1} = -\Delta_0 \frac{\omega_\perp}{\theta(0)} \exp\left[2\int_{-\infty}^{0} d\tau(\omega_\perp - \theta)\right] \tag{76}$$

The effect of transversal excitation on the splitting depends on the behavior of $\theta(\tau)$ and is not simple as in the longitudinal case. If $\theta(\tau)$ grows (decreases) monotonically along the instanton path, then the excitation of the transversal mode suppresses (promotes) the tunneling splitting. Besides, as will be shown later, this 2D model is still not good enough, since the θ cannot be well approximated by only one mode perpendicular to the instanton path.

The above formulation can be generalized to a general multidimensional case in the form invariant under any coordinate transformation, as was done before for the ground-state case. We consider the general Hamiltonian given by Eq. (32). The formulation can be carried out in the same way as before. The equation for the additional term w is given by

$$g^{ij}\frac{\partial W_0}{\partial q_i}\frac{\partial w}{\partial q_j} + \Delta E = 0 \tag{77}$$

the solution can be obtained as

$$w = \int_{-\infty}^{\tau}[\theta(\tau) - \Delta E]d\tau - \ln(\mathbf{U}^T(\tau)\Delta\mathbf{x}) \tag{78}$$

where $\mathbf{U}$ and θ are defined as

$$\dot{U}_k = \theta(\tau)U_k - \left[g^{ij}\tilde{A}_{ik} + \frac{\partial g^{ij}}{\partial q_k}p_{0i}\right]U_j \tag{79}$$

and

$$\theta(\tau) = \sum_{ij} U^i U^j \mathbf{A}_{ij} \tag{80}$$

where $U^i = g^{ik}U_k$ and the normalization condition should be satisfied as

$$\sum_{ij} g^{ij}U_iU_j = 1 \tag{81}$$

The tunneling splitting is finally given by

$$\Delta_{n_{\gamma=1}} = \Delta_0\omega_\gamma \left(\mathbf{U}^T\left[\mathbf{A}^{-1} + \frac{(\mathbf{A}^{-1}\mathbf{p}_0)\cdot(\mathbf{p}_0^T\mathbf{A}^{-1})}{(\mathbf{p}_0^T\mathbf{A}^{-1}\mathbf{p}_0)}\right]\mathbf{U}\right)_\Sigma \exp[-\Delta S_1] \tag{82}$$

where ΔS_1 is the correction due to the first term of w,

$$\Delta S_1 = 2\int_{-\infty}^0 [\theta(\tau) - \omega_\gamma]d\tau \tag{83}$$

The matrix $\mathbf{A}$ is the same as before.

Finally, the tunneling splittings are given by Eq. (69) and Eq. (82) for the longitudinal and transversal single mode excitation, respectively. Generalization to the M-mode excitation ($n_{\gamma_k} = 1, k = 1, 2, \cdots, M, n_{\gamma'} = 0, \gamma' \neq \gamma_k$) is straightforward. This is not discussed here [32].

The above theory has been applied to HO_2 and the vinyl radical C_2H_3 [32, 104]. Only the final numerical results are shown here. Table III gives the results for HO_2. The quantum numbers n_1, n_2, and n_3 correspond to HO stretch, HO_2 bend, and O_2 stretch. As can be seen, the effect of vibrational excitation varies from mode to mode and the peculiarity is well reproduced by the present semiclassical theory. Figure 16 shows the change of θ along the instanton path and clearly demonstrates the insufficiency of the adiabatic approximation. There are strong interactions among transversal vibrations and these features cannot be reproduced by any 2D model. Table IV shows the results for the vinyl radical. Again the effect of the excitation varies from mode to mode. In this case, the lowest excitation, that is, the rocking vibration, corresponds to the longitudinal one and the enhancement is much larger than the others. The absolute values of splitting cannot be well reproduced by the low level of quantum chemical calculations, such as MP2, but the ratio Δ_n/Δ_0 can be relatively well reproduced by them.

TABLE III
Tunneling Splitting of the Low Excited States of HO_2[a,b]

(n_1, n_2, n_3)	Splitting in Excited States	
	Δ_n/Δ_0(exact)	Δ_n/Δ_0(SC)
(0,0,0)	1	1
(0,0,1)	1.82	1.62
(0,1,0)	0.77×10^2	1.02×10^2
(0,1,1)	1.34×10^2	1.65×10^2
(1,0,0)	1.12×10^3	0.9×10^3
(1,0,1)	2.32×10^3	2.0×10^3
(1,1,0)	5.76×10^4	8.7×10^4

[a] The exact value and the semiclassical value of the ground state are Δ_0(exact) $= 0.77 \times 10^{-12}$eV and Δ_0(SC) $= 0.78 \times 10^{-12}$eV. The parameter n_1, n_2, and n_3 are quantum numbers corresponding to HO stretch, HO_2 bend, and O_2 stretch, respectively.

B. Decay of Metastable State through Tunneling (Predissociation)

The theory developed for tunneling splitting can be easily extended to the decay of the metastable state through multidimensional tunneling, namely, tunneling predissociation of polyatomic molecules. In the case of predissociation, however, the instanton trajectory cannot be fixed at both ends, but one end should be free (see Fig. 17). The boundary conditions are

$$\mathbf{q}_0(\tau = -\infty) = \mathbf{q}_m \tag{84}$$

and

$$V(\mathbf{q}(\tau = 0)) = 0 \tag{85}$$

The basis functions $\{\phi_n(z)\}$ used to define the instanton trajectory is

$$\phi_n(z) = z^n \tag{86}$$

where $z \in [0, 1]$ and $\phi_n(z = 0) = 0$. The functionality of $z(\tau)$ is the same as before, and $z(\tau = -\infty) = 0, \mathbf{q}_0(z = 0) = \mathbf{q}_m$, and $V(\mathbf{q}_0(\alpha)) = 0$, where α is a scaling factor introduced below. Thus the instanton path is given by

$$q_0^i(z) = q_m^i + \sum_{n=1}^{N_b} C^{in} z^n \tag{87}$$

where the expansion coefficients $\{C^{in}\}$ are determined iteratively by minimizing the action in the same way as before. However, the old path cannot be used as the

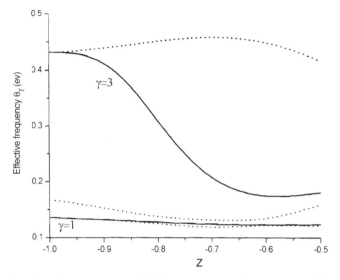

Figure 16. The effective frequency $\theta(\tau)$ for two transversal excitations in HO_2. Solid line shows the results of numerical solution of Eqs. (79) and (80). Dotted line represents the frequency in the adiabatic approximation. $\gamma = 1[3]$ corresponds to the mode $(0, 0, 1)$ $[(1, 0, 0)]$ (see Fig. 14). Taken from Ref. [32].

initial guess for the next iteration, since the second boundary condition Eq. (85) is not satisfied. Thus one more step to renormalize the parameter z, that is, the scaling $z \rightarrow \alpha z$, is introduced so that Eq. (85) is satisfied. This scaling simply corresponds to the transformation

$$C^{in} \rightarrow \alpha^{in} C^{in} \tag{88}$$

TABLE IV
Normal Frequencies in (cm^{-1}) and Corresponding
Tunneling Splitting for the First Excited States of Vinyl Radical[a,b]

		Frequency and Tunneling Splitting			
n	Type of the Motion	ω_n[CCSD(T)]	ω_n[MP2]	Δ/Δ_0[CCSD(T)]	Δ/Δ_0[MP2]
1	C_2H_1 rocking vibration	711	771	36.0	41.1
2	Wagging (out-of-plane)	813	996	2.20	1.76
3	Wagging (out-of-plane)	923	1063	1.28	1.12
4	plane distortion	1062	1129	3.0	2.3
5	$H_4C_3H_5$ bending	1390	1465	17.0	1.5
6	C_2C_3 stretching	1632	1863	200.0	70.0

[a]See Fig. 14.
[b]The C_3 atom in the wagging mode $n = 2$ shifts more than the atom C_2 in the wagging mode $n = 3$.

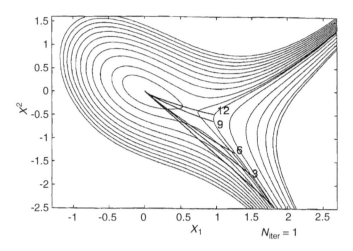

Figure 17. Iteration process of the calculation of instanton trajectory in the cubic potential for $N = 2$ and $C_2 = 0.6$ in Eq. (98). The parameter N_{iter} is the number of iteration. Taken from Ref. [31].

Except for this scaling, the other procedures are the same as in the case of tunneling splitting. The canonically invariant formula for the decay rate k can be finally obtained as

$$k = \sqrt{\frac{\det \mathbf{A}_m}{\pi g_m \det \mathbf{A}(0)}} \frac{\mathbf{p}^T \mathbf{g} \mathbf{p}_0}{\sqrt{(\mathbf{p}^T \mathbf{A}^{-1} \mathbf{p})_0}} \exp[-S_0 - S_1] \tag{89}$$

where S_0 is the action integral along the instanton path,

$$S_1 = \int_{-\infty}^{0} d\tau [\text{Tr}(\mathbf{A}(\tau) - \mathbf{A}_m) + \mathbf{p}^T \boldsymbol{\Lambda}] \tag{90}$$

$$\Lambda^i = \frac{\partial}{\sqrt{g} \partial q_j} (\sqrt{g} g^{ij}) \tag{91}$$

where $\text{Tr}(\mathbf{A}) \equiv A_i^i = g^{ij} A_{ji}$. The symmetric matrix $\mathbf{A}(\tau)$ is the solution of of the equation

$$\dot{\mathbf{A}} = -H_{\mathbf{qq}} - H_{\mathbf{qp}} \mathbf{A} - \mathbf{A} H_{\mathbf{pq}} - \mathbf{A} H_{\mathbf{pp}} \mathbf{A} \tag{92}$$

Note that there are removable singularities at both ends of the instanton path, namely, at $z = 0$ and $z = 1$. Thus, in the practical numerical computations some modifications to remove the singularities are necessary. The details are not given here and the reader should refer to Ref. [30, 31]. Finally, the numerically stable formula is obtained as follows from Eq.(89):

$$k = \omega_z \sqrt{\frac{g_0 \det \mathbf{A}_m}{\pi g_m \det \mathbf{A}_0}} \frac{(\tilde{\mathbf{p}}^T \mathbf{g} \tilde{\mathbf{p}})_0}{\sqrt{(\tilde{\mathbf{p}}^T \mathbf{A}^{-1} \tilde{\mathbf{p}})_0}}$$

$$\times \exp\left(-S_0 - \int_0^1 dz \left[I(z) - \frac{d\dot{z}}{\dot{z} dz} + \frac{1}{z} \right]\right) \tag{93}$$

where

$$I(z) = \frac{1}{z}[\mathrm{Tr}(\mathbf{A}(z) - \mathbf{A}_m) + \mathbf{p}^T \Lambda] \tag{94}$$

$$\tilde{p}_i(z) = \frac{p_i(z)}{\dot{z}(z)} \tag{95}$$

$$\dot{z}(z) = \omega_z \exp\left(\int_0^z dz \left[\frac{d\dot{z}}{\dot{z} dz} - \frac{1}{z} \right]\right) \tag{96}$$

$$\omega_z = \frac{\dot{z}}{z}\Big|_{z=0} \tag{97}$$

In Eq.(94), everything is convergent.

Numerical calculations were carried out in order to test the whole numerical algorithm and accuracy of the rate calculation. The potential system employed is a nonlinearly transformed model of the separable case [31]. That is

$$V(\mathbf{x}) = \frac{\mu \omega_0^2 x_1^2}{2}\left(1 - \frac{x_1}{x_0}\right) + \sum_{i=2}^N \frac{\mu \omega_i^2 x_i^2}{2} \tag{98}$$

where μ is taken to be the mass of proton, $\omega_0 = 650\,\mathrm{cm}^{-1}$, $\omega_i = 1000\,\mathrm{cm}^{-1}$ $(i = 1, 2, \cdots, N)$, $x_0 = 1.5$ a.u., and N represents the dimension of the system. Since this is a separable system, the integrals can be performed analytically and the the rate constant is given by [59]

$$k = 4\frac{\omega_0^2 x_0 \mu}{\pi} \exp[-S_0] \tag{99}$$

with

$$S_0 = \frac{8}{15}\mu \omega_0 x_0^2 \tag{100}$$

Numerical calculations were performed after applying the following nonlinear transformation,

$$q_k = T_{k1}x_1 + \sum_{i=2}^N T_{ki}(x_i + 1 + x_1^2)^2 \qquad k = 1, 2, \cdots, N \tag{101}$$

where $T = \{T_{ki}\}$ is an orthogonal matrix defined by

$$T = T^{(1N)} T^{(1(N-1))} \cdots T^{(12)} \tag{102}$$

and $T^{(1j)}$ is the matrix of rotation by angle $\gamma = 0.1$ in (x_1, x_j) plane. The calculations were carried out for $N = 2, 4, 10, 20$. The accuracy was found to be the same in all these cases. The number of basis functions used is $N_b = 10$, which is good enough to obtain the converged results in all cases, and the number of iteration (N_{iter}) is 17, when the straight line is used for the initial guess of the instanton path. The final result of the rate in the present semiclassical theory is $k = 3.611 \times 10^{-5}$ in comparison with the analytical answer $k = 3.608 \times 10^{-5}$.

C. Tunneling in Reactions

As mentioned before, in order to take into account quantum mechanical tunneling effects in the semiclassical methods, it is necessary to detect the boundary between the classically accessible region and the classically forbidden region. This is nothing but the well-known turning point in the ordinary 1D problems and is called *caustics* in general multidimensional case. The caustics are an envelope of the turning points of each classical trajectory and makes a multidimensional surface. In this section, an efficient method to detect the caustics is explained together with some numerical applications.

1. How to Detect Caustics

It is well known that in a $2N$-dimensional phase space, N-dimensional Lagrange manifold is generated by a continuous set of the map of coordinates and momenta in time, namely, classical trajectories, $\{q(t), p(t)\}$. On projecting this manifold onto the configuration space, some points show up as singularities, namely, caustics. These singularities can be mathematically expressed in two forms: either $\partial q(t)/\partial q(0) = 0$ or $\partial p(t)/\partial q(t) = \infty$. This suggests two approaches to determine the location of caustics. The first approach to propagate $\partial q(t)/\partial q(0)$ is based on the solution of a certain coupled linear differential equations [105, 106]. This system of coupled equations can be reduced to a second-order differential equation and in the long time propagation the solution may become unstable due to exponentially growing and decreasing terms. The second approach we have proposed is to propagate $\partial p(t)/\partial q(t)$, which is a solution of the nonlinear Riccati-type differential equation [30]. The high degree of numerical stability common to this type of differential equation makes this approach particularly appealing. The solution diverges at the caustics that are an intrinsic property of this solution. We have devised a method to avoid this divergence and to efficiently detect caustics [29].

In a Lagrange manifold, the matrix

$$A_{ij} = \frac{\partial p_i(t)}{\partial q_j(t)} \qquad (i,j) = 1, 2, \cdots, N \tag{103}$$

satisfies, along the classical trajectory, the following Riccati-type differential equation [30]

$$\frac{dA}{dt} = -H_{qq} - H_{qp}A - AH_{pq} - AH_{pp}A \qquad (104)$$

where $H_{\alpha\beta}(\alpha, \beta = p, q)$ are the matrices of second dreivative of the Hamiltonian taken along the classical trajectory, namely, $\partial^2 H/\partial q \partial p$, and so on. During the propagation, the solution of Eq. (104) diverges at the caustics as

$$\text{Det}|A(t_{\text{caustics}})| = \infty \qquad (105)$$

Once this divergence happens, further solution of the differential equation is not possible beyond this point, and we have to reformulate the problem. To clarify our idea, let us consider the 1D problem. At the turning point, $p(q) = 0$ and A diverges. If we invert A to $\tilde{A} = \partial q/\partial p$, the divergence is removed and the propagation of $\tilde{A}$ proceeds smoothly through the caustics. This inversion is equivalent to the canonical transformation, $(p, q) \rightarrow (-\tilde{q}, \tilde{p})$. It can be easily shown that Eq.(104) does not change under this transformation. In a general N-dimensional case, it is recommended to invert only the diverging element(s) selectively. If the diverging element is assumed to be A_{NN}, the transformation $(p_N, q_N) \rightarrow (-\tilde{q}_N, \tilde{p}_N)$ eliminates the divergence and mixes the coordinates and momenta. The new matrix $\tilde{A}$ formed becomes free of any diverging element and its propagation proceeds smoothly.

In the general $N \times N$ matrix case, this procedure can be summarized as follows: (1) Irrespective of the position of the diverging element, a simple rotation ensures the repositioning of the diverging element as the (N, N) element. This rotation is a canonical transformation and can be achieved best by using the orthogonal matrix, which diagonalizes the matrix A. Denoting this diagonalizing matrix as S, the rotationally transformed matrix is obtained by

$$p' = Sp \qquad \text{and} \qquad q' = Sq \qquad (106)$$

so that

$$A' = SAS^T \qquad (107)$$

(2) By invoking the transformation $(p'_N, q'_N) \rightarrow (-\tilde{q}_N, \tilde{p}_N)$, we obtain $A' \rightarrow \tilde{A}$, where

$$\begin{aligned} \tilde{p}_i &= p'_i & i &= 1, \ldots, N-1 & (108) \\ \tilde{q}_i &= q'_i, & i &= 1, 2 \ldots, N-1 & (109) \\ p'_N &= -\tilde{q}_N \quad \text{and} \quad q'_N = \tilde{p}_N & & & (110) \end{aligned}$$

The second derivative coefficients of the Hamiltonian in the new representation are derived by rotating the old coefficients. For example,

$$(H_{\tilde{q}\tilde{q}})_{NN} = (H_{pp})_{NN} \tag{111}$$

The new matrix $\tilde{A}$ satisfies Eq. (104) using the modified second derivatives as coefficients. The propagation of the new matrix $\tilde{A}$ runs smoothly through the hitherto divergent region and the caustics can be accurately detected from the solution of Eq. (104) in the new representation. Well beyond the point of divergence, the inverse transformation is carried out in exactly reverse order to revert to the matrix A and the original propagation is performed. In some cases, it may happen that two or more eigenvalues of the matrix A almost simultaneouly diverge around the same propagation time, especially when the multidimensional potential has a deep well or the dynamics becomes chaotic. This requires consecutive multiple canonical transformations in a short time step. Some care should be taken not to miss this kind of closely occurring caustics by using small time steps. Since the numerical procedure itself is quite stable, it is enough to use small time steps in such a case.

2. Numerical Examples

a. *Henon–Heiles System.* To test the ideas presented above, we have applied the method to a 2D Henon–Heiles Hamiltonian (in au)

$$H = \frac{1}{2}(p_x^2 + p_y^2) + \frac{1}{2}(x^2 + y^2) + (x^2 y - \frac{1}{3}y^3) \tag{112}$$

The classical trajectory was generated from the turning point $(\vec{p}(0) = 0)$, which corresponds to the initial condition $A^{-1} = 0$ for Eq. (104). After a short time propagation of A^{-1}, the further propagation is made with the matrix A. The actual initial conditions for (x, y) are $x_0 = -0.43$ with $y_0 = -0.37, -0.39$ and -0.41, which correspond to the regular, partially chaotic and fully chaotic regimes, respectively. The trajectories (lines) and caustics (crosses) are shown in Figs. 18–20. In the case of regular regime (Fig. 18), the caustics clearly provide the envelope of the family of trajectories. As the system becomes chaotic, the caustics is no longer discernible with the tori partially (Fig. 19) or fully (Fig. 20) destroyed. As can be seen in these figures, the present method works well in all these cases.

b. *Reaction Dynamics.* The second example is a triatomic chemical reaction. As was mentioned in Section II, in order to run tunneling trajectories in the classically forbidden region, it is inevitable to detect efficiently the caustics along each trajectory in the classically allowed region. We have employed the ground

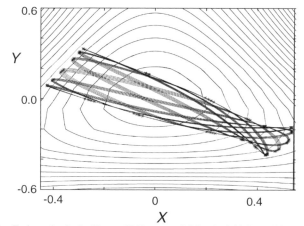

Figure 18. Trajectories in the Henon–Heiles potential for the initial condition: $x_0 = -0.43$ and $y_0 = -0.37$. The marks (*) indicate the location of caustics. Taken from Ref. [29].

adiabatic potential energy surface obtained from the DIM matrix mimicking the CH_2 molecule [51]. The DIM potential matrix is given by

$$V = \begin{pmatrix} G_1 & (g_3 - h_3)/2 & (g_2 - h_2)/2 \\ (g_3 - h_3)/2 & G_2 & (g_1 - h_1)/2 \\ (g_2 - h_2)/2 & (g_1 - h_1)/2 & G_3 \end{pmatrix} \tag{113}$$

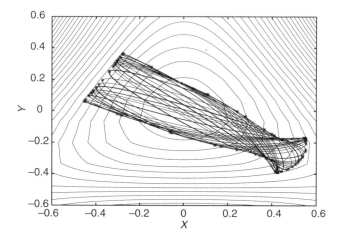

Figure 19. The same as in Fig. 18 for the initial condition: $x_0 = -0.43$ and $y_0 = -0.39$. Partial destruction of regular caustics is seen. Taken from Ref. [29].

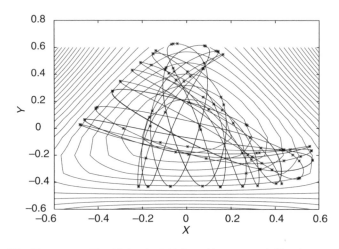

Figure 20.   The same as Fig. 18 for the initial condtion: $x_0 = -0.43$ and $y_0 = -0.41$. Total destruction of regular caustics is seen. Taken from Ref. [29].

where $G_i = g_i + (g_j + g_k + h_i + h_k)/2 (i = 1, 2, 3; j, k \neq i)$, and (h_1, h_2, h_3) and (g_1, g_2, g_3) represnt the ground and excited states of H_2, CH_a, and CH_b, respectively. The functional forms of them are given by

$$h_1(r) = -D_h[1 + \alpha_1(r - r_{H_2}) + \alpha_2(r - r_{H_2})^2$$

$$+ \alpha_3(r - r_{H_2})^3] \exp[-\alpha_4(r - r_{H_2})] \tag{114}$$

$$g_1(r) = D_g[1 + \beta_1 r + \beta_2 r^2] \exp[-\beta r] \tag{115}$$

$$h_{2,3}(r) = B_h[\exp[-\gamma_h(r - r_{CH})] - 2] \exp[-\gamma_h(r - r_{CH})] \tag{116}$$

$$g_{2,3}(r) = B_g[\exp[-\gamma_g(r - r_{CH})] - 2] \exp[-\gamma_g(r - r_{CH})] \tag{117}$$

where the parameters r_{CH}, r_{H_2} are 2 and 1.401 a.u., $\alpha_i = 2.1977034$, $1.2932502, 0.64375666, 2.835071 (i = 1 - 4)$, $\beta_i = -1.3874149, 0.9098728$, $2.181301 (i = 1 - 3)$, $D_j = 0.15796326, 4.502447 (j = h, j)$, $B_j = 0.13, 0.10$ $(j = h, g)$, and $\gamma_j = 1.3, 1.5 (j = h, g)$. The total angular mometum J is taken to be zero, the collision energy and the initial rovibrational states are assumed to be 1.2 eV and $(v = 0, j = 0)$. Note that this potential energy surface has an attractive deep well of depth ~ 2.3 eV. Thus many trajectories are trapped in the well region for long time. The four coordinates to describe the triatomic system with $J = 0$ are denoted as $X, Y, x,$ and y. The initail condition for the matrix A is derived from the energy and momentum conservations. Differentiating each of these conservation equations partially with respect to the

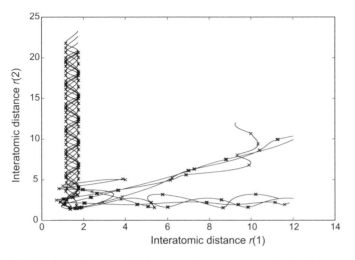

Figure 21. Family of reactive trajectories in the ground adiabatic potential energy surface determined by Eq. (13). Crosses indicate the caustics. Taken from Ref. [29].

four coordinates, we can obtain the analytical expressions for the Jacobian matrix $D(P_X, P_Y, p_x, p_y)/D(X, Y, x, y)$ in the asymptotic region. The initial condition so obtained is

$$A = \begin{pmatrix} AA & 0 \\ 0 & aa \end{pmatrix} \tag{118}$$

where

$$AA = \frac{1}{\mathbf{P_X X} + \mathbf{P_Y Y}} \begin{pmatrix} P_Y^2 & -P_X P_Y \\ -P_X P_Y & P_X^2 \end{pmatrix} \tag{119}$$

and

$$aa = \frac{1}{\mathbf{p_x x} + \mathbf{p_y y}} \begin{pmatrix} -\mu x \frac{\partial v}{\partial x} + p_y^2 & -\mu x \frac{\partial V}{\partial y} - p_x p_y \\ -\mu y \frac{\partial V}{\partial x} - p_x p_y & -\mu y \frac{\partial V}{\partial y} + p_x^2 \end{pmatrix} \tag{120}$$

An example of a family of classical trajectories together with the caustics are shown in Fig. 21. The caustics in the asymptotic region appear periodically as turning points of the vibrational motion. In the rearrangement region, the trajectories are no more periodic and the caustics appear rather randomly. As is seen, the present method is demonstrated to work well.

Figure 22 shows an application of the present method to the H_3 reaction system and the thermal rate constant is calculated. The final result with tunneling effects included agree well with the quantum mechanical transition state theory calculations, although the latter is not shown here.

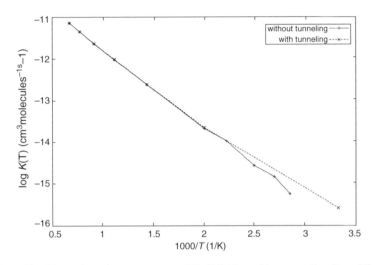

Figure 22. Thermal reaction rate constant of H_3. Solid line-without tunneling. Dotted line-with tunneling. Taken from Ref. [9].

IV. ELECTRON TRANSFER: IMPROVEMENT OF THE MARCUS THEORY

The electron-transfer reaction from donor to acceptor molecule plays a fundamental role in chemical and biological systems and numerous theoretical works have been done since the pioneering work by Marcus (see, e.g., [107–109]). This is nothing but a nonadiabatic process due to the potential energy surface crossing and thus is a good target for the ZN theory to be applied. For the nonadiabatic transition, the simple perturbation theory or the LZ formula has been employed in most of the theoretical works. In this sense, the ZN formulas can be usefully utilized in order to improve the theory to cover from the intermediate to strong electronic coupling regime. Besides, one can treat the classically forbidden transitions that cannot be dealt with by the above two methods except for perturabtion theory in the weak coupling regime. In many cases, the nonadiabatic transition and the nuclear tunneling are treated separately. As the ZN formulas in the NT case clearly indicate, this is not allowed, that is, both processes are coupled and cannot be separated unless the energy is very low compared to the barrier top of the lower adiabatic potential.

The nonadiabatic transition state theory given in the Section II.C, namely, Eq. (17), can be applied to the electron-transfer problem [28]. Since the electron transfer theory should be formulated in the free energy space, we introduce the free energy profile $F(\xi)$ as

$$\exp[-\beta F(\xi)] = \int d\mathbf{Q} \exp[-\beta V(\mathbf{Q})] |\nabla(S(\mathbf{Q}))\delta(S(\mathbf{Q}) - \xi) \tag{121}$$

Then the rate constant k can be rewitten as

$$k = (Z_r^q)^{-1} \sqrt{\frac{1}{2\pi\beta}} \int d\xi \delta(\xi - \xi_0) \exp[-\beta F(\xi)] \bar{P}(\beta, \xi) \tag{122}$$

The average transition probability $\bar{P}(\beta, \xi)$ is defined by

$$\bar{P}(\beta, \xi) = \frac{\int d\mathbf{Q} \exp[-\beta V_1(\mathbf{Q})] |\nabla S(\mathbf{Q})| \delta(\xi - S(\mathbf{Q})) P(\beta, \mathbf{Q})}{\int d\mathbf{Q} \exp[-\beta V_1(\mathbf{Q})] |\nabla S(\mathbf{Q})| \delta(\xi - S(\mathbf{Q}))} \tag{123}$$

where the transition probability $P(\beta, \mathbf{Q})$ at a give temperature and at the position $\mathbf{Q}$ on the seam surface is given by Eq. (18). The crossing seam surface $\xi_0 = S(\mathbf{Q})$ of the potentials of donor and acceptor is taken to be the nonadiabatic transition state. It can be shown that Eqs. (122) and (123) are essentially the same as those in Ref. [110], when the 1D reaction coordinate is assumed and the nuclear tunneling effect is neglected. In our treatment,the multidimensionality is taken into account and the instantaneous normal mode analysis is made at the transition point to determine the normal direction to the seam, that is, the reaction coordinate. In this way, the reaction coordinate has the maximum mean free path and thus the theory is supposed to be applicable in a wide range of friction.

In order to find the relation between Eq. (122) and the Marcus theory, we employ the linear response approximation. In this case, the free energies $F_j(\xi)(j = 1, 2)$ for the donor and acceptor become a parabolic function of ξ as

$$F_1(\xi) = -\frac{1}{\beta} \ln \left[\int d\mathbf{Q} \exp[-\beta V_1(\mathbf{Q})] |\nabla S(\mathbf{Q})| \delta(\xi - S(\mathbf{Q})) \right]$$
$$= \frac{1}{2} \omega^2 (\xi - \xi_{01})^2 \tag{124}$$

and

$$F_2(\xi) = -\frac{1}{\beta} \ln \left[\int d\mathbf{Q} \exp[-\beta V_2(\mathbf{Q})] |\nabla S(\mathbf{Q})| \delta(\xi - S(\mathbf{Q})) \right]$$
$$= \frac{1}{2} \omega^2 (\xi - \xi_{02})^2 + \Delta G \tag{125}$$

where $\xi_{0i}(i = 1, 2)$ are the positions of donor and acceptor free energy minima, respectively, and ΔG represents the exothermicity of the reaction that is determined by setting $\xi = \xi_{02}$ in Eq. (125).

Then, finally we can have

$$k = \kappa k_{\text{Marcus}} \tag{126}$$

with

$$\kappa = \frac{\hbar\omega}{2\pi H_{AD}^2} \sqrt{\frac{\lambda}{\pi\beta}} \bar{P}(\beta, \xi_0) \qquad (127)$$

where k_{Marcus} is the Marcus formula defined by

$$k_{Marcus} = \frac{H_{AD}^2}{\hbar} \sqrt{\frac{\pi\beta}{\lambda}} \exp\left[-\frac{\beta(\lambda + \Delta G)^2}{4\lambda}\right] \qquad (128)$$

and H_{AD} is the electronic coupling between acceptor and donor (AD). The reorganization energy λ is defined by

$$\lambda = \frac{1}{2}\omega^2(\xi_{02} - \xi_{01})^2 \qquad (129)$$

The factor κ takes into acount the effects of nonadiabatic transition and tunneling properly. Also note that the electronic coupling H_{AD} is assumed to be constant in the Marcus formula, but this is not necessary in the present formulation. The coupling H_{AD} cancels out in k of Eq. (126) and the ZN probability can be calculated from the information of adiabatic potentials.

The present formula Eq. (126) is tested in comparison with the Bixon–Jortner perturbation theory in the weak electronic coupling regime [109]. The Arrhenius plot is shown in Fig. 23, where the electronic coupling H_{AD} is taken

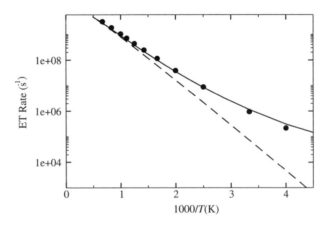

Figure 23. Arrhenius plot of the electron transfer rate. The electronic coupling strength is $H_{AD} = 0.0001$ a.u. Solid line-Bixon–Jortner perturbation theory Ref. [109]. Full-circle:present results of Eq. (26}). Dashed line-results of Marcus's high temperature theory [Eq.(129)]. Taken from Ref. [28].

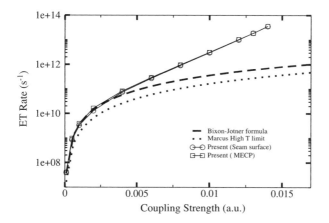

Figure 24. Electron-transfer rate versus electronic coupling strength. The temperature is $T = 500$ K. Solid line with circle-present results from Eq. (126) with the transition probability averaged over the seam surface. Solid line with square-present results with the transition probability taken at the minimum energy crossing point (MECP). Dashed line-Bixon–Jortner theory Ref. [109]. Dotted line-Marcus's high temperature theory. Taken from Ref. [28].

to be 0.0001 a.u. and the results of the Marcus high temperature theory is also presented. It is clearly seen that the present formula is accurate enough and thus expected to be applicable at stronger couplings. The model used is the symmetric $(\Delta G = 0)$ 12-dimensional harmonic oscillators, the parameters of which are given in [28]. Figure 24 shows the rate against the coupling strength at $T = 500$ K in comparison with the Bixon–Jortner and the Marcus high temperature theories. The latter two naturally do not work well when the electronic coupling increases. The square in the figure shows the present results in the framework of the (minimum energy crossing point MECP) approximation in which this crossing point is taken to be the representative point of hopping. This seems to work well, but this is simply because the electronic coupling is assumed to be constant here. In general, it is important to take into account the whole geometry of crossing seam surface.

One might think that the present model is rather special, since the potentials used are symmetric. This is not the case, however, since the ZN theory holds irrespective of the potential symmetry. Figure 25 shows the results of numerical applications to an asymmetric case [28]. This is the thermal electron-transfer rate constant at $T = 500$ K for the system of 12 harmonic oscillators with $\Delta G = -2247\,\mathrm{cm}^{-1}$. The potential parameters are given in [28]. The following three interesting features should be noted: (1) The perturabtion theory (Bixon–Jortner theory [109]) starts to break down with increasing the coupling strength as is seen from the comparison between the solid and the dotted lines. (2) The

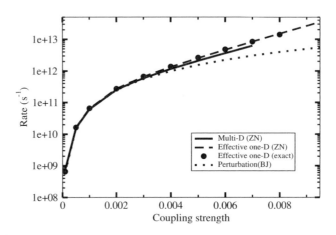

Figure 25. Electron-transfer rate the electronic coupling strength at $T = 500\,\text{K}$ for the asymmetric reaction ($\Delta G = -3\omega_2, \omega_2 = 749\,\text{cm}^{-1}$). Solid line-present full dimensional results with use of the ZN formulas. Dotted line-full dimensional results obtained from the Bixon–Jortner formula. Filled dotts-effective 1D results of the quantum mechanical flux–flux correlation function. Dashed line-effective 1D results with use of the ZN formulas. Taken from Ref. [28].

effective 1D model (filled dots) with use of the effective one frequency based on the Dogonadze model [111] works well in the relatively weak coupling regime, although the model does not work well in the strong coupling regime (not shown here). (3) The good agreement between the filled dots and the dash line confirms the accuracy of the present theory [Eq. (126)] with use of the ZN formulas. The 1D exact results (filled dots) have been calculated by using the flux–flux correlation function method.

The present approach has been applied to the experiment done by Nelsen et al., [112], which is a measurement of the intramolecular electron transfer of 2,7-dinitronaphthalene in three kinds of solvents. Since the solvent dynamics effect is supposed to be unimportant in these cases, we can use the present theory within the effective 1D model approach. The basic parameters are taken from the above reference except for the effective frequency. The results are shown in Fig. 26, which shows an excellent agreement with the experiment. The electronic coupling is quite strong and the perturbative treatment cannot work. The effective frequencies used are 1200, 950, and $800\,\text{cm}^{-1}$ for CH_3CN, dimethylformamide (DMF), and PrCN [113].

The electron transfer discussed above corresponds to the so-called normal case in which the NT type of nonadiabatic transition plays the essential role. There is another important case called inverted case, in which the LZ type of nonadiabatic transition plays a role. Since the ZN theory can describe this type of transition also, the corresponding electron-transfer theory can be formulated [114]. On the other hand, the realistic electron transfer occurs in solution and

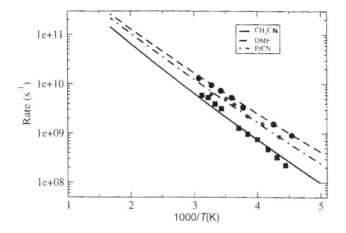

Figure 26. Electron-transfer rates for 2,7-dinitronaphthalene as a function of temperature for three solvents. Symbols are the experimental results by Nelsen et al. [112]. Lines are the present theoreical results. Solid line and square CH_3CN. Dot–dash line and *: PrCN. Dashed line and circle: N, N-dimethylforamide (DMF). Taken from Ref. [113].

the solvent dynamics effects are known to play crucial roles in many cases. Incorporation of the ZN formulas into this case presents an important and intriguing study in the future.

V.   CONTROL OF CHEMICAL DYNAMICS BY EXTERNAL FIELDS

Controlling chemical dynamics as we wish has been a dream of chemists for a long time. Thanks to the remarkable progress of laser technology, this is not just an unrealizable dream anymore. Once the mechanisms of chemical dynamics are understood to a good extent, then we can think of controlling them by using external fields. Catalytic reactions in organic chemistry are an example of reaction control, but external fields can be a kind of new catalysis and open a new dimension for reaction control. Although photochemistry has a long history, the technological developments of high power lasers have created various new possibilities in the field. New theoretical and experimental methods, such as the coherent control, the OCT, and the experiment with use of the genetic algorithm have been developed [33, 34, 115]. From the viewpoint of theoretical development, the OCT is a powerful method, being applicable virtually to any types of dynamical processes. The quantum mechanical version of that is, however, unfortunately not feasible for systems of more than three degrees of freedom. It is definitely desirable to develop such methods and/or theories that can be applied to realistic large systems and at the same time can be useful to comprehend the mechanisms. By doing that, the efficiency of control can also be enhanced. There are two basic elements in chemical dynamics and we try to develop new methods that are efficient for controlling them. They are (1) electronic transition of wavepacket between two adiabatic potential energy surfaces and (2) motion of wavepacket on a single adiabatic potential energy surface. For (1), we proposed the method of periodic chirping of laser frequency that is discussed in the next Section V.A. For (2) a new semiclassical *guided* optimal control theory applicable to multidimensional realistic systems has been formulated and is discussed in Section V.A.1. In addition to these ideas, it would also be nice to use any peculiar intriguing phenomena, if any. The complete reflection phenomenon can be such an example and will be discussed in Section V.C.

A. Periodic Chirping

By introducing the Floquet (or dressed) state formalism [116], various dynamic processes in periodic external fields can be regarded as a sequence of nonadiabatic transitions among the adiabatic dressed states. Energy levels and potential curves are shifted up and down by the amount of photon energy and thus potential curve crossings are created. The key idea of the present periodic chirping method is to create curve crossings and to directly control the nonadiabatic transitions by using the interference effects. This makes the essential difference from the various methods based on the linear chirping and the adiabatic rapid passage [33]. The nonadiabatic transitions in this context are the time-dependent ones and the time-dependent version of the ZN theory is usefully utilized.

1. Selective and Complete Excitation of Energy Levels

First, our general theory of periodic chirping is explained [36]. In the case of two-level problem, the energy levels vary as a function of the time-dependent field parameter F, laser frequency, for example, and the curve crossing is created, as shown in Fig. 27. If the field–matter interaction is taken into account and is diagonalized, the adiabatic states (solid lines) depict the avoided crossing. When the field $F(t)$ is swept as a function of time, the nonadiabatic transitions are induced at this avoided crossing. The transition matrix $T_{a \to b}$ that describes the transition from F_a to F_b(see Fig. 27) is given by

$$T_{a \to b} = \begin{pmatrix} \sqrt{1-p}\exp[i(\phi+\sigma_1/2+\sigma_2/2)] & \sqrt{p}\exp[i(\sigma_0-\sigma_1/2+\sigma_2/2)] \\ -\sqrt{p}\exp[-i(\sigma_0-\sigma_1/2+\sigma_2/2)] & \sqrt{1-p}\exp[-i(\phi+\sigma_1/2+\sigma_2/2)] \end{pmatrix}$$

$$(130)$$

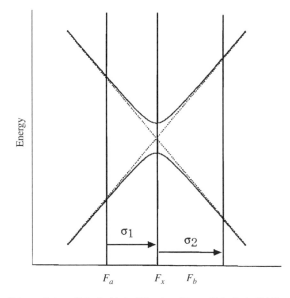

Figure 27. Schematic two diabatic (dotted lines) and two adiabatic (solid lines) potentials in an external field. The external field parameter F oscillates between F_a and F_b, striding the avoided crossing point F_x. σ_1 and σ_2 represent the phases that can be controlled by changing F_a and F_b. Taken from Ref. [36].

where p is the nonadiabtic transition probability by one passage of the crossing point F_x. The various phases are defined as

$$\sigma_0 = \mathrm{Re}\left(\int_{F_x}^{F_*} \Delta E(F)dt\right) = \mathrm{Re}\left(\int_{t_x}^{t_*} \Delta E(t)dt\right) \tag{131}$$

$$\sigma_1 = \int_{t_a}^{t_x} \Delta E(t)dt \tag{132}$$

$$\sigma_2 = \int_{t_x}^{t_b} \Delta E(t)dt \tag{133}$$

$$\phi = -\frac{\delta}{\pi} + \frac{\delta}{\pi}\ln\left(\frac{\delta}{\pi}\right) - \arg\Gamma\left(\frac{\delta}{\pi}\right) - \frac{\pi}{4} \tag{134}$$

where $\Delta E(F)$ is the adiabatic energy difference at the field strength F, the time $t_\gamma(\gamma = a, b, x, *)$ is the time at which $F(t_\gamma) = F_\gamma$ is satisfied, and F_* is the complex solution of $\Delta E(F_*) = 0$. The phase σ_0 can be rewritten as the real part of the following complex integral based on the time-dependent quadratic potential model:

$$\sigma_0 + \delta = \frac{1}{2\sqrt{\alpha}}\int_{-\beta}^{i}\left(\frac{1+\zeta^2}{\zeta+\beta}\right)^{1/2}d\zeta \tag{135}$$

where α and β are the same as a^2 and b^2 in the time-independent ZN theory of nonadiabatic transition (see the appendix).

Now we try to control the transition by sweeping the field parameter F as a function of time striding the avoided crossing position F_x. First, let us consider n periods of oscillation between F_a and F_b. The final overall transition amplitude T_n is given by

$$T_n = T^n = (T_{a \to b}^t T_{a \to b})^n \tag{136}$$

where T^t is the transpose of T. Roughly speaking, the transition probability p, the Stokes phase ϕ, and the phsae σ_0 are dependent on the local functionality of the adiabatic potentials around the crossing point and the sweeping speed dF/dt of the external field. The phase factors σ_1 and σ_2, on the other hand, depend on the global functionality of the adiabatic potentials in the range (F_a, F_b). We want to find appropriate parameters that satisfy

$$P_{12}^{(n)} = |(T_n)_{12}|^2 = 0 \qquad \text{or} \qquad 1 \tag{137}$$

Using the Lagrange–Sylvester formula, we obtain

$$T_n = T^n = \frac{\lambda_+ \lambda_- (\lambda_-^{n-1} - \lambda_+^{n-1})}{\lambda_+ - \lambda_-} E + \frac{\lambda_+^n - \lambda_-^n}{\lambda_+ - \lambda_-} T \tag{138}$$

where E is the unit matrix and $\lambda_\pm$ are the eigenvalues of T, which are defined as

$$\lambda_\pm = \exp[\pm i\xi] \tag{139}$$

with

$$\cos \xi = (1 - p)\cos(2\psi - \sigma) + p \cos(\sigma) \tag{140}$$

where

$$\psi = \phi + \sigma_0 + \sigma_1 \tag{141}$$
$$\sigma = 2\sigma_0 + \sigma_1 - \sigma_2 \tag{142}$$

The unitarity of T requires ξ to be real, and thus the probability p should satisfy

$$\frac{1 - |\cos \xi|}{2} \leq p \leq \frac{1 + |\cos \xi|}{2} \tag{143}$$

Then, Eq. (137), leads to

$$P_{12}^{(n)} = 4 \frac{\sin^2(n\xi)}{\sin^2 \xi} p(1 - p) \sin^2 \psi = 0 \tag{144}$$

or

$$P^{(n)}_{12} = 4\frac{\sin^2(n\xi)}{\sin^2\xi}p(1-p)\sin^2\psi = 1 \tag{145}$$

In the case of Eq. (144), we simply have the condition $\sin(n\xi) = 0$ or $\sin\psi = 0$. On the other hand, Eq. (145) is equivalent to $P^{(2n)}_{12} = 0$ and thus provides the following two consitions:

$$\sin(2n\xi) = 0 \qquad \text{namely} \qquad \sin^2(n\xi) = 1 \tag{146}$$

and

$$4p(1-p)\sin^2\psi = \sin^2\xi \tag{147}$$

Equation (146) determines ξ for a given n and Eq. (147) gives a condition for p and ψ for a given ξ. The phase σ can be determined from Eq. (140). The phases ψ and σ can be adjusted by changing σ_1 and σ_2, respectively.

In the case of n and half periods of oscillation of the field, the following condition should first be satisfied

$$\sin^2[(2n+1)\xi] = 0 \tag{148}$$

The additional conditions are obtained as follows:

$$P^{(n+1/2)}_{12} = 0 \rightarrow 4(1-p)\sin^2(\psi - \sigma) = \frac{\sin^2\xi}{\sin^2(n\xi)} \tag{149}$$

and

$$P^{(n+1/2)}_{12} = 1 \rightarrow 4p\sin^2(\psi - \sigma) = \frac{\sin^2\xi}{\sin^2(n\xi)} \tag{150}$$

The above set of conditions are complete in the sense that a transition from any initial state to any final state can be controlled perfectly. This idea can also be applied to multilevel problems. In the practical applications, the quadratic chirping, that is, one-period oscillation, is quite useful, as demonstrated by numerical applications given below.

First, let us consider a selective and complete excitation in a three-level problem by quadratically chirping the laser frequency as shown in Fig. 28 [42] (the field parameter F is the laser frequency ω). The energy separation ω_{23}

(a)

(b)

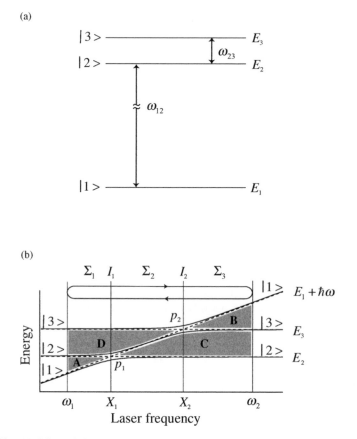

Figure 28. (a) Schematic level structure of a three-level model. (b) Floquet diagram of the three-level model shown in (a) as a function of laser frequency. Taken from Ref. [42].

between the two excited states is assumed to be much smaller than the separation between the ground state and the level $|2>$ or $|3>$, namely, $\omega_{12} \gg \omega_{23}$, where $\omega_{ij} = (E_j - E_i)/\hbar$. Since the applied laser frequency ω is close to ω_{12}, the transition between the states $|2>$ and $|3>$ are negligible and the Floquet Hamiltonian is expressed as

$$H_{\text{Floquet}} = \begin{pmatrix} E_1 + \hbar\omega(t) & -\mu_{12}\epsilon(t)/2 & -\mu_{13}\epsilon(t)/2 \\ -\mu_{12}\epsilon(t)/2 & E_2 & 0 \\ -\mu_{13}\epsilon(t)/2 & 0 & E_3 \end{pmatrix} \quad (151)$$

where μ_{ij} is the transition dipole moment between $|i>$ and $|j>$ and $\epsilon(t)$ is an envelope function of the laser field. When the laser frequency is swept from ω_1 to ω_2, the excitation process is divided into two parts: adiabatic propagation along the adiabatic Floquet states and nonadiabatic transitions at the avoided crossings X_1 and X_2. The transition matrix corresponding to the half period of frequency sweeping $\omega_1 \rightarrow \omega_2$ is gven by

$$T_{\omega_1 \rightarrow \omega_2} = \Sigma_3 I_2 \Sigma_2 I_1 \Sigma_1 \tag{152}$$

where Σ_j represents the adiabatic propagation from X_{j-1} to X_j, and I_j represents the nonadiabatic transition at the crossing X_j. They are explicitly given by

$$(\Sigma_j)_{pq} = \exp[-i\sigma_j^{(p)}]\delta_{pq} \tag{153}$$

$$\sigma_j^{(k)} = \frac{1}{\hbar} \int_{X_{j-i}}^{X_j} E_k^{(ad)}(\omega)dt \tag{154}$$

$$I_1 = \begin{pmatrix} \sqrt{1-p_1}\exp[i\phi_1] & \sqrt{p_1}\exp[i\psi_1] & 0 \\ -\sqrt{p_1}\exp[-i\psi_1] & \sqrt{1-p_1}\exp[-i\phi_1] & 0 \\ 0 & 0 & 1 \end{pmatrix} \tag{155}$$

$$I_2 = \begin{pmatrix} 1 & 0 & 0 \\ 0 & \sqrt{1-p_2}\exp[i\phi_2] & \sqrt{p_2}\exp[i\psi_2] \\ 0 & -\sqrt{p_2}\exp[-i\psi_2] & \sqrt{1-p_2}\exp[-i\phi_2] \end{pmatrix} \tag{156}$$

Here, p_j denotes the nonadiabatic transition probability for one passage of the avoided crossing X_j, ϕ_j and ψ_j are the dynamical phases due to the nonadiabatic transition at X_j, $E_k^{(ad)}$ is the kth adiabatic Floquet state, $X_0 = \omega_1$ and $X_3 = \omega_2$. The transition amplitude Eq. (152) can be explicitly expressed as

$$T_{\omega_1 \rightarrow \omega_2} =$$

$$\begin{pmatrix} \sqrt{1-p_1} & \sqrt{p_1}e^{-iA} & 0 \\ -\sqrt{p_1(1-p_2)}e^{-iC} & \sqrt{(1-p_1)(1-p_2)}e^{-i(A+C)} & \sqrt{p_2}e^{-i(A+C+D)} \\ \sqrt{p_1 p_2}e^{-i(B+C)} & -\sqrt{(1-p_1)p_2}e^{-i(A+B+C)} & \sqrt{1-p_2}e^{-i(A+B+C+D)} \end{pmatrix}$$

$$\tag{157}$$

where

$$A = \phi_1 - \psi_1 + \Delta\sigma_1^{(2,1)} \tag{158}$$

$$B = \phi_2 + \psi_2 + \Delta\sigma_3^{(3,2)} \tag{159}$$

$$C = \phi_1 + \psi_1 - \phi_2 + \Delta\sigma_2^{(2,1)} + \Delta\sigma_3^{(2,1)} \tag{160}$$

$$D = -\phi_1 + \phi_2 - \psi_2 + \Delta\sigma_1^{(3,2)} + \Delta\sigma_2^{(3,2)} \tag{161}$$

$$\Delta\sigma_j^{(k,l)} = \sigma_j^{(k)} - \sigma_j^{(l)} \tag{162}$$

These phases roughly correspond to the areas shown in Fig. 28(b). The overall transition matrix $T^{(n=1)}$ for one period of oscillation is given by

$$T^{(n=1)} = (T_{\omega_1 \to \omega_2})^t T_{\omega_1 \to \omega_2} \tag{163}$$

and the transition probability from $|1 >$ to $|2 >$ is explicitly given by

$$P_{12}^{(n=1)} = |(T^{(n=1)})_{21}|^2 = p_1(1-p_1)|\exp[2iC] - 1 + p_2(1-\exp[-2iB])|^2 \tag{164}$$

The condition of the complete excitation to $|2 > (P_{12}^{(n=1)} = 1)$ is expressed as

$$p_1 = 1/2, \quad B = m\pi \quad \text{and} \quad C = (n+1/2)\pi \quad (m,n = \text{integer}) \tag{165}$$

The physical meaning of $B = m\pi$ is that no bifurcation into the diabatic state $|3 >$ occurs at X_2 on the second half of the sweep whatever the probability p_2 is. The conditions of p_1 and C guarantee that the interference between $|1 >$ and $|2 >$ at X_1 on the way back leads to the complete excitation to $|2 >$. The complete excitation to $|3 >$ can be achieved by one period of sweeping, if we start from ω_2. The condition is given by

$$p_2 = 1/2, \quad A = m\pi \quad \text{and} \quad D = (n+1/2)\pi \quad (m,n = \text{integer}) \tag{166}$$

Numerical examples are shown in Figs. 29–31. The parameters used are

$$\omega_{12} = 500\,\text{cm}^{-1}, \omega_{23} = 10\,\text{cm}^{-1}, \mu_{12} = \mu_{13} = 1.0 \text{ a.u.} \tag{167}$$

The laser frequency is swept quadratically as a function of time as

$$\omega(t) = -a(t - t_0)^2 + b - E_1 \tag{168}$$

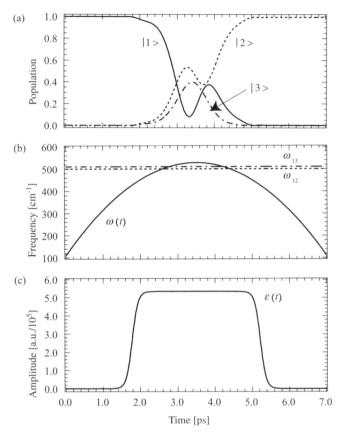

Figure 29. Complete excitation from $|1>$ to $|2>$ by one period of frequency chirping in the case of the three-level model. Upper part-time variation of the population. Middle part-time variation of laser frequency. Bottom part-envelope of the laser pulse. Taken from Ref. [42].

where

$$a = 8(V_{13})^3 \alpha_f / \hbar^2, \quad b = E_3 - 2V_{13}\beta_f, \quad V_{13} = -\mu_{13}E_0/2 \qquad (169)$$

with E_0 being the amplitude at the pulse peak. The laser pulse shape is taken as

$$\epsilon(t) = \begin{cases} E_0[1 + \tanh(\beta_e(t - t_{0e}))]/2 & \text{for } t \le t_0 \\ E_0[1 - \tanh(\beta_e(t - t_{1e}))]/2 & \text{for } t > t_0 \end{cases} \qquad (170)$$

The parameters α_f, β_f, and the peak intensity are determined to be 0.6005, 1.58142, and $0.1\,\text{GW cm}^{-2}$ from the conditions given above and the ZN formulas. The other parameters of laser field are $t_0 = 3.5\,\text{ps}$, $t_{0e} = 1.7775\,\text{ps}$,

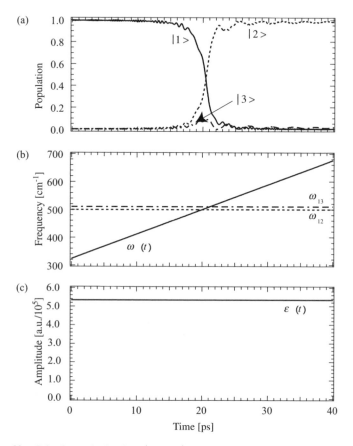

Figure 30. Selective excitation from $|1>$ to $|2>$ by the adiabatic rapid passage (ARP) in the case of three-level model. Upper part time variation of the population. Middle part-time variation of laser frequency. Bottom part-envelope of the laser pulse. Taken from Ref. [42].

$t_{1e} = 2t_0 - t_{0e}$, and $\beta_e = 6.515\,ps^{-1}$. As seen from Fig. 29, the transition time is $\sim 3\,ps$, which is very close to the time $\Delta t = 2\pi/\Delta E \approx 3.3\,ps$ determined from the uncertainty principle. Thus the present scheme gives the shortest possible time of transition. Figure 30 shows the excitation by using the conventional ARP (adiabatic rapid passage) method. The laser frequency is linearly chirped, $\omega(t) = \omega_{12} + c(t - 20\,ps)$. The chirp rate c and the laser intensity are determined to be $8.816\,cm^{-1}\,ps^{-1}$ and $0.1\,GW\,cm^{-2}$ so that the excitation probability becomes 0.99. Selective excitation is possible, but it takes quite a long time ($\sim 20\,ps$) and complete excitation is not possible. The case of π-pulse is shown in Fig. 31. The pulse shape is hyperbolic-secant $= \epsilon(t) = E_0 \text{sech}[\beta_e(t - t_0)]$. The parameters are $t_0 = 10\,ps$, the peak intensity

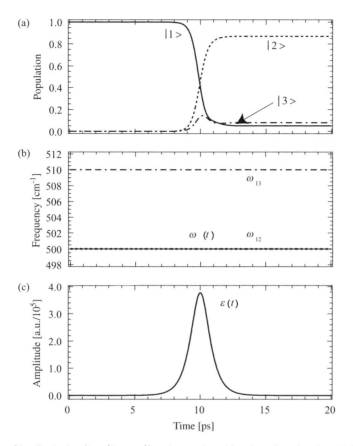

Figure 31. Excitation from $|1 >$ to $|2 >$ by π-pulse with a short time duration. (a) Part-time variation of the population. (b) Part-time variation of laser frequency. (c) Part-envelope of the laser pulse. Taken from Ref. [42].

is $0.05\,\mathrm{GW\,cm^{-2}}$, and $\beta_e = 1.56022\,\mathrm{ps^{-1}}$. These are determined so that the transition time is comparable to the case of one-period sweeping ($\sim 3\,\mathrm{ps}$). As is clearly seen, the complete excitation is not possible.

Next, let us consider a four-level problem and try to excite the middle level $|3 >$ among the three excited states (see Fig. 32). The laser pulse is the same as Eq. (170) with $t_{0e} = 981.2\,\mathrm{fs}$ and $\beta_e = 4.257\,\mathrm{ps^{-1}}$. The system parameters are

$$\omega_{12} = 500\,\mathrm{cm^{-1}} \qquad \omega_{23} = 10\,\mathrm{cm^{-1}} \qquad \omega_{34} = 10\,\mathrm{cm^{-1}}$$
$$\mu_{ij}(ij = 12, 13, 14) = 1.0\,\mathrm{a.u.} \tag{171}$$

(a)

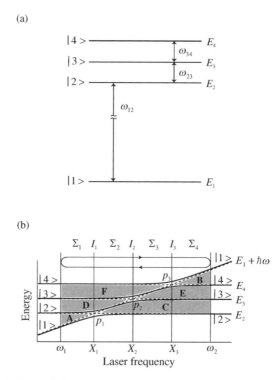

(b)

Figure 32. (a) Schematic level structure of a four-level model. (b) Floquet diagram of the four-level model as a function of laser frequency. Taken from Ref. [42].

The laser frequency is a combination of three linear chirping:

$$\omega(t) = \omega_{14} + c(t - t_0) \qquad \text{for} \qquad (t \le t_1)$$
$$= \omega_{14} - c(t - t_2) \qquad \text{for} \qquad (t_1 < t \le t_3)$$
$$= \omega_{14} + c(t - t_4) \qquad \text{for} \qquad (t > t_3) \qquad (172)$$

where $t_1 = t_0 + \Delta t_1, t_2 = t_0 + 2\Delta t_1, t_3 = t_2 + \omega_{24}/c + \Delta t_2, t_4 = 2t_3 - t_2$, $c = 346.6 \text{ cm}^{-1} \text{ ps}^{-1}$, $\Delta t_1 = 726.3 \text{ fs}$, $\Delta t_2 = 90.08 \text{ fs}$, and the peak intensity is $0.5917 \text{ GW cm}^{-2}$. The results are shown in Fig. 33. The transition time is ~ 3.0 ps and is close to the limit of the uncertainty principle. The excitation by $\pi-$ pulse and by ARP take one order of magnitude longer time compared to the present periodic chirping method. The numerical results are not shown here. Please refer to Ref. [42].

In the above numerical examples the field parameter F is taken to be the laser frequency and the nonadiabatic transition used is the Landau–Zener type of curve-crossing. The periodic chirping method, however, can actually be more

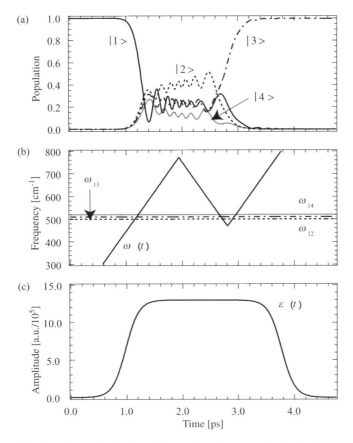

Figure 33. Complete excitation from $|1>$ to $|3>$ by one and a half period of frequency chirping in the case of four-level model. (a) Patart-time variation of the population. (b) Part-time variataion of laser frequency. (c) Part-envelope of the laser pulse. Taken from Ref. [42].

general [43]. First of all, the external field is not necessarily laser field, but can be any other field such as magnetic field or electric field [36]. The field parameter F can also be other than field frequency, such as the field strength. Other types of nonadiabatic transitions can also be utilized, such as the Rosen–Zener type and the exponential potential model, although the Landau–Zener-type curve crossing transition is probably most efficient. These examples can be found in Refs. [1, 36, 43].

2.   *Electronic Transition of Wave Packet*

In order to control chemical dynamics, it is crucial to control wave packet motions. Here we consider efficient electronic excitation of wave packet by ultrashort broadband laser pulses, which has fundamental importance in chemical dynamics. The quadratic chirping method formulated in Section V.A.1 is demonstrated to be effective and useful for that also [37]. Although it is possible to use two linearly chirped pulses to control wave packet dynamics [42], it is more efficient to use quadratically chirped pulses. Here we assume that the nuclear configuration does not change during the electronic transition and regard the system as a coordinate-dependent energy level problem with the effect of the kinetic energy operator taken into account as a perturbation. Then tha laser parametrs can be designed again analytically from the ZN theory. The present approach can be used not only for ordinary electronic excitation but also for many other parocesses such as the pump–dump method and selective bond breaking. The efficiency can be up to ~ 80% for a relatively weak laser intensity.

The total excitation probability from the ground state is approximated as

$$\mathcal{P} = \int P_{12}(x) |\Psi_g(x, t = 0)|^2 dx \tag{173}$$

where $\Psi_g(x, t = 0)$ is the initial wave packet and the nonadiabatic transition probability $P_{12}(x)$ is calculated from the corresponding two-level problem that depends on x parametrically. By taking into account the kinetic energy operator as a perturbation, the Floquet Hamiltonian is given by [37, 39]

$$H_0 = \frac{1}{2} \begin{pmatrix} \hbar\omega(t) - \tilde{\Delta}(x) & -\mu\epsilon(t) \\ -\mu\epsilon(t) & \tilde{\Delta}(x) - \hbar\omega(t) \end{pmatrix} \tag{174}$$

where

$$\tilde{\Delta}(x) = \Delta(x) + \Delta t \vec{v} \cdot \nabla\Delta(x) \tag{175}$$

with

$$\Delta(x) = V_e(x) - V_g(x) \tag{176}$$

where $V_j(X)(j = e, g)$ are the ground- and excited-state-potentials, $\vec{v}$ is the mean velocity of the wave packet, μ is the transition dipole moment, ϵ is the laser pulse envelope, and Δt is the time delay measured from the pulse center (see the appendix of Ref. [37]). In the case of quadratically chirped pulse the time-dependent laser frequency satisfies the resonance condition twice and the nearly complete excitation can be achieved by the constructive interference. The nonadiabatic transition probability $P_{12}(x)$ is expressed as usual from the ZN theory as

$$P_{12}(x) = 4p_{ZN}(1 - p_{ZN}) \sin^2 \Psi_{ZN} \tag{177}$$

with

$$p_{ZN} = \exp\left[-\frac{\pi}{4\sqrt{\alpha\beta}} \left(\frac{2}{1 + \sqrt{1 + \beta^{-2}(0.4\alpha + 0.7)}} \right)^{1/2} \right] \tag{178}$$

$$\Psi_{ZN} = \sigma_{ZN} + \phi_S \tag{179}$$

$$\phi_S = -\frac{\delta_{ZN}}{\pi} + \frac{\delta_{ZN}}{\pi} \ln\left(\frac{\delta_{ZN}}{\pi}\right) - \arg\Gamma\left(\frac{\delta_{ZN}}{\pi}\right) - \frac{\pi}{4} \tag{180}$$

$$\sigma_{ZN} + i\delta_{ZN} = \frac{1}{2\sqrt{\alpha}} \int_{-\beta}^{i} \left(\frac{1 + \zeta^2}{\zeta + \beta}\right)^{1/2} d\zeta \tag{181}$$

where the two basic parameters of the nonadiabatic transition, α and β, are defined as

$$\alpha = \frac{\hbar\alpha_\omega}{(\mu\epsilon)^3} \tag{182}$$

and

$$\beta = \frac{\Delta(x) - \hbar\beta_\omega + \frac{(\vec{v}\cdot\nabla\Delta)^2}{4\hbar\alpha_\omega}}{\mu\epsilon} \tag{183}$$

The parameters α_ω and β_ω are the chirping rate and the carrier frequency, namely, the laser frequency is chirped quadratically as

$$\omega(t) = \alpha_\omega(t - t_p)^2 + \beta_\omega \tag{184}$$

The laser parameters should be chosen so that α and β can make the nonadiabatic transition probability $\mathcal{P}$ as close to unity as possible. Figure 34 depicts the probability P_{12} as a function of α and β. There are some areas in which the probabilty is larger than 0.9, such as those around $(\alpha = 1.20, \beta = 0.85)$, $(\alpha = 0.53, \beta = 2.40)$, $(\alpha = 0.38, \beta = 3.31)$, and so on. Due to the coordinate dependence of the potential difference $\Delta(x)$ and the transition dipole moment $\mu(x)$, it is generally impossible to achieve perfect excitation of the wave packet by a single quadratically chirped laser pulse. However, a very high efficiency of the population transfer is possible without significant deformation of the shape of the wave packet, if we locate the wave packet parameters inside one of these islands. The biggest, thus the most useful island, is around $\alpha = 1.20, \beta = 0.85$. The transition probability P_{12} is > 0.9, if

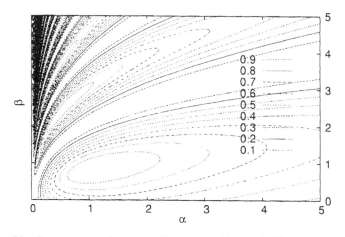

Figure 34. Contour map of the nonadiabatic transition probability P_{12} induced by a quadratically chirped pulse as a function of the two basic parameters α and β. Taken from Ref. [37].

$\alpha \in (0.62, 2.21)$ and $\beta \in (0.45, 1.30)$. Thus the condition for the "nearly complete" excitation is expressed as

$$0.70 \lesssim \alpha \lesssim 2.0 \tag{185}$$

and

$$0.50 \lesssim \beta \lesssim 1.20 \tag{186}$$

If the potential difference $\Delta(x)$ can be approximated by a linear function of x within the range of the wave packet, the range of the parameter β is estimated as

$$\beta(x_0) - \frac{\sigma_{xx}|\nabla\Delta|}{\mu\epsilon} \lesssim \beta \lesssim \beta(x_0) + \frac{\sigma_{xx}|\nabla\Delta|}{\mu\epsilon} \tag{187}$$

where x_0 is the center of the wave packet and σ_{xx} is the variance of the wave packet in the coordinate space. If we take $\beta(x_0) \sim 0.85$, we have

$$\mu\epsilon \gtrsim 2.86\sigma_{xx}|\Delta| \tag{188}$$

The single inequality is used, simply because we want to reduce the necessary laser intensity as much as possible. Equation (188) now provides the "nearly complete" excitation condition.

The following three numerical applications are shown below [37]: (1) electronic excitation of a wave packet from a nonequilibrium displaced position,

(2) creation of a localized wave packet on the ground potential energy surface by the quadratically chirped pump–dump method, and (3) selective bond breaking of a triatomic molecule. The envelope of the laser pulse employed is the same tangent hyperbolic type as before,

$$\epsilon(t) = \frac{\epsilon_0}{2} \left(\tanh\left[\frac{t - t_c + \tau/2}{s} \right] - \tanh\left[\frac{t - t_c - \tau/2}{s} \right] \right) \qquad (189)$$

where t_c, s, τ, and ϵ_0 are the center time, switching time, duration, and maximum amplitude, respectively. The center time is taken to be equal to the frequency center time, that is, $t_c = t_p$.

First, by taking the diatomic molecule LiH, we try to excite a displaced wave packet on the ground-state $X^1\Sigma^+$ at $R = 6.0$ a.u. to the $B^1\Pi$ excited state (see Fig. 35). The potential energy curves and the transition dipole moment are taken from [117]. The time evolution of the populations on the ground and excited states is shown in Fig. 36 More than 86% of the initial state is excited to the B state within the period shorter than a few femtoseconds. The integrated total transition probability $\mathcal{P}$ given by Eq. (173) is $\mathcal{P} = 0.879$, which is in good agreement with the value 0.864 obtained by numerical solution of the original coupled Schroedinger equations. This means that the population deviation from 100% is not due to the approximation, but comes from the intrinsic reason, that is, from the spread of the wavepacket. Note that the LiH molecule is one of the

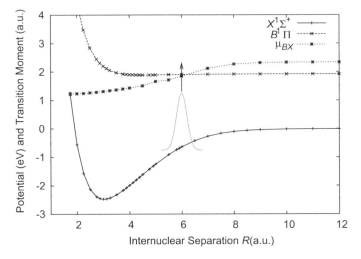

Figure 35. Electronic excitation of a LiH wave packet from the outer classical turning point ($\sim 6a_0$) of the ground $X^1\Sigma^+$ state. The $X \to B$ transition is considered. The initial wavepacket is the shifted ground vibrational state. Taken from Ref. [37].

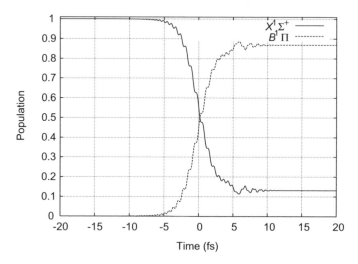

Figure 36. Time variation of the wave packet population on the ground X and excited B states of LiH. The system is excited by a single quadratically chirped pulse with parameters: $\alpha_\omega = 5.84 \times 10^{-2}\,\mathrm{eV\,fs^{-2}}$, $\beta_\omega = 2.319\,\mathrm{eV}$, and $I = 1.00\,\mathrm{TWcm^{-2}}$. The pulse is centered at $t = 0$ and has a temporal width $\tau = 20\,\mathrm{fs}$. Taken from Ref. [37].

most difficult systems for the present method, since the mass is very light and the gradient of potential difference is relatively large. These difficulties have been overcome by using the quick quadratic chirping, which demonstrates the usefulness of the method. The initial displaced localized wavepacket can be prepared by a sequence of vibrational excitations by quadratically chirped pulses, which is not explained here, or more directly by using the semiclassical optimal control theory explained in the Section V.B. As another extreme example, the NaK molecule is employed. As shown in Fig. 37, the initail wave packet is a less localized one at the inner turning point on the ground state $X^1\Sigma^+$ prepared by the quadratically chirped pump–dump method explained below. The potential energy curve and the transition dipole moment are taken from [118]. The total excitation probabiltiy $\mathcal{P}$ is as high as $\mathcal{P} = 0.905$ with the laser intensity of only $0.2\,\mathrm{TW\,cm^{-2}}$ (see Fig. 38). If we use a well-localized initial wave packet, then the total excitation efficiency can be higher. Note that the above method can be applied even to the wave packet moving away from the turning point.

The second example is the quadratically chirped pump–dump scheme. Since the pioneering work by Tannor and Rice [119], the pump–dump method has been widely used to control various processes. However, since it is not possible to transfer a wave packet from one potential energy surface to another nearly completely by using the ordinary transform limited or linear chirped pulses, the

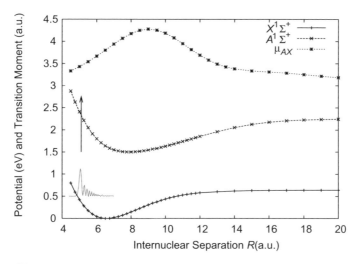

Figure 37. Electronic excitation of the NaK wavepacket from the inner turning point of the ground X state. The $X \rightarrow A$ transition is considered. The initial wave packet is prepared by two quadratically chirped pulses within the pump–dump mechanism. Taken from Ref. [37].

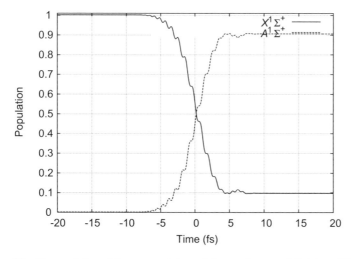

Figure 38. Time variation of the wavepacket population on the ground X state and the excited A state of NaK. The system is excited by a quadratically chirped pulse with parameters: $\alpha_\omega = 3.13 \times 10^2 \, \text{eV} \, \text{fs}^{-2}$, $\beta_\omega = 1.76 \, \text{eV}$, and $I = 0.20 \, \text{TW} \, \text{cm}^{-2}$. The pulse is centered at $t = 0$ and has a temporal width $\tau = 20 \, \text{fs}$. Taken from Ref. [37].

efficiency of the ordinary pump–dump method cannot be high. The nearly complete pump–dump now becomes possible by using the present quadratically chirped pulses. As an example, starting from the ground vibrational state $(v = 0)$ on the ground electronic state of NaK, we try to prepare the wave packet at the inner turning point on the ground state that was used in the above example as the initial packet for the electronic excitation (see Fig. 37). A quadratically chirped pulse is applied to this ground vibrational state to excite to the excited state $A^1\Sigma^+$ (see Fig. 37) and the wave packet is dumped down to the ground electronic state when it arrives at the right turning point on the excited state. The time variation of the population is shown in Fig. 39. The overall pump–dump probability is found to be > 0.981. The final wave packet that arrived at the left turning point is nothing but the one shown in Fig. 37. As mentioned above, this pump–dump procedure can be applied even to the wave packet moving in between the turning points, if the velocity of the wave packet is not too high. An example is shown in Fig. 40, in which the same wave packet in Fig. 37 is dumped at the minimum potential position on its way to the right turning point. The overall pump–dump probability is about as high as 0.958.

The present excitation scheme of quadratic chirping can be applied to higher dimensional systems easily. As an example, we consider the bond-selective

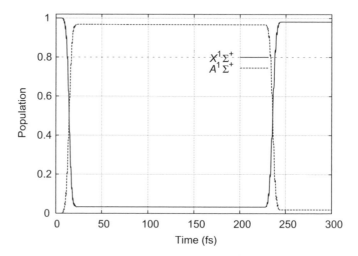

Figure 39. Pump–dump control of NaK molecule by using two quadratically chirped pulses. The initial state taken as the ground vibrational eigenstate of the ground state X is excited by a quadratically chirped pulse to the excited state A. This excited wavepacket is dumped at the outer turning point at $t \simeq 230$ fs by the second quadratically chirped pulse. The laser parameters used are $\alpha_\omega = 2.75(1.972) \times 10^{-2}$ eV fs^{-2}, $\beta_\omega = 1.441(1.031)$ eV, and $I = 0.15(0.10)$ TW cm^{-2} for the first (second) pulse. The two pulses are centered at $t_1 = 14.5$ fs and $t_2 = 235.8$ fs, respectively. Both of them have a temporal width $\tau = 20$ fs. (See color insert.) Taken from Ref. [37].

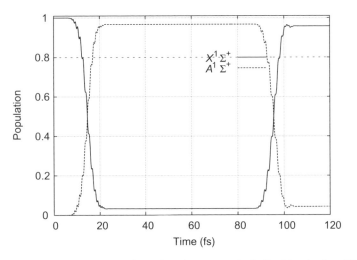

Figure 40. Pump–dump control of NaK by using two quadratically chirped pulses. The initial state and the first step of pump are the same as in Fig. 39. The excited wave packet is now dumped at $R \simeq 6.5a_0$ on the way to the outer turning point. The parameters of the second pulse are $\alpha_\omega = 1.929 \times 10^{-2}\, \text{eV fs}^{-2}$, $\beta_\omega = 1.224\, \text{eV}$, and $I = 0.10 \text{TW cm}^{-2}$. The second pulse is centered at $t = 95.5\, \text{fs}$ and has a temporal width $\tau = 20\, \text{fs}$. (See color insert.) Taken from Ref. [37].

photodissociation of a 2D model of H_2O on the $\tilde{A}$ excited potential energy surface. The 2D model is adapted from [120]. The initial wave packet on the ground electronic state is a 2D Gaussian wave packet in the form

$$\Phi = \phi_1 \phi_2 \qquad (190)$$

with

$$\phi_i = (\pi \sigma_i^2)^{-1/4} \exp\left[-\frac{(R_i - R_{ie})^2}{2\sigma_i^2} + \frac{i}{\hbar} P_{i0}(R_i - R_{ie}) \right] \qquad (191)$$

where $i(= 1, 2)$ is the index of the two OH bonds, R_{ie} and P_{i0} are the central position and the initial momentum of the wave packet, respectively. The overall total transition probability $\mathcal{P}$ are calculated from Eq. (173) and compared with the full numerical solution $\mathcal{P}_{num}$ of the original coupled Schroedinger equations. The following three kinds of initial wave packets are considered:(a)wave packet at the equilibrium position with zero momentum, (b)wave packet slightly shifted in the x direction with zero momentum, and (c)wave packet at the equilibrium position with a finite momentum directed along the x direction. The laser parameters and the final transition probabilities ($\mathcal{P}$ and $\mathcal{P}_{num}$) are listed below.

Case (a)

$$R_{ie} = 1.82 \, \text{a.u.} \, (i = 1, 2), \quad P_{i0} = 0.0, \quad \sigma_{xx} = 0.20$$
$$\alpha_\omega(\text{eV fs}^{-2}) = 1.34, \quad \beta_\omega(\text{eV}) = 6.17, \quad I(\text{TW cm}^{-2}) = 20.0$$
$$\mathcal{P} = 81.7\%, \quad \mathcal{P}_{num} = 80.2\% \tag{192}$$

Case (b)

$$R_{1e} = 2.32 \, \text{a.u.}, \quad R_{2e} = 1.82 \, \text{a.u.} \quad P_{i0} = 0.0 \quad \sigma_{xx} = 0.20$$
$$\alpha_\omega(\text{eV fs}^{-2}) = 2.89, \quad \beta_\omega(\text{eV}) = 4.34, \quad I(\text{TW cm}^{-2}) = 57.8$$
$$\mathcal{P} = 74.4\%, \quad \mathcal{P}_{num} = 73.0\% \tag{193}$$

Case (c)

$$R_{ie} = 1.82 \, \text{a.u.}, \quad P_{10} = 12.0 \, \text{a.u.}, \quad P_{20} = 0.0, \quad \sigma_{xx} = 0.20$$
$$\alpha_\omega(\text{eV fs}^{-2}) = 1.34, \quad \beta_\omega(\text{eV}) = 6.61, \quad I(\text{TW cm}^{-2}) = 20.0$$
$$\mathcal{P} = 81.5\%, \quad \mathcal{P}_{num} = 78.8\% \tag{194}$$

The laser intensities are taken to be the possible lowest. The intensity in case (b) is almost three times larger than the others. This is simply due to the fact that the transition dipole moment exponentially decays from the equilibrium position and also the potential energy difference increases. Note again that the coordinate-dependent level approximation works well. In order to demonstrate the selectivity the time evolution of the wave packets on the excited state are shown in Fig 41. As a measure of the selectivity, we have calculated the target yield Y_e by

$$Y_e = \int_D |\Psi_e(t)|^2 dR_1 dR_2 \tag{195}$$

where $\Psi_e(t)$ is the wave packet at time t on the excited state and the integration domain D is taken to be $R_1 \in (2.5, 7.5)$ and $R_2 \in (1.0, 3.0)$ in au. In case (a) (top panels of Fig. 41), the wave packet naturally dissociates equally into the two directions with $Y_e = 0.495$. In the case of a shifted wave packet [case (b)], Ψ_e stands on the slope of the valley at $R_1 > R_2$, where the force is directed toward $R_1(x - \text{direction})$. Because of the small mass of H_2O, a small portion of Ψ_e goes beyond the barrier and appears in the $R_2(y)$ direction and the yield is $Y_e = 0.72$. Case (c) is the most efficient in the present system. The target yield is $Y_e = 0.95$

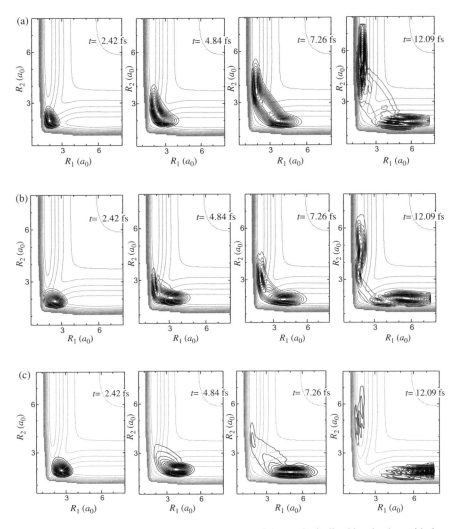

Figure 41. Selective bond breaking of H_2O by means of the quadratically chirped pulses with the initial wave packets described in the text. The dynamics of the wavepacket moving on the excited potential energy surface is illustrated by the density. (a) The initail wave packet is the ground vibrational eigen state at the equilibrium position. (b) The initial wave packet has the same shape as that of (a), but shifted to the right. (c) The initail wave packet is at the equilibrium position but with a directed momentum toward x direction. Taken from Ref. [37]. (See color insert.)

and this high yield is attained by the rapid motion of Ψ_e on the excited potential energy surface within a very short period ($\sim 2\,\text{fs}$). The preparation of the wave packet with directed momentum can be achieved by using the semiclassical optimal control theory explained in Section V.B.

Note again that the parameters of the optimal laser pulses can be estimated from the ZN theory of nonadiabatic transition regardless of the dimensionality of the system.

B. Semiclassical Guided Optimal Control Theory

As demonstrated in the Section V.A.1, the quadratic chirping method is very powerful for controlling transitions among energy levels, for example, among energy levels of atoms or among vibrational states of molecules. The efficiency can be theoretically 100%. But this is not effective at all, unfortunately, in the case of vibrational levels of realistic multidimensional systems, since the vibrational energy level calculation itself is already very time consuming and actually not feasible. In this sense, the OCT is more useful. We can control wave packet motion on a single potential energy surface by OCT [33]. Electronic transitions can also be treated by OCT, but the quadratic chirping method is more efficient, as demonstrated in Section V.A.2. The general idea of the OCT is to design such a laser field that the initial wave packet driven by the field becomes as close as possible to the desired target wave packet by solving the appropriate equations iteratively to optimize a certain functional. There have been formulated various versions of quantum and classical OCT [33]. The quantum mechanical versions, however, are essentially based on the grid method and can be applied only to low dimensional systems, at the highest three dimensions at present, unfortunately. The classical mechanical versions, on the other hand, can be easily used for higher dimensional systems. However, they cannot be reliable, since the various phases play crucial roles in laser control of dynamics [121, 122]. This is quite different from the ordinary chemical dynamics, in which some sort of averaged physical quantities can be obtained without phases. We have developed a semiclassical *guided* optimal control theory [38, 40, 41] with use of the Herman–Kluk type frozen Gaussian wave packets [17].

1. Formulation

In the general procedure of OCT, the optimal laser field $E_k(t)$ is obtained by iteratively solving a set of Schroedinger equations starting from a certain initial guess (typically zero filed). If we use the idea of the quantum gradient search method [123], the correction to the controlling field is given by

$$\delta E_k(t) \sim \hbar^{-1} \langle \Phi_t | \phi(T) \rangle \text{Im}[\Theta_k(t)] \tag{196}$$

with

$$\Theta_k(t) = \langle \phi(t) | \mu_k(\mathbf{r}) | \chi(t) \rangle \tag{197}$$

where $\phi(t)$ and $\chi(t)$ are the wave packets in the previous step in the iteration and $\mu_k(\mathbf{r})$ is the dipole moment with the index k denoting the polarization vector component. The time-dependent wavepackets $\phi(t)$ and $\chi(t)$ are the solutions of the time-dependent Schroedinger equation on a single potential energy surface in a laser field given by

$$\left[i\hbar \frac{\partial}{\partial t} + \frac{\hbar^2}{2m} \Delta_\mathbf{r} - V(\mathbf{r}) + \mu(\mathbf{r})\mathbf{E}(t) \right] \psi(t) = 0 \tag{198}$$

The solution $\psi = \phi(t)$ is propagated forward in time from $t = 0$ to $t = T$ and satisfies the initial condition $\phi(t = 0) = \Phi_i$. On the other hand, $\psi = \chi(t)$ is propagated backward from $t = T$ to $t = 0$ with the initial condition $\chi(t = T) = \Phi_t$. Since the wavepacket propagations should be performed at each iteration, the numerical costs of quantum computations become huge for multidimensional systems. In our formulation, we have introduced the following two ideas in order to make the theoretical framework applicable to large systems. One is to incorporate the semiclassical Herman–Kluk-type frozen Gaussian wavepacket propagation into the above OCT. The second is the idea of *guiding* the control process by introducing a set of intermediate target states. The outline of the formulation is given below.

The wave packets $\phi(t)$ and $\chi(t)$ to be propagated forward and backward, respectively, are expanded in terms of the frozen Gaussian wave packets as (see also Section II.B)

$$\phi(t) = \int \frac{d\mathbf{q}_0 d\mathbf{p}_0}{(2\pi\hbar)^N} g_{\gamma,\mathbf{q}_t,\mathbf{p}_t} C_{\gamma,\mathbf{q}_t,\mathbf{p}_t} \exp[iS_{\gamma,\mathbf{q}_t,\mathbf{p}_t}/\hbar] \langle g_{\gamma,\mathbf{q}_t,\mathbf{p}_t} | \phi(0) \rangle \tag{199}$$

The similar expansion applies to $\chi(t)$. The frozen Gaussian wave packets $g_{\gamma,\mathbf{q}_t,\mathbf{p}_t}$ are explicitly given by

$$g_{\gamma,\mathbf{q},\mathbf{p}} = \Pi_{j=1}^N \left(\frac{2\gamma_j}{\pi} \right)^{1/4} \exp[-\gamma_j(r_j - q_j)^2 + ip_j(r_j - q_j)/\hbar] \tag{200}$$

where $S_{\mathbf{q}_0,\mathbf{p}_0,t}$ and $C_{\gamma,\mathbf{q}_0,\mathbf{p}_0}$ are the action and the preexponential factor at time t along the trajectory starting from $(\mathbf{q}_0, \mathbf{p}_0)$ at $t = 0$. The parameters γ_j are constants in time and are taken to be common for all the frozen Gaussians. Then

the correlation function defined by Eq. (197) becomes in the semiclassical
approximation

$$\Theta_k(t) = \int \frac{d\mathbf{q}_0 d\mathbf{p}_0}{(2\pi\hbar)^N} C^*_{\gamma \mathbf{q}_0 \mathbf{p}_0, t} \exp[-iS_{\mathbf{q}_0, \mathbf{p}_0, t}/\hbar]$$

$$\times \langle \phi(0) | g_{\gamma, \mathbf{q}_0 \mathbf{p}_0} \rangle \Omega_k(\gamma, \mathbf{q}_t, \mathbf{p}_t) \tag{201}$$

where

$$\Omega_k(\gamma, \mathbf{q}_t, \mathbf{p}_t) = \int \frac{d\mathbf{q}'_0 d\mathbf{p}'_0}{(2\pi\hbar)^N} \langle g_{\gamma, \mathbf{q}_t, \mathbf{p}_t} | \mu_k(\mathbf{r}) | g_{\gamma', \mathbf{q}'_t, \mathbf{p}'_t} \rangle$$

$$\times C_{\gamma, \mathbf{q}'_0, \mathbf{p}'_0, t} \exp[iS_{\mathbf{q}'_0, \mathbf{p}'_0, t}/\hbar] \langle g_{\gamma', \mathbf{q}'_0, \mathbf{p}'_0} | \chi(0) \rangle \tag{202}$$

Without loss of generality $\gamma' = \gamma$ can be assumed. If the dipole moment can be
assumed to be a linear function of coordinate within the spread of the frozen
Gaussian wave packet, the matrix element $\langle g_{\gamma, \mathbf{q}_t, \mathbf{p}_t} | \mu_k(\mathbf{r}) | g_{\gamma, \mathbf{q}'_t, \mathbf{p}'_t} \rangle$ can be evaluated
analytically. Since the integrand in Eq. (201) has distinct maxima usually, we can
introduce the linearization approximation around these maxima. Namely, the
Taylor expansion with respect to $\delta \mathbf{q}_0 = \mathbf{q}'_0 - \mathbf{q}_0$ and $\delta \mathbf{p}_0 = \mathbf{p}'_0 - \mathbf{p}_0$ is made,
where $\mathbf{q}'_0$ and $\mathbf{p}'_0$ represent the maximum positions. The classical action $S_{\mathbf{q}'_0, \mathbf{p}'_0, t}$ is
expanded up to the second order, the final phase-space point $(\mathbf{q}_t, \mathbf{p}_t)$ to the first
order, and the Herman–Kluk preexponential factor $C_{\gamma, \mathbf{q}'_0, \mathbf{p}'_0}$ to the zeroth order.
This approximation is the same as the cellularization procedure used in Ref. [18].
Under the above assumptions, various integrations in $\Omega_k(\gamma, \mathbf{q}_t, \mathbf{p}_t)$ can be carried
out analytically and we have

$$\Theta_k(t) = \int \frac{d\mathbf{q}_0 d\mathbf{p}_0}{(2\pi\hbar)^N} \langle \phi(0) | g_{\gamma, \mathbf{q}_0, \mathbf{p}_0} \rangle \langle g_{\gamma, \mathbf{q}_0, \mathbf{p}_0} | \chi(0) \rangle$$

$$\times [\mu_k(\mathbf{q}_t) - \nabla \mu_k(\mathbf{q}_t) \mathbf{F}_{\gamma, \mathbf{q}_0 \mathbf{p}_0, t}] G_{\gamma, \mathbf{q}_0, \mathbf{p}_0, t} \tag{203}$$

where $\mathbf{F}_{\gamma, \mathbf{q}_0 \mathbf{p}_0, t}$ and $G_{\gamma, \mathbf{q}_0, \mathbf{p}_0, t}$ are functions of the parameters of the trajectories
$(\mathbf{q}_t, \mathbf{p}_t)$. The details of the derivation and the expressions are given in
Ref. [38, 40].

Now, consider the special case of $\mu(\mathbf{r}) = 1$ for the moment. Then the
correlatoion function Eq. (197) is just a simple overlap of $\phi(t)$ and $\chi(t)$. With
use of the orthogonality and completeness of frozen Gaussian basis, this overlap
can be simply expressed as

$$\langle \phi(t) | \chi(t) \rangle = \langle \phi(0) | \chi(0) \rangle = \int \frac{d\mathbf{q}_0 d\mathbf{p}_0}{(2\pi\hbar)^N} \langle \phi(0) | g_{\gamma, \mathbf{q}_0, \mathbf{p}_0} \rangle \langle g_{\gamma, \mathbf{q}_0, \mathbf{p}_0} | \chi(0) \rangle. \tag{204}$$

Equations (203) and (204) should be equal for any couple of wavepackets, which
is possible only if $G_{\gamma \mathbf{q}_0 \mathbf{p}_0} = 1$. Thus, if the dipole moment does not significantly

change within the width of frozen Gaussians, then the correlation function can be very much simplified as

$$
\Theta_k(t) \simeq \int \frac{d\mathbf{q}_0 d\mathbf{p}_0}{(2\pi\hbar)^N} \langle \phi(0) | g_{\gamma,\mathbf{q}_0,\mathbf{p}_0} \rangle \mu_k(\mathbf{q}_t) \langle g_{\gamma,\mathbf{q}_0,\mathbf{p}_0} | \chi(0) \rangle \tag{205}
$$

This equation requires propagation of classical trajectories only and simplifies the calculation of optimal field very much. This semiclassical theory has been confirmed to work well by numerical applications for 2D systems [38, 40] and is applicable to realistic multidimensional systems for obtaining optimal laser field. In the practical applications, there is another problem, though. The final target state is usually far from the initial state and the overlap between the target wave packet and the zeroth order propagated initial wave packet is too small to carry out further iterations efficiently. To overcome this difficulty, we divide the whole process into a sequence of a certain number of steps and prepare an intermediate target state in each step. The intermediate target states cannot be defined uniquely, of course; but that is not so crucial, since we can easily physically guess what the intermediate state should roughly be like. Besides, the efficiency at the intermediate steps are not necessarily so high. The final efficiency at the final step is the most important quantity, naturally.

2. Numerical Examples

The above mentioned method has been applied to the four-dimensional model of isomerization of the HCN molecule, namely, isomerization in a plane spanned by the two 2D Jacobi vectors. The molecule is assumed to be initially linear in X direction. The potential energy surface and the dipole moment are taken from [124]. The whole process is divided into the following three steps: (1) acceleration of the initial wave packet which is taken to be the ground vibrational state in the HCN configuration so that the energy exceeds the top of the interstate barrier, (2) field-free propagation toward the target (CNH) region, and (3) deceleration so that the wave packet stays in the target region. Figure 42 shows the potential contours for the motion of H when N and C are fixed. Figure 43 depicts the probability densities of the wavepacket after 30 fs of field-free propagation as a function of (a) proton coordinate, (b) N—C bond coordinate, (c) proton momentum, and (d) N—C bond momentum. Figure 44 shows the efficiency and the final optimal field strength as a function of time. The efficiency is calculated according to the formula

$$
P_{\text{iso}} = \left\langle \phi(t) \Big| h \left(-\frac{\mathbf{R}_{\text{N=C}} \cdot \mathbf{R}_{\text{H}}}{R_{\text{N=C}} R_{\text{H}}} \right) \Big| \phi(t) \right\rangle \tag{206}
$$

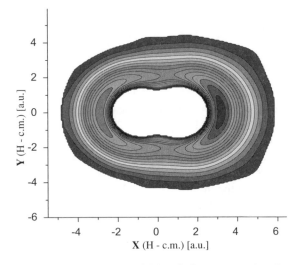

Figure 42. The potential energy contour felt by a hydrogen atom when the two atoms N (left side) and C (right side) are fixed. (See color insert.)

where $\mathbf{R}_{N=C}$ and $\mathbf{R}_H$ are the vector from N to C and the vector from the center of mass to H, and h is the step function. The final efficiency achieved here is $\sim 81\%$, even though the efficiencies at the intermerdiate steps are lower. The peak intensity of the field is as high as $10^{14}\,\mathrm{W\,cm^{-2}}$, but this is only for 15–20 fs and multiphoton ionization would not totally destroy the present process.

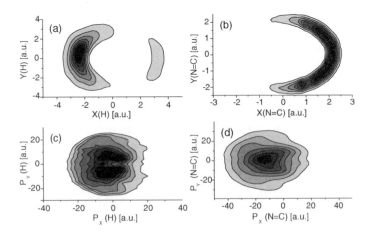

Figure 43. Probability density of the wavepacket after 30 fs of field-free propagation of the accelerated wavepacket as a function of (a) proton coordinate, (b) N=C bond length, (c) proton momentum, and (d) N=C bond momentum. Taken from Ref. [41].

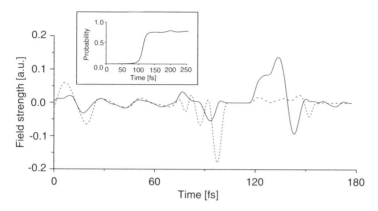

Figure 44. Optimal field calculated for the HCN–CNH isomerization control. Solid line-component along the x axis. Dashed line-component along the y axis. Inset is the time variation of the isomerization probability. Taken from Ref. [41].

C. Utilization of Complete Reflection Phenomenon

Zhu and Nakamura proved that the intriguing phenomenon of complete reflection occurs in the 1D NT type potential curve crossing [1, 14]. At certain discrete energies higher than the bottom of the upper adiabatic potential, the particle cannot transmit through the potential from right to left or vice versa. The overall transmission probability P (see Fig. 45) is given by

$$P = \frac{4\cos^2 \Psi(E)}{4\cos^2 \Psi(E) + \frac{p^2}{1-p}} \qquad (207)$$

where p is the nonadiabatic transition probability for one passage of the crossing point and Ψ is the phase along the upper adiabatic potential curve at energy E plus the contribution from the dynamical phases. This indicates that when the phase Ψ satisfies the condition,

$$\Psi(E) = (n + 1/2)\pi \qquad (n = 0, 1, 2, \cdots) \qquad (208)$$

the transition probability P becomes exactly zero. This phenomenon is due to the quantum mechanical interference between the wave trapped on the upper adiabatic potential and the wave transmitting along the lower adibataic potential without any transition to the upper state, and occurs whatever the potential shape and the coupling strength are. The above equations are the expressions in the semiclassical approximation, but the phenomenon itself is proved to happen quantum mechanically exactly [14, 125]. This phenomenon has been found in some other types of two-state potential systems [126]. The cases found are

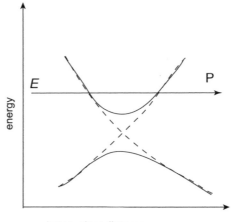

internuclear distance

Figure 45. Schematic picture representing the nonadiabatic tunneling-type transition.

(1) diabatically avoided crossing case, (2) double crossing NT type potentials, and (3) two-state three-channel system with tunneling. The semiclassical conditions have been formulated. Since this is a very unique and interesting phenomenon, we can think of various applications. Although the "completeness" is destroyed in multidimensional systems naturally, in some cases the reflection dips survive and we can use them to control molecular processes. In a periodic system or a system with a finite number of potential units, the complete transmission can occur. Thus we can think of molecular switching by using the complete reflection and transmission phenomena. Both 1D and 2D models have been discussed in Refs. [44, 46]. Laser control of photodissociation branching with use of this complete reflection phenomenon is also possible, since a CW laser can create NT type of potential crossings easily. The models of 2D HOD and CH_3SH molecules have been considered and it is numerically demonstrated that the selective dissociation is possible to some extent [45]. Since these models including the molecular switching are discussed already in the book [1], another example is presented here [127].

The selective photodissociation of an HI molecule is considered in the enrgy range $\hbar\omega = 3-6\,eV$. In the case of a diatomic molecule, predissociation can be stopped, if the condition, Eq. (208), is satisfied in the region designated in Fig. 46. In the energy range $\hbar\omega = 3-6\,eV$, there are three electronically excited states, $^1\Pi_1, ^3\Pi_{0+}$, and $^3\Pi_1$ (see Fig. 47). The ground state $^1\Sigma$ is coupled to these excited states through the transition dipole moments (see Fig. 48). The couplings among the excited states can be neglected, because the corresponding transitions are off-resonant. Thus we have the following potential matrix:

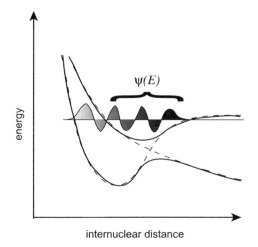

Figure 46. Schematic picture representing the complete reflection condition in a diatomic molecule. Taken from Ref. [126].

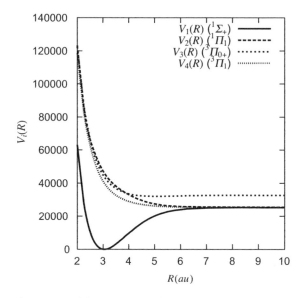

Figure 47. *Ab initio* potential energy curves of the HI molecule. The unit of *y* axis is reaprocal centemeters.

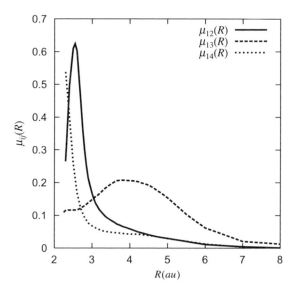

Figure 48. *Ab initio* transition dipole moments between the electronically ground and excited states. Taken from Ref. [126].

$$V(R,t) = \begin{pmatrix} V_1(R) & -\mu_{12}(R)E(t) & -\mu_{13}(R)E(t) & -\mu_{14}E(t) \\ -\mu_{12}(R)E(t) & V_2(R) & 0 & 0 \\ -\mu_{13}E(t) & V_3(R) & 0 & 0 \\ -\mu_{14}E(t) & 0 & 0 & V_4(R) \end{pmatrix} \quad (209)$$

where $i = 1, 2, 3, 4$ correspond to $^1\Sigma, ^1\Pi_1, ^3\Pi_{0+}, ^3\Pi_1$, respectively. The potentials and transition dipole moments are taken from the *ab initio* data and spline fitting is made to them [128, 129]. The CW laser field $E(t)$ is taken as

$$E(t) = E_0 \cos(\omega t)\theta(t) \quad (210)$$

where $\theta(t)$ is the envelope of the laser pulse that should be wide and smooth enough so that unnecessary transitions are not induced due to the sudden switching of the field. The actual intensity used in the calculations is $1\,\mathrm{TW\,cm^{-2}}$. The two excited states $V_2 = ^1\Pi_1$ and $V_4 = ^3\Pi_1$ correlate to the ground-state inodine I and the state $V_3 = ^3\Pi_{0+}$ correlates to the excited iodine I*. Thus, in order to selectively produce the excited state I* we have to be able to stop the dissociation through the other two excited states. Numerical solutions were made by solving the time-dependent coupled equations

$$i\hbar\frac{\partial}{\partial t}\phi(R,t) = \left(-\frac{\hbar^2 d^2}{2m dR^2} + V(R,t)\right)\phi(R,t) \quad (211)$$

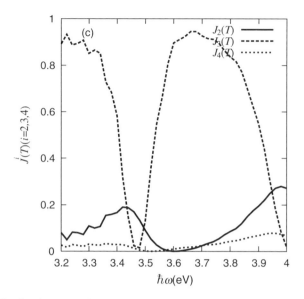

Figure 49. The time-integrated fluxes at $t = 3.5$ ps as a function of $\hbar\omega$ for $v = 4$. Taken from Ref. [126].

The step sizes used are $\Delta R = 7.8 \times 10^{-3}$ a.u. and $\Delta t = 0.043$ fs, and the absorbing potential is put at $R = 9-10$ a.u. We have first found the complete reflection manifold in which Eq. (208) is satisfied and roughly estimate the appropriate energy region and vibrational states. One example of the results is shown in Figs. 49 and 50 for $v = 4$. Figure 49 shows the time-integrated dissociation flux at $t = 3.5$ ps. The time-integrated flux is defined as

$$J_i(t) = \int_0^t dt \, \frac{\hbar}{m} \text{Im} \left[\phi_i^*(R, t) \frac{d}{dR} \phi_i(R, t) \right] |_{R=6} \qquad (212)$$

where $\phi_i(R, t)$ is the wave function on the potential i. It can be seen that the two dissociation channels through the states $i = 2$ and 4 are almost stopped at the photon energy $\hbar\omega \simeq 3.58$ eV. Figure 50 shows the time varation of the time-integrated flux to confirm the highly selective production of I^*. The condition for the inverse case to produce I selectively can be found easily. For example, $\hbar\omega \simeq 3.47$ eV with $v = 4$ meets the condition. In the present treatment, the rotational degree of freedom has been neglected. In order to compare with any real experiment, it is required to take into account the effects of initial rotational state distribution that depends on the experimental condition. The completeness would be deteriorated to some extent, but the control may be achieved to a good extent. One defect of the method based on the complete

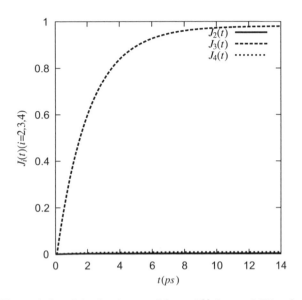

Figure 50. Time variation of the time-integrated fluxes $J_i(t)$ for $v = 4$. Taken from Ref. [126].

reflection phenomenon is that the initial state should be prepared in a certain excited vibrational state.

VI. MANIFESTATION AND CONTROL OF MOLECULAR FUNCTIONS

Nonadiabatic transitions definitely play crucial roles for molecules to manifest various functions. The theory of nonadiabatic transition is very helpful not only to comprehend the mechanisms, but also to design new molecular functions and enhance their efficiencies. The photochromism that is expected to be applicable to molecular switches and memories is a good example [130]. Photoisomerization of retinal is well known to be a basic mechanism of vision. In these processes, the NT type of nonadiabatic transitions play essential roles. There must be many other similar examples. Utilization of the complete reflection phenomenon can also be another candidate, as discussed in Section V.C. In this section, the following two examples are cosidered (1) photochromism due to photoisomerization between cyclohexadiene (CHD) and hexatriene (HT) as an example of photoswitching molecular functions, and (2) hydrogen transmission through a five-membered carbon ring.

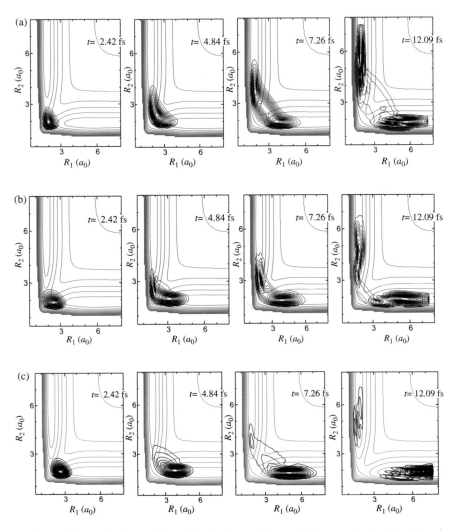

Figure 41 (from the chapter "Nonadiabatic Chemical Dynamics"). Selective bond breaking of H_2O by means of the quadratically chirped pulses with the initial wave packets described in the text. The dynamics of the wavepacket moving on the excited potential energy surface is illustrated by the density. (a) The initail wave packet is the ground vibrational eigen state at the equilibrium position. (b) The initial wave packet has the same shape as that of (a), but shifted to the right. (c) The initail wave packet is at the equilibrium position but with a directed momentum toward x direction. Taken from Ref. [37].

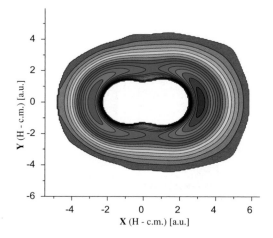

Figure 42 (from the chapter "Nonadiabatic Chemical Dynamics"). The potential energy contour felt by a hydrogen atom when the two atoms N (left side) and C (right side) are fixed.

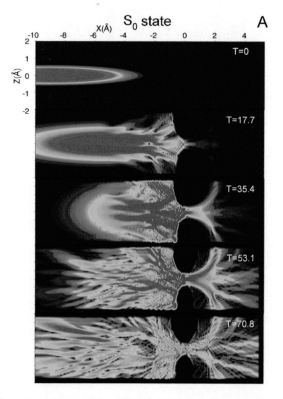

Figure 63 (from the chapter "Nonadiabatic Chemical Dynamics"). Two-dimensional snapshots $(y = 0)$ of wave packet on the ground S_0 state. The symbol T means the time duration in fs. Taken from Ref. [47].

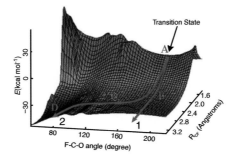

Figure 12 (from the chapter "Exploring Multiple Reaction Paths to a Single Product Channel"). Two-dimensional cut through the potential surface for fragmentation of the transition state $[OH \cdots CH_3 \cdots F]^-$ complex as a function of the CF bond length and the FCO angle. All other coordinates are optimized at each point of this PES. Pathway 1 is the direct dissociation, while pathway 2 leads to the hydrogen-bonded $[CH_3OH \cdots F^-]$ structure. The letter symbols correspond to configurations shown in Fig. 11. Reprinted from [63] with permission from the American Association for the Advancement of Science.

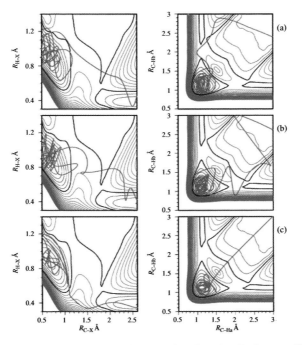

Figure 16 (from the chapter "Exploring Multiple Reaction Paths to a Single Product Channel"). Projections onto 2D surfaces of trajectories (in green) of $CH_3O \rightarrow H_2 + HCO$. The left column is a projection onto the surface of Fig. 15. The right column is a projection onto the surface of Fig. 14. The black contour represents the saddle point energy for the $H + H_2CO \rightarrow H_2 + HCO$ reaction. Blue contours are lower in energy; red contours are higher. Reprinted with permission from [67]. Copyright 2001 American Chemical Society.

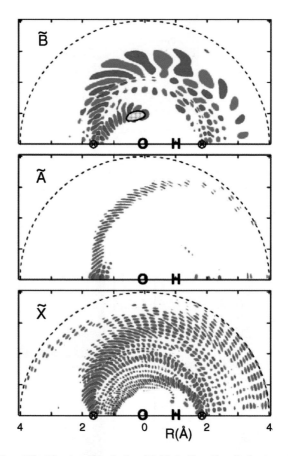

Figure 19 (from the chapter "Exploring Multiple Reaction Paths to a Single Product Channel"). Outgoing waves on the first three electronic states of H_2O following excitation of the $(\tilde{B}^1A_1) \leftarrow (\tilde{X}^1A_1)$ transition. Green and red lobes have positive and negative amplitudes, respectively. The gray-filled contour on the $\tilde{B}$ surface represents the Franck–Condon region of excitation from the ground state. Reprinted from [75] with permission from the American Association for the Advancement of Science.

A. Photochromism: Photoisomerization between Cyclohexadiene and Hexatriene

1. Potential Energy Surfaces and Reaction Scheme

Photoinduced reversible ring-opening–closure of photochromic molecules, such as diarylethenes and fulgides, are applicable to molecular switches and memories [130–133]. Murakami et al. [132] recently found that the reaction yield of the photoinduced ring opening of diarylethene strongly depends on the time duration of laser irradiation. The control of the reaction yield of photochromism by the optimal laser is thought to be particularly important for the "gated-function", which is crucial to achieve low fatigue and nondestructive read-out capability in practical photoswitches and optical rewritable memories (i.e., efficiently isomerized by the specific optimal laser, but not isomerized by sunlight or read-out light). The clarification of the ultrafast photochemical dynamics enables us to build up reaction control strategies to achieve high isomerization yield, quick response, and gated functions, which are required in practical photochromic systems. Recent advances in experimental and theoretical techniques have made it possible to reveal time-dependent pictures of ultrafast reactions in detail [134], which provides a wide perspective for reaction controls and molecular designs. Photoisomerization between 1,3-cyclohexadiene (CHD) and all-cis-hexatriene (HT) (Fig. 51) has been attracting great interest [135–139] not only as a prototype of ultrafast photochemistry, but also as a model system to understand universal reaction mechanisms of photochromism, because CHD/HT is the reaction center of various photochromic molecules. The ground-state CHD equilibrium geometry is of C_2 symmetry, at which the electronic ground (S_0), first (S_1) and second (S_2) excited states have 1^1A, 1^1B and 2^1A characters, respectively, and the photoexcitation to S_1 (1^1B) is much stronger than that to S_2

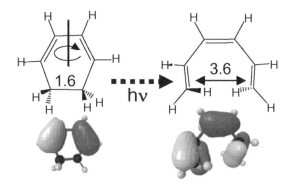

Figure 51. 1,3-cyclohexadiene (a) and all-cis-hexatriene (b) isomers of C_6H_8 with the respective third active orbitals (highest occupied π-orbital) and breaking C–C distances in angetroms. The bold line indicates the C_2 rotation axis. Taken from Ref. [48].

(2^1A) because of the large $1^1A - 1^1B$ transition dipole moment. Fus et al. [135] experimentally observed the CHD/HT photoisomerization in the vapor phase in femtosecond (fs) time-resolution as follows (with time durations from photo-excitation): (1) photoexcitation of CHD to S_1 (1^1B) in the Franck–Condon (FC) region (0 fs). (2) State character change of S_1 from 1^1B to 2^1A (53 fs). (3) Radiationless decay from S_1 to S_0 (130 fs). (4) CHD or HT formation on S_0 (200 fs).

We have theoretically studied this CHD/HT photoisomerization process as a prototype system of photochromism [48, 140]. The overall reaction scheme is shown in Fig. 52. The *ab initio* potential energy surface calculations are performed using the MOLPRO 2002 [100] and the GAMESS [141] codes. The 6-31G basis set with d and p polarization functions $[6 - 31G(d,p)]$ is used for all the calculations. The molecular orbitals are optimized using the complete active space self-consisitent field (CASSCF) method. The active space comprises six active orbitals and six active electrons [CASSCF(6,6)]; the active orbitals are composed of the $\pi(\pi^*)$ and the breaking $\sigma(\sigma^*)$ orbitals. The equilibrium molecular geometries on the respective states and the minima on the CoIn(conical intersection) hypersurface are optimized using the (state averaged) CASSCF energy gradient without symmetry constraint. The notations $1^1A, 2^1A$, and 1^1B in the C_2 symmetry are used to indicate the dominant electronic state characters, even though the symmetry broken structures are considered. Robb and co-workers carried out the pioneering theoretical studies using the CASSCF calculations [136]. In their calculations, however, the 1^1B

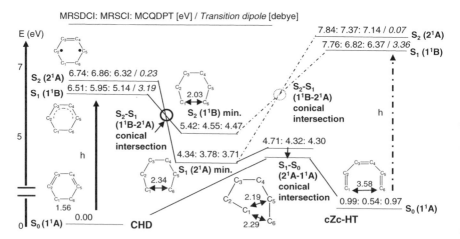

Figure 52. Diagram of CHD/HT photochemical interconversion. The MRSDCI, MRSCI, and MCQDPT energies are relative values from those of the ground-state CHD. Transition dipole moments and equilibrium geometries are also shown. Taken from Ref. [139].

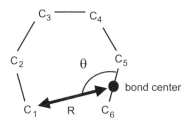

Figure 53. Two-dimensional Jacobi coordinates employed. Taken from Ref. [48].

and 2^1A states were separately calculated within the constraint of C_2 symmetry and the transition between 1^1B and 2^1A was not elucidated. Besides, they have taken the minimum energy path through the conical intersection based on the assumption that the thermal equilibrium is reached before decaying to the ground state. However, the experiment [135] implies that the excited state decays to the ground state before it reaches the thermal equilibrium. We have calculated the *ab initio* potential energy surfaces along the 2D reduced coordinates shown in Fig. 53. The other nonreactive coordinates are optimized. The single point energies of the respective states at the optimized geometries are calculated using the multireference configuration interaction (MRCI) method with single excitations to take into account dynamic correlation energies. The potential energy scheme is shown in Fig. 54. First, the photoexcitation occurs

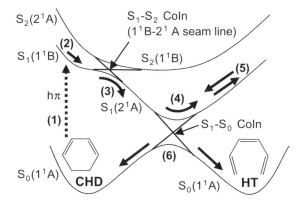

Figure 54. Reaction scheme of CHD/HT photoisomerization. (1) CHD is photoexcited to $S_1(1^1B)$, (2) ring opening proceeds descending the $S_1(1^1B)$ potentiaql energy surface, (3) electronic state character of S_1 changes from 1^1B to 2^1A at the first conical intersection, (4) the $S_1(2^1A)$ wavepacket ascends the potential energy surface toward the open ring direction, where the wave packet is still compact, (5) the wavepacket turns back toward the closed-ring direction, and (6) radiationless decay to S_0 occurs through the second conical intersection located along the direction toward the five-membered ring (5MR). Taken from Ref. [49].

from the ground state $S_0(1^1A)$ to $S_1(1^1B)$, since the transition dipole moment to the $S_1(1^1B)$ state is much larger than that to $S_2(2^1A)$. There are two important CoIns on the way to the conversion, one is between S_1 and S_2 and the other between $S_1(2^1A)$ and $S_0(1^1A)$. At the first CoIn, the state character changes as $S_1(1^1B) \rightarrow S_1(2^1A)$ and $S_2(2^1A) \rightarrow S_2(1^1B)$. This occurs by breaking the C_2 symmetry.

On the potential energy surfaces thus obtained 2D wavepacket dynamics calculations have been performed in the diabatic state representation. The reduced massses are regarded as those of CH_2–ethylene system. The validity was examined by using on-the-fly *ab initio* molecular dynamics that were supplementarily performed. The dynamics calculations performed are composed of the following steps:

1. The eigenfunction of the vibrational ground state is calculated on the *ab initio* 2D S_0 potential energy surface by solving the eigenvalue problem.

2. The initial wave packet on the excited $S_1(1^1B)$ state is prepared by the product of the S_0 eigenfunction and the $S_0 - S_1$ transition dipole moment at each grid point.

3. The wave packet is propagated from the FC region on the S_1 and S_2 surfaces (on which the geometries are optimized by the 1^1B CASSCF energy gradient.

4. After the transition to $S_1(2^1A)$, the wave packet is projected onto the $S_1 - S_0$ surfaces on which geometries are optimized by the S_1 CASSCF energy gradient.

5. When the wave packet reaches the CHD or HT product regions on S_0, the product populations are calculated at respective regions.

6. Wave packet absorptions are imposed in order to exclude the effect of secondary thermal isomerization.

The reaction scheme clarified by the present studies is depicted in Fig. 54. Note that the real dynamics goes on the 2D space. The wavepacket bifurcates at the first CoIn into $S_1(2^1A)$ and $S_2(1^1B)$ after 15–25 fs from the first photoexcitation. Here it turns out that most of the wave packet ($\sim 80\%$) goes to the $S_1(2^1A)$. Since the 2D consideration is not good enough for this nonadiabatic transition, the full dimensional analysis was made with use of the ZN formula under the harmonic normal mode approximation and the final transition probability was estimated. The subsequent dynamics on the $S_1 - S_0$ coupled surfaces is made as follows: (1) The wave packet ascends the S_1 surface toward the open-ring direction due to the excess kinetic energy from the FC region (~ 45 fs), (2) turns around to the close-ring direction (70–80 fs), (3) is refelected by the steep ascending slope at the close-ring region (90–120 fs), (4)

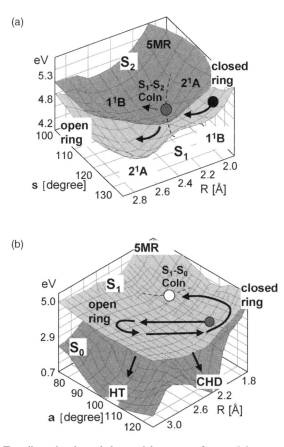

Figure 55. Two-dimensional coupled potential energy surfaces and the wavepacket motion. (a) $S_1 - S_2$ surfaces and (b) $S_1 - S_0$ surfaces. The black, gray, and white circles and dotted lines indicate the locations of the FC region, $S_1 - S_2$ conical intersection minimum, 5MR $S_1 - S_0$ conical intersection minimum, and seam lines, respectively. The solid arrows indicate the schematic wavepacket pathway in the case of natural photoisomerization starting from the vibrational ground state. Taken from Ref. [49].

reaches the $S_1 - S_0$ CoIn region owing to the shallow potential toward the 5MR (120–130 fs), and (5) the first decay occurs due to the nonadiabatic transition across the seam line toward the 5MR (130–180 fs). This overall time duration is in good agreement with the experiment [135]. Figure 55 depicts the motion of the wave packet schematically. Figure 56 shows the wavepacket population on the respective states as a function of the elapsed time. This indicates ∼ 45% of the S_1 wave packet decays to S_0 during the first major seam line crossing toward the 5MR at 130–180 fs. The nonadiabatic transition at this second CoIn is also corrected by considering the multidimensionality. The final CHD/HT branching

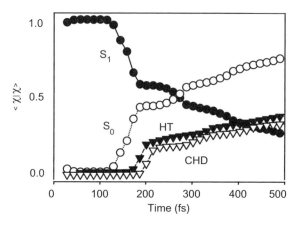

Figure 56. Changes of the wavepacket populations on the $S_1 - S_0$ coupled potential energy surfaces as a function of time duration from the FC region. Taken from Ref. [48].

ration is found to be $\sim$5:5. The experimental value for the CHD/HT ratio in solution is 6:4 [142]. As is clear from the above analysis, the second CoIn between $S_1(2^1A)$ and $S_0(1^1A)$ is crucial for the final branching.

2. Laser Control of Photoswitching Functions

Once the overall mechanisms of molecular functions dictated by nonadiabatic transitions are understood, we can think of controlling them. Here, we consider possible control schemes generally applicable to the photoswitching molecules by taking the CHD/HT photoisomerization as an example. We investigate the following two ideas to control the molecular functions [49]: (1) nearly complete electronic transitions by using quadratically chirped pulses and (2) preparation of the initial wavepacket on the ground state with appropriately directed momentum. These two methods have already been demonstrated to be effective in practical examples of laser control, as discussed in Section IV.

In the case of quadratic chirping, the following scheme is proposed (see Fig. 57): (1) The nearly complete initial excitation can be achieved to enhance the total population involved in the photoisomerization process, and (2) the branching ratio of HT can be increased by the laser-induced dumping when the wavepacket reaches above the ground state HT basin before the natural radiationless decay occurs through the $S_1 - S_0$ CoIn. The $S_1(2^1A)$ wave packet can be efficiently dumped to $S_0(1^1A)$ via pumping to $S_2(1^1B)$ because of the large transition dipole memontes. The effiency of these pump–dump processes can be very much enhanced by using the quadratic chirping method. Figure 58 shows the changes of the wave packet populations on the $S_0(1^1A)$ and $S_1(1^1B)$ states under the quadratically chirped pulses at $3.5\,\mathrm{TW\,cm^{-2}}$ intensity. At the

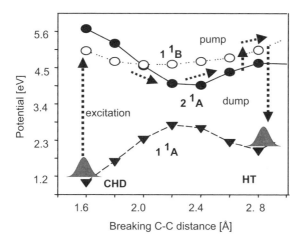

Figure 57. Schematic potential energy profiles and the pump-dump scheme. Taken from Ref. [49].

first photoexcitation process $\sim 90\%$ efficiency is achieved [see Fig. 58(a)]. Although the wave packet on the $S_1(2^1A)$ state passes over the ground state HT region, it is not efficient in the present system to dump it directly to the ground HT state because of the small transition dipole moment bewteen 2^1A and 1^1A.

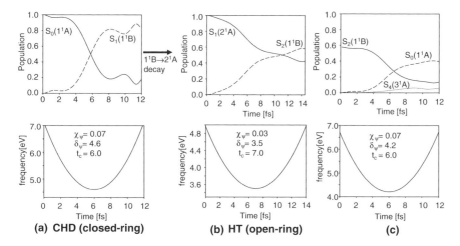

Figure 58. Changes of the wavepacket populations on the respective states (upper panels) under the 3.5 TW cm^{-2} quadratically chirped pulses (lower panels) during the sequential pump-dump scheme via the (a) $1^1A \rightarrow 1^1B$ pumping at CHD and [(b) and (c)] $2^1A \rightarrow 1^1B \rightarrow 1^1A$ pump dump at HT as a function of time duration. Taken from Ref. [49].

Thus, as mentioned above, the pump–dump process $2^1A \rightarrow 1^1B \rightarrow 1^1A$ is used. Fig. 58(b) shows the population changes of the 2^1A and 1^1B states under the quadratically chirped pump pulse. The $S_2(1^1B)$ population can be $\sim 60, 56$, and 80% at 3.5, 8.0, and 14.0 TW cm^{-2}, respectively. Finally, the wave packet on the $S_2(1^1B)$ state is dumped to $S_0(1^1A)$. Since the wave packet is spread a bit already, it is not easy to attain a high efficiency. Approximately 40 [see Fig. 58(c)], 46%, and 55% populations are dumped at 3.5, 8.0, and 14.0 TW cm^{-2}, respectively. In this pump–dump process we have to be careful about the deterioration by the other states (see Fig. 57), such as the processes of $2^1A \rightarrow 2^1B \rightarrow 1^1A$ and $S_2(1^1B) \rightarrow S_4(3^1A)$. These processes are found, however, not to deteriorate the present pump–dump scheme (see Ref. [49] for more details). The total overall reaction yield of HT becomes 70 [see Fig. 58(c)], 73, and 77%, respectively, at 3.5, 8.0, and 14.0 TW cm^{-2} (it is $\sim 50\%$ without the pump–dump laser).

Let us next consider the control by the directed momentum method. The initial wave packet with any directed momentum can be easily prepared by the semiclassical optimal control theory explained in Section V.B. Since the direct pathway to the 5MR region is far away from the $S_1 - S_2$ CoIn, we do not have to worry about the transition to $S_2(1^1B)$ through that CoIn, if we give the momentum directed to 5MR. We have rather arbitrarily chosen the kinetic energy ~ 6 kcal mol^{-1} and the direction toward 5MR and run the wave packet in the same way as before. This wave packet directly goes to the $S_1 - S_0$ CoIn region without any excursion around the open-ring region and $\sim 24\%$ of that makes a transition to S_0 (see Fig. 59). The multidimensionality correction is carried out in the same way as before. Most part of this S_0 wave packet finally goes to the product HT. Finally, the total HT branching ratio is $\sim 62\%$. One may

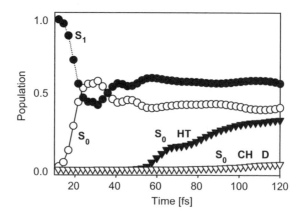

Figure 59. Changes of wave packet populations on S_1 and S_0 after the initial excitation of the wavepacket accelerated toward the 5MR. Taken from Ref. [49].

claim that this efficiency is not high enough comapred to 50% in the case of natural decay. Note, however, that the magnitude and direction of the wavepacket are not optimally designed in this calculation and thus the total yield can be much higher by using a more appropriate design. An important point here is that we can generally design appropriate laser fields to enhance the desired total yield and branching ratio. The final efficiency naturally depends on the potential energy surface topography and the transition dipole moment of the system.

B.   Hydrogen Transmission through Five-membered Carbon Ring and Hydrogen Encapsulation

It has been found that atomic hydrogen penetrates through a five-membered carbon ring with the help of the nonadiabatic tunneling phenomenon [47]. The example of pentaboron-substituted corannulenyl radical ($C_{15}H_{10}B_5$) is explained here (see Fig. 60). In the case of the original corannulene molecule, the hydrogen atom penetration is not possible, as may be guessed from its stable electronic structure. Thus boron-substitution is introduced so that the molecule attains the electronegativity. The five carbon atoms in the second layer are replaced by borons. Then the molecule abstracts an electron from the incident hydrogen atom to gain a stable ionic structure. It would be easy for thus

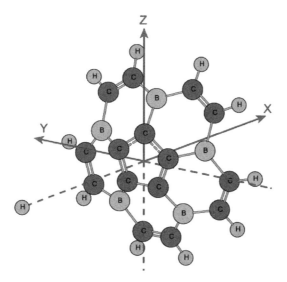

Figure 60.   Pentaboron-subsituted corannulenyl radical ($C_{15}H_{10}B_5$) and the coordinate system used. The origin is put at the center of the five-membered ring and the x axis is set to be perpendicular to the molecular plane. Taken from Ref. [47].

produced proton to transmit through the ring. The present study could be a model to think about any possibility of hydrogen encapsulation by fullerenes [143]. The ground and excited electronic potential energy surfaces are calculated by using the multireference (MR) polarization (POL) CI method for $C_{15}H_{10}B_5 + H$. The following excitations from the reference configuration have been generated: single excitations of [core to active], [core to external], and [active to external]. The molecular orbitals are determined by using the CASSCF method. The calculations have been made at ~ 500 configurations that cover the whole necessary range for 3D wavepacket calculations. During the transmission of H the geometry of the $C_{15}H_{10}B_5$ moiety is fixed, that is, the sudden approximation is used. This is considered not to be so bad, since the transmission time is as short as 100 fs. The molecular structure is quite flat, actually flatter than the original $C_{15}H_{15}$, and thus the symmetry for the transmission is higher.

The quantum dynamics calculations are carried out in the following two ways: (1) 1D calculations in order to explore the possibility of the hydrogen transmission and to confirm our idea, and (2) 3D wave packet dynamics to find out the actual transmission probability. The potential energy curves for the 1D case, namely, the potential curves when the hydrogen atom impacts the molecule perpendicularly to the molecular plane aiming at the center, are shown in Fig. 61. There are two nonadiabatic tunneling type avoided curve crossings, which play the essential role for the resonant-type transmission. The steric potential barrier is also found in the the middle in between the two crossings. Since we are interested in the energy region lower than the

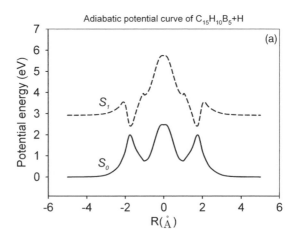

Figure 61. Adiabatic potential energy curves for the atomic hydrogen transmission through the five-membered ring in the case of center approach. Taken from Ref. [47].

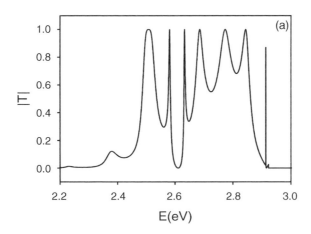

Figure 62. Hydrogen transmission probability as a function of total energy (eV) for the center approach. Taken from Ref. [47].

dissociation limit of the first excited state, higher excited states are not shown. The quantum mechanical R-matrix calculations are performed to calculate the transmission probability as a function of energy. The result is depicted in Fig. 62. Some resonant-type complete transmission peaks are revealed. This confirms our idea. Then the 3D quantum mechanical wave packet dynamics calculations are carried out by shooting an appropriate wave packet perpendicularly to the molecular plane. The initial wavepacket used has the half-widths of $\Delta q_x = 1.67$ Å and $\Delta q_y = \Delta q_z = 0.34$ Å in the coorinate space and $\Delta p_x = 0.63$ a.u. and $\Delta p_y = \Delta p_z = 2.82$ a.u. in the mometum space, where x is the direction perpendicular to the molecular plane. The incident central energy is $E \simeq 2.79$ eV. Figure 63 shows the two-dimensional snapshots of the wave packet motion on the ground electronic surface. $x = 0$ corresponds to the molecular center where the potential barrier exists. The violent interference structure appears when the packet reaches the crossing region, but after that a substantial portion of the packet transmits through the five-membered ring. The transmission probability in the present case reaches $\sim 27\%$.

As is well known, many experimental studies have been made extensively to search for a possibility of encapsulation of atoms by hollow fullerenes since the discovery of C_{60} by Kroto et al. [143]. These methods, however, usually require high tempratures and high pressures, or ion implantation. The yields are also as low as $0.4-10^{-5}\%$. In this sense, the efficiency in our case is much higher and the required conditions are much milder with collison energy of ~ 2 eV. However, the boron substitution is a bottle neck, although Smalley and co-workers successfully synthesized boron-doped fullerenes [144].

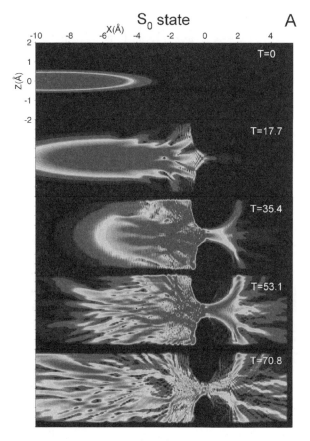

Figure 63. Two-dimensional snapshots $(y = 0)$ of wave packet on the ground S_0 state. The symbol T means the time duration in fs. Taken from Ref. [47]. (See color insert.)

Recently, a hydrogen molecule was encapsulated by fullerene by a synthetic approach [145]. This, however, requires several steps, such as opening an orifice, inserting a hydrogen molecule, and closing the orifice. Our approach would be a one step, although the boron substitution is necessary and the hydrogen atom has to be used. In any case, the presently proposed mechanism with use of the nonadiabatic transitions would be worthwhile to be further investigated.

VII. CONCLUDING REMARKS

Nonadiabatic transitions due to potential energy surface crossings definitely play crucial roles in chemical dynamics. They (1) are important to comprehend the mechanisms of chemical reactions, (2) provide us guiding principles how to improve the efficiencies of the various dynamics, (3) present useful tools to control the dynamics by using lasers, and (4) play key roles to manifest and create new functions of molecules. The Zhu–Nakamura (ZN) theory presents a set of analytical formulas to describe the nonadiabatic transitions in the potential curve crossing problems, and can play basic roles in these studies. Although the theory is a one-dimensional (1D) one, it works well because the nonadiabatic coupling is a vector defined in multidimensional space.

Comprehension of reaction dynamics in large systems can be done by using the semiclassical frozen wave packet dynamics or by using the generalized trajectory surafce hopping (TSH) method in which the ZN formulas can be usefully utilized, as explained in Section II. Especially, the generalized TSH method can be applied easily to high dimensional systems with effects of nonadiabatic transitions and tunneling included. The commonly used molecular dynamics (MD) method can be very much improved. The multi-dimensional tunneling, splitting, and tunneling decay presented in Section III can now be applied to virtually any large systems as far as high quality quantum chemical calculations can be carried out to obtain potential energy surfaces. The efficient detection of caustics along classical trajectories also discussed in Section III can be combined with the generalized TSH method, as mentioned above. The newly developed theory of electron transfer can be further strengthened and applied to real systems by combining the quantum chemical calculations of electronic structures and the statistical theory, such as the RISM theory [146], although the effects of solvent dynamics should be properly incorporated. Laser control of various chemical dynamics can be thought of by combining the semiclassical *guided* optimal control theory for the wave packet motion on a single adiabatic potential energy surface and the quadratic chirping method for electronic transitions. Now, we can deal with realistic multi-dimensional systems. There is no doubt that molecular functions are also dictated by nonadiabatic transitions in many cases. The manifestation and control of new molecular functions can be thought of from the viewpoint of nonadiabatic transition. The viewpoint of efficient utilization of nonadiabatic transition to manifest and control molecular functions would be useful for synthesizing new molecules.

In all these studies mentioned above, the ZN theory of nonadiabatic transition can be very useful, as discussed in this chapter.

VIII.   APPENDIX: ZHU-NAKAMURA THEORY AND SUMMARY OF THE FORMULAS

Zhu and Nakamura treated the two-state time-independent linear potential model (two linear diabatic potentials coupled by a constant diabatic coupling) quantum mechanically and obtained the quantum mechanically exact expressions of the reduced scattering matrix in terms of one Stokes constant that is a function of the two basic parameters a^2 and b^2 [14]. This was successfully done for the two types of transitions:(1) Landau–Zener (LZ) type of transition in which the two diabatic potentials have the same sign of slopes, and (2) nonadiabatic tunneling (NT) type of transition in which the two diabatic potentials have opposite signs of slopes.These are Eqs. (A.7–A.11) in the LZ case and Eqs.(A.41–A.42) in the NT case. The existence of the phenomenon of complete reflection was thus quantum mechanically exactly proved by this solution. Actual ZN formulas recommended for practical use were derived from these expressions so as to be applicable to general curved potentials.

Below, the final expressions of the ZN formulas that can be directly applied to practical problems are summarized. Note that there are some typographical errors in the expressions given in Refs. [1] and [13]. The necessary corrections are explained in Ref. [2]. The whole set of correct expressions are provided here.

Note that the formulas presented here contain some empirical corrections that have been introduced so that the analytical solutions can cover some small limiting regimes [1, 13, 14].

The time-dependent problems are generally simpler than the time-independent ones and the corresponding version of the ZN formulas is presented in Section C. This corresponds to the LZ type, since the NT type does not occur in the time-dependent case.

A. LANDAU–ZENER TYPE OF TRANSITION

The two-dimensionless basic parameters a^2 and b^2 in terms of the diabatic potentials are defined as

$$a^2 = \frac{\hbar^2}{2m} \frac{F(F_1 - F_2)}{8V_X^3} \tag{A.1}$$

and

$$b^2 = (E - E_X) \frac{F_1 - F_2}{2FV_X} \tag{A.2}$$

with $F = \sqrt{|F_1 F_2|}$, where $F_j (j = 1, 2)$, V_X, and E_X are the slopes of diabataic potentials, the diabatic coupling, and the energy at the potential crossing, respectively. These parameters can be reexpressed in terms of the adiabatic potentials (E_1 and E_2 with $E_2 > E_1$) as given below. This means that the diabatization of adiabatic potentials are not necessary; besides the transition probabilities can be estimated more accurately with use of these parameters expressed in terms of adiabatic potentials:

$$a^2 = \sqrt{d^2 - 1}\, \frac{\hbar^2}{m(T_2^0 - T_1^0)^2 [E_2(R_0) - E_1(R_0)]} \tag{A.3}$$

and

$$b^2 = \sqrt{d^2 - 1}\, \frac{2E - [E_2(R_0) + E_1(R_0)]}{[E_2(R_0) - E_1(R_0)]} \tag{A.4}$$

where

$$d^2 = \frac{[E_2(T_1^0) - E_1(T_1^0)][E_2(T_2^0) - E_1(T_2^0)]}{[E_2(R_0) - E_1(R_0)]^2} \tag{A.5}$$

The position R_0 corresponds to the minimum separation of the two adiabatic potentials, and T_1^0 and T_2^0 are defined as (see Fig. A.1)

$$E_X = [E_1(R_0) + E_2(R_0)]/2 = E_1(T_1^0) = E_2(T_2^0) \tag{A.6}$$

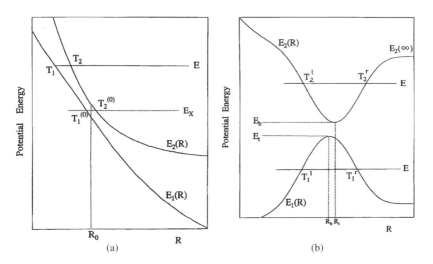

Figure A.1. Schematic adiabatic potentials and various parameters used in the ZN formulas. (a) Landau–Zener type. (b) Nonadiabatic tunneling type. Taken from Ref. [9].

In terms of the Stokes constant U_1, the reduced scattering matrix S^R can be quantum mechanically exactly given by

$$S^R = \begin{pmatrix} (1 + U_1 U_2)\exp(-2i\sigma) & -U_2 \\ -U_2 & (1 - U_1^* U_2)\exp(2i\sigma) \end{pmatrix} \qquad (A.7)$$

where

$$U_2 = \frac{U_1 - U_1^*}{1 + |U_1|^2} \qquad (A.8)$$

The overall nonadiabatic transition probability between the two adiabatic states is given by

$$P_{12} = |S_{12}^R|^2 = \frac{4(\mathrm{Im}U_1)^2}{(1 + |U_1|^2)^2} = 4p(1 - p)\sin^2\psi \qquad (A.9)$$

with

$$\psi = \arg(U_1) \qquad (A.10)$$

and

$$p = \frac{1}{1 + |U_1|^2} \qquad (A.11)$$

where p represents the nonadiabatic transition probability for one passage of the crossing point. Note that the above expressions are quantum mechanically *exact*, as far as the Stokes constant U_1 is exact. Below the semiclassical expressions in the ZN theory are given.

1.   Case A: $E \geq E_X$

The Stokes constant U_1, which is actually a function of the parameters, is given as

$$U_1 = \sqrt{\frac{1}{p} - 1}\, \exp(i\psi) \qquad (A.12)$$

where

$$p = \exp\left[-\frac{\pi}{4a}\left(\frac{2}{b^2 + \sqrt{b^4 + 0.4a^2 + 0.7}} \right)^{1/2} \right] \qquad (A.13)$$

and

$$\psi = \sigma + \phi_S = \sigma - \frac{\delta_\psi}{\pi} + \frac{\delta_\psi}{\pi} \ln\left(\frac{\delta_\psi}{\pi}\right) - \arg \Gamma\left(i\frac{\delta_\psi}{\pi}\right) - \frac{\pi}{4} \qquad (A.14)$$

The parameters σ and δ are defined in Section A.3 The nonadiabatic transition amplitude that connects the wave function just before and right after the transition at the avoided crossing is given by

$$I_X = \begin{pmatrix} \sqrt{1-p}\,\exp(i\phi_S) & -\sqrt{p}\,\exp(i\sigma_0) \\ \sqrt{p}\,\exp(-i\sigma_0) & \sqrt{1-p}\,\exp(-i\phi_S) \end{pmatrix} \qquad (A.15)$$

2. Case B: $E \leq E_X$

The Stokes constant U_1 is given by

$$\mathrm{Re}U_1 = \cos(\sigma)\left\{ \sqrt{B(\sigma/\pi)}\,\exp(\delta) - g_1 \sin^2(\sigma)\frac{\exp(-\delta)}{\sqrt{B(\sigma/\pi)}} \right\} \qquad (A.16)$$

and

$$\mathrm{Im}U_1 = \sin(\sigma)\left\{ B(\sigma/\pi)\exp(2\delta) - g_1^2 \sin^2(\sigma)\cos^2(\sigma)\frac{\exp(-2\delta)}{B(\sigma/\pi)} \right.$$
$$\left. + 2g_1 \cos^2(\sigma) - g_2 \right\}^{1/2} \qquad (A.17)$$

The probability p is given by

$$p = [1 + B(\sigma/\pi)\exp(2\delta) - g_2 \sin^2(\sigma)]^{-1} \qquad (A.18)$$

where

$$g_1 = 1.8(a^2)^{0.23}\exp(-\delta) \qquad (A.19)$$

$$g_2 = \frac{3\sigma}{\pi\delta}\ln(1.2 + a^2) - \frac{1}{a^2} \qquad (A.20)$$

and

$$B(X) = \frac{2\pi X^{2X}\exp(-2X)}{X\Gamma^2(X)} \qquad (A.21)$$

3. Definitions of σ, δ, and δ_ψ

These parameters introduced above are defined here. The expressions of σ and δ are dependent on the energy as described below. Since σ_0 and δ_0, which appear below, and the parameter δ_ψ are common in the all energy regions, these are defined first.

$$\delta_\psi = \delta\left(1 + \frac{5a^{1/2}}{a^{1/2} + 0.8}10^{-\sigma}\right) \tag{A.22}$$

$$\sigma_0 + i\delta_0 \equiv \int_{R_0}^{R_*}[K_1(R) - K_2(R)]dR \simeq \frac{\sqrt{2}\pi}{4a}\frac{F_-^c + iF_+^c}{F_+^2 + F_-^2} \tag{A.23}$$

where

$$F_\pm = \sqrt{\sqrt{(b^2 + \gamma_1)^2 + \gamma_2} \pm (b^2 + \gamma_1)}$$

$$+ \sqrt{\sqrt{(b^2 - \gamma_1)^2 + \gamma_2} \pm (b^2 - \gamma_1)} \tag{A.24}$$

$$F_+^c = F_+[b^2 \to (b^2 - b_c^2)] \tag{A.25}$$

$$F_-^c = F_-(\gamma_2 \to \gamma_2') \tag{A.26}$$

$$\gamma_1 = 0.9\sqrt{d^2 - 1} \tag{A.27}$$

and

$$\gamma_2 = \frac{7}{16}\sqrt{d^2} \tag{A.28}$$

with

$$b_c^2 = \frac{0.16b_x}{\sqrt{1 + b^4}} \tag{A.29}$$

$$\gamma_2' = \frac{0.45\sqrt{d^2}}{1 + 1.5\exp(2.2b_x|b_x|^{0.57})} \tag{A.30}$$

and

$$b_x = b^2 - 0.9553 \tag{A.31}$$

Now, σ and δ are given below:

(a) When $E \geq E_2(R_0)$,

$$\sigma = \int_{T_1}^{R_0}K_1(R)dR - \int_{T_2}^{R_0}K_2(R)dR + \sigma_0 \tag{A.32}$$

and

$$\delta = \delta_0 \tag{A.33}$$

(b) When $E \leq E_1(R_0)$,

$$\sigma = \sigma_0 \tag{A.34}$$

and

$$\delta = -\int_{R_0}^{T_1} |K_1(R)| dR + \int_{R_0}^{T_2} |K_2(R)| dR + \delta_0 \tag{A.35}$$

(c) When $E_1(R_0) < E < E_2(R_0)$

$$\sigma = \int_{T_1}^{R_0} K_1(R) dR + \sigma_0 \tag{A.36}$$

and

$$\delta = \int_{R_0}^{T_2} K_2(R) dR + \delta_0 \tag{A.37}$$

B.   NONADIABATIC TUNNELING-TYPE OF TRANSITION

The two parameters a^2 and b^2 in terms of diabatic potentials are the same as Eqs.(A.1) and (A.2). In terms of adiabatic potentials, however, they are differently defined as

$$a^2 = \frac{(1 - \gamma^2)\hbar^2}{m(R_b - R_t)^2 (E_b - E_t)} \tag{A.38}$$

and

$$b^2 = \frac{2E - (E_b + E_t)}{E_b - E_t} \tag{A.39}$$

where

$$\gamma = \frac{E_b - E_t}{E_2\left(\frac{R_b + R_t}{2}\right) - E_1\left(\frac{R_b + R_t}{2}\right)} \tag{A.40}$$

The reduced scattering matrix in terms of the Stokes constant U_1 is given quantum mechanically exactly as

$$S^R = \frac{1}{1 + U_1 U_2} \begin{pmatrix} \exp(i\Delta_{11}) & U_2 \exp(i\Delta_{12}) \\ U_2 \exp(i\Delta_{12}) & \exp(i\Delta_{22}) \end{pmatrix} \tag{A.41}$$

where

$$U_2 = \frac{U_1 - U_1^*}{|U_1|^2 - 1} \tag{A.42}$$

Note that the Stokes constants are naturally different from those in the LZ type of transition. Both R_t and E_t (R_b and E_b) represent the position and energy of the top (bottom) of the lower (upper) adiabatic potential (see Fig. A.1). The suffices 1 and 2 of the S matrix designate not the adiabatic potential, but the regions in coordinate space with respect to the potential crossing. Namely, the off-diagonal element S_{12} represents the transmission through the crossing region. When the adiabatic potentials are symmetric around the crossing point and $R_t = R_b$ is satisfied, γ becomes unity and the appropriate limit should be taken to define the parameter a^2, which gives

$$a^2 = \frac{\hbar^2}{4m(E_b - E_t)^2} \left[\frac{\partial^2 E_2(R)}{\partial R^2} \Big|_{R=R_b} - \frac{\partial^2 E_1(R)}{\partial R^2} \Big|_{R=R_t} \right] \tag{A.43}$$

The semiclassical expressions in the ZN theory are given below for the Stokes constant and other imporatant physical quantities.

1.   Case A: $E \geq E_b$

The Stokes cosntant U_1 is given by

$$U_1 = i\sqrt{1 - p} \exp(i\psi) \tag{A.44}$$

where the nonadiabatic transition probability p for one passage of the crossing point and the phase ψ are defined as

$$p = \exp\left[-\frac{\pi}{4a} \left(\frac{2}{b^2 + \sqrt{b^4 - 0.72 + 0.62a^{1.43}}} \right) \right] \tag{A.45}$$

and

$$\psi = \sigma - \phi_S = \sigma + \frac{\delta}{\pi} - \frac{\delta}{\pi} \ln\left(\frac{\delta}{\pi}\right) + \arg\Gamma\left(i\frac{\delta}{\pi}\right) + \frac{\pi}{4} - g_7 \tag{A.46}$$

with

$$\sigma = \int_{T_2^l}^{T_2^r} K_2(R)dR \tag{A.47}$$

$$\delta = \frac{\pi}{16ab} \frac{\sqrt{6 + 10\sqrt{1 - b^{-4}}}}{1 + \sqrt{1 - b^{-4}}} \tag{A.48}$$

and

$$g_7 = \frac{0.23a^{1/2}}{a^{1/2} + 0.75} 40^{-\sigma} \tag{A.49}$$

The phases appearing in the definition of S matrix are given as

$$\Delta_{11} = 2 \int_{T_2^l}^{R_b} K_2(R)dR - 2\sigma_0 \tag{A.50}$$

$$\Delta_{22} = 2 \int_{R_b}^{T_2^r} K_2(R)dR + 2\sigma_0 \tag{A.51}$$

$$\Delta_{12} = \sigma \tag{A.52}$$

with

$$\sigma_0 = \frac{R_b - R_t}{2} \left[K_1(R_t) + K_2(R_b) + \frac{[K_1(R_t) - K_2(R_b)]^2}{3[K_1(R_t) + K_2(R_b)]} \right] \tag{A.53}$$

T_2^r and T_2^l are the turning points on $E_2(R)$ (see Fig. A1). The nonadiabatic transition amplitude to connect the wave functions between the right and the left sides of the crossing point is given by

$$I_X = \begin{pmatrix} \sqrt{1-p}\exp(i\phi_S) & \sqrt{p}\exp(i\sigma_0) \\ -\sqrt{p}\exp(-i\sigma_0) & \sqrt{1-p}\exp(-i\phi_S) \end{pmatrix} \tag{A.54}$$

The overall transmission probability from the left to the right or the vice versa is given by

$$P_{12} = \frac{4\cos^2(\psi)}{4\cos^2(\psi) + p^2/(1-p)} \tag{A.55}$$

2.   Case B: $E_b \geq E \geq E_t$

The Stokes constant U_1 is given by

$$U_1 = i[\sqrt{1 + W^2} \exp(i\phi) - 1]/W \qquad (A.56)$$

with

$$W = \frac{1 + g_5}{a^{2/3}} \int_0^\infty \cos\left[\frac{t^3}{3} - \frac{b^2}{a^{2/3}}t - \frac{g_4}{a^{2/3}}\frac{t}{0.61\sqrt{2 + b^2} + a^{1/3}t}\right] dt \qquad (A.57)$$

and

$$\phi = \sigma + \arg \Gamma(1/2 + i\delta/\pi) - \frac{\delta}{\pi}\ln(\frac{\delta}{\pi}) + \frac{\delta}{\pi} - g_3 \qquad (A.58)$$

where

$$\sigma = -\frac{(1 - b^2)\sqrt{5 + 3b^2}}{\sqrt{a^2}}\left[0.057(1 + b^2)^{0.25} + \frac{1}{3}\right] \qquad (A.59)$$

$$\delta = \frac{(1 + b^2)\sqrt{5 - 3b^2}}{\sqrt{a^2}}\left[0.057(1 - b^2)^{0.25} + \frac{1}{3}\right] \qquad (A.60)$$

$$g_3 = \frac{0.34a^{0.7}(a^{0.7} + 0.35)(0.42 + b^2)}{a^{2.1} + 0.73}\left(2 + \frac{100b^2}{100 + a^2}\right)^{0.25} \qquad (A.61)$$

$$g_4 = \frac{\sqrt{a^2} - 3b^2}{\sqrt{a^2} + 3}\sqrt{1.23 + b^2} \qquad (A.62)$$

and

$$g_5 = 0.38(1 + b^2)^{1.2 - 0.4b^2}/a^2 \qquad (A.63)$$

The phases appearing in the definition of S-matrix are defined as

$$\Delta_{11} = \sigma - 2\sigma_0 \qquad (A.64)$$
$$\Delta_{22} = \sigma + 2\sigma_0 \qquad (A.65)$$
$$\Delta_{12} = \sigma \qquad (A.66)$$

and

$$\sigma_0 = -\frac{1}{3}(R_t - R_b)K_1(R_t)(1 + b^2) \qquad (A.67)$$

The overall transmission probability takes the form

$$P_{12} = \frac{W^2}{1 + W^2} \tag{A.68}$$

3. Case C: $E \leq E_t$

The Stokes constant U_1 is given by

$$\mathrm{Re}U_1 = \sin(2\sigma)\left[\frac{0.5\sqrt{a^2}}{1 + \sqrt{a^2}}\sqrt{B(\sigma_c/\pi)}\exp(-\delta) + \frac{\exp(\delta)}{\sqrt{B(\sigma_c/\pi)}}\right] \tag{A.69}$$

and

$$\mathrm{Im}U_1 = \cos(2\sigma_c)\sqrt{\frac{(\mathrm{Re}U_1)^2}{\sin^2(2\sigma_c)} + \frac{1}{\cos^2(2\sigma_c)}}$$
$$-\frac{1}{2\sin(\sigma_c)}\left|\frac{\mathrm{Re}U_1}{\cos(\sigma_c)}\right| \tag{A.70}$$

where

$$\sigma_c = \sigma(1 - g_6) \tag{A.71}$$

$$\sigma = \frac{\pi}{16a|b|}\frac{\sqrt{6 + 10\sqrt{1 - b^{-4}}}}{1 + \sqrt{1 - b^{-4}}} \tag{A.72}$$

$$\delta = \int_{T_1^l}^{T_1^r} |K_1(R)|dR \tag{A.73}$$

and

$$g_6 = 0.32 \times 10^{-2/a^2}\exp(-\delta) \tag{A.74}$$

T_1^r and T_1^l are the turning points on $E_1(R)$ (see Fig. A1). The phases appearing in the definition of S-matrix are

$$\Delta_{11} = \Delta_{22} = \Delta_{12} = -2\sigma \tag{A.75}$$

In this energy region, physically meaningful quantities are the overall transmission and reflection probabilities. The transmission probability is given by

$$P_{12} = \frac{B(\sigma_c/\pi)\exp(-2\delta)}{\left[1 + \frac{0.5\sqrt{a^2}}{1+\sqrt{a^2}}B(\sigma_c/\pi)\exp(-2\delta)\right]^2 + B(\sigma_c/\pi)\exp(-2\delta)} \tag{A.76}$$

This expression contains both effects of quantum mechanical tunneling $(\exp[-2\delta]$ is the Gamov factor) and nonadiabatic transition that is represented by the factor σ_c. This transmission probability is always smaller than the ordinary tunneling probability through the lower adiabatic potential with the nonadiabatic coupling effect neglected. When the diabatic coupling is infinitely strong, namely, $a^2 \to 0$, the above transmission probability goes to the ordinary potential penetration probability $= \exp(-2\delta)/[1 + \exp(-2\delta)]$.

C. TIME-DEPENDENT VERSION OF THE ZHU–NAKAMURA FORMULAS

The formulas derived in the time-independent framework can be easily transferred into the corresponding time-dependent solutions. The formulas in the time-independent linear potential model, for example, provide the formulas in the time-dependent quadratic potential model in which the two time-dependent diabatic quadratic potentials are coupled by a constant diabatic coupling [1, 13, 147]. The classically forbidden transitions in the time-independent framework correspond to the diabatically avoided crossing case in the time-dependent framework. One more thing to note is that the nonadiabatic tunneling (NT) type of transition does not show up and only the LZ type appears in the time-dependent problems, since time is unidirectional.

The following replacements of the parameters in the time-independent ZN formulas are good enough for the transfer.

$$a^2 \Leftrightarrow \alpha = \frac{\sqrt{d^2 - 1}\hbar^2}{2V_0^2(t_t^2 - t_b^2)} \tag{A.77}$$

$$b^2 \Leftrightarrow \beta = -\sqrt{d^2 - 1}\frac{t_t^2 + t_b^2}{t_t^2 - t_b^2} \tag{A.78}$$

and

$$\sigma + i\delta = \frac{1}{\hbar}\left[\int_0^{t_b} E_+(t)dt - \int_0^{t_t} E_-(t)dt + \sqrt{\frac{\beta}{\alpha}} + \Delta_1\right] \tag{A.79}$$

with

$$\Delta_1 = \frac{t_0 - (t_b + t_t)/2}{\sqrt{\alpha(\beta^2 + i)(t_b - t_t)}}\sqrt{\frac{d^2}{d^2 - 1}} + \frac{1}{2\sqrt{\alpha}}\int_0^i \left(\frac{1 + t^2}{t + \beta}\right)^{1/2} dt \tag{A.80}$$

$$V_0 = \frac{1}{2}[E_+(t_0) - E_-(t_0)] \tag{A.81}$$

and

$$d^2 = \frac{[E_+(t_b) - E_-(t_b)][E_+(t_t) - E_-(t_t)]}{E_+(t_0) - E_-(t_0)} \tag{A.82}$$

where $E_\pm(t)$ are the adiabatic potentials with $E_+ > E_-$, $t_b(t_t)$ is the bottom (top) of the adiabatic potential $E_+(t)(E_-(t))$, t_0 is the position at which the adiabatic energy difference becomes minimum. When the complex integral is annoying, then the following formula can be used

$$\sigma + i\delta = \frac{1}{\hbar}\left[\int_0^{t_0} E_+(t)dt - \int_0^{t_0} E_-(t)dt + \Delta\right] \tag{A.83}$$

with

$$\Delta = \frac{\sqrt{2}\pi}{4\sqrt{\alpha}} \frac{F_-^c + iF_+^c}{F_+^2 + F_-^2} \tag{A.84}$$

where $F_\pm^c$ are the same as those defined by Eqs. (A.24–A.31).

Acknowledgments

The author would like to thank all the group members in the past and present who carried out all the researches discussed in this chapter: Drs. C. Zhu, G. V. Mil'nikov, Y. Teranishi, K. Nagaya, A. Kondorskiy, H. Fujisaki, S. Zou, H. Tamura, and P. Oloyede. He is indebted to Professors S. Nanbu and T. Ishida for their contributions, especially on molecular functions and electronic structure calculations. He also thanks Professor Y. Zhao for his work on the nonadiabatic transition state theory and electron transfer. The work was supported by a Grant-in-Aid for Specially Promoted Research on "Studies of Nonadiabatic Chemical Dynamics based on the Zhu–Nakamura Theory" from MEXT of Japan.

References

1. H. Nakamura, *Nonadiabatic Transition: Concepts, Basic Theories and Applications*, World Scientific, Singapore, 2002.

2. H. Nakamura, *J. The. Comp. Chem.* **4**, 127 (2005).

3. E. E. Nikitin and S.Ya. Umanskii, *Theory of Slow Atomic Collisions*, Springer, Berlin, 1984.

4. M. S. Child, *Molecular Collision Theory*, Academic Press, London, 1974; *Semiclassical Mechanics with Molecular Applications*, Clarendon, London, 1991.

5. E. S. Medevedev and V. I. Osherov, *Radiationless Transitions in Polyatomic Molecules*, Springrer Series in Chemical Physics, Springer, Berlin, 1994, vol. 57.

6. J. C. Tully, In *Dynamics of Molecular Collisions*, Part B, W. H. Miller (ed.), Plenum, New York, 1976.

7. J. Michl and V. Bonacic-Koutecky, *Electronic Aspects of Organic Photochemistry*, (John Wiley & Sons, Inc., New York, 1990.

8. J. R. Bolton, N. Mataga, and G. MvLendon, *Electron Transfer in Inorganic, Organic, and Biological Systems* Advances in Chemistry Series 228, American Chemical Society, Washington, D.C., 1991.

9. H. Nakamura, *J. Phys. Chem.* **A110**, 10929 (2006).

10. H. Nakamura, in *Dynamics of Molecules and Chemical Reactions*, R. E. Wyatt and J. Z. H. Zhang (eds.) Marcel Dekkaer, New York, 1996.

11. H. Nakamura and C. Zhu, *Comm. Atom. Molec. Phys.* **32**, 249 (1996).

12. C. Zhu, G. Mil'nikov, and H. Nakamura, in *Modern Trends in Chemical Reaction Dynamics*, K. Liu and X. Yang (eds.), World Scientific, Singapore 2004.

13. C. Zhu, Y. Teranishi, and H. Nakamura, *Adv. Chem. Phys.* **117**, 127 (2001).

14. C. Zhu and H. Nakamura, *J. Math. Phys.* **33**, 2697 (1992); C. Zhu, H. Nakamura, N. Re, and V. Aquilanti, *J. Chem. Phys.* **97**,1892 (1992); C. Zhu and H. Nakamura, *J. Chem. Phys.* **97**, 8497 (1992); **98**, 6208 (1993); **101**,4855 (1994); **108**, 7501 (1998); **101**, 10630(1994): **102**, 7448 (1995); **109**, 4689 (1998); **106**, 2599 (1997): **107**, 7839 (1997); *Chem. Phys. Lett.* **258**, 342 (1996); **274**, 205 (1997); *Comp. Phys. Comm.* **74**, 9 (1993).

15. N. Makri and W. H. Miller, *J. Chem. Phys.* **89**, 2170 (1988); W. H. Miller, *J. Chem. Phys.* **53**, 3578 (1970); X. Sun and W. H. Miller, *J. Chem. Phys.* **106**, 6346 (1997); W. H. Miller, *J. Phys. Chem.* **105**, 2942 (2001).

16. T. V. Voorhis and E. J. Heller, *Phys. Rev.* **A66**, 050501 (2002); E. J. Heller, *J. Chem. Phys.* **75**, 2923 (1981).

17. M. F. Herman and E. Kluk, *Chem. Phys.* **91**, 27 (1984); E. Kluk, M. F. Herman, and H. D. Davis, *J. Chem. Phys.* **84**, 326 (1986); M. F. Herman, *Annu Rev. Phys. Chem.* **45**, 83 (1994).

18. A. R. Walton and D. E. Manolopoulos, *Molec. Phys.* **87**, 961 (1996), M. L. Brewer, J. S. Hulme and D. E. Manolopoulos, *J. Chem. Phys.* **106**, 4832 (1997).

19. A. Kondorskiy and H. Nakamura, *J. Chem. Phys.* **120**, 8937 (2004).

20. A. Bjerre and E. E. Nikitin, *Chem. Phys. Lett.* **1**, 179 (1967).

21. J. C. Tully and R. Preston, *J. Chem. Phys.* **54**, 4297 (1970).

22. J. C. Tully, *J. Chem. Phys.* **93**, 1061 (1990).

23. M. S. Toplar, T. C. Allison, D. W. Schwenke,and D. G. Truhlar, *J. Phys. Chem.* **A102**, 1666 (1998); Y. L. Volonuev, M. D. Hack, and D. G. Truhlar, *J. Phys. Chem.* **A103**, 6309 (1999); M. D. hack, A. Jasper, Y. L. Volobuev, D. Schwenke, and *J. Phys. Chem.* **A104**, 217 (2000); Y. L. Volobuev, M. D. Hack, M. Toplara, and D. G. Truhlar, *J. Chem. Phys.* **112**, 9716 (2000); A. Jasper, M. D. Hack, and D. G. Truhlar, *J. Chem. Phys.* **115**, 1804 (2001); M. D. Hack and D. G. Truhlar, *J. Chem. Phys.* **114**, 2894 (2001).

24. J. C. Tully, in *Modern Methods for Multidimensional Dynamics Computations in Chemistry*, D. L. Thompson (ed.), World Scientific, Singapore, 1998.

25. W. Domcke, D. Yarkony, and H. Koppel, *Conical Intersections, Electronic Structures, Dynamics and Spectroscopy*, World Scientific, Singapore, 2004.

26. A. Jasper, B. K. Kendrick, C. A. Mead, and D. G. Truhlar, in *Modern Trends in Chemical Reaction Dynamics: Experiment and Theory*, K. Liu and X. Yang (eds.), World Scientific, Singapore, 2004.

27. Y. Zhao, G. V. Mil'nikov, and H. Nakamura, *J. Chem. Phys.* **121**, 8854 (2004).

28. Y. Zhao, W. Lia, and H. Nakamura, *J. Phys. Chem.* **A110**, 8204 (2006).

29. P. Oloyede, G. V. Mil'nikov, and H. Nakamura, *J. Theo. Comp. Chem.* **3**, 91 (2004).

30. G. V. Mil'nikov and H. Nakamura, *J. Chem. Phys.* **115**, 6881 (2001).

31. G. V. Mil'nikov and H. Nakamura, *J. Chem. Phys.* **117**, 10081 (2002).

32. G. V. Mil'nikov and H. Nakamura, *J. Chem. Phys.* **122**, 124311 (2005).

33. S. A. Rice and M. Zhao, *Optical Control of Molecular Dynamics*, John Wiley & Sons, Inc., New York, 2002.

34. A. Bandrauk, *Molecules in Lase Field*, Mardel Dekker, New York, 1994.

35. P. Brumer and M. Shapiro, *Annu. Rev. Phys. Chem.* **48**, 601 (1997).

36. Y. Teranishi and H. Nakamura, *Phys. Rev. Lett.* **81**, 2032 (1998).

37. S. Zou, A. Kondorskiy, G. V. Mil'nilkov, and H. Nakamura, *J. Chem. Phys.* **122**, 084112 (2005).

38. A. Kondorskiy and H. Nakamura, *J. Theor. Comp. Chem.* **4**, 75,89 (2005).

39. S. Zou, A. Kondorskiy, G. Mil'nikov, and H. Nakamura, in *Progress in Ultra Fast Intense Laser Science*, Springer, New York, 2006, vol. II, Sect. 5.

40. A. Kondorskiy, G. V. Mil'nikov, and H. Nakamura, in *Progress in Ultrafasy Laser Science*, Springer-Verlag, Berlin, 2006, vol. II, Sect. 6.

41. A. Kondorskiy, G. V. Mil'nikov, and H. Nakamura, *Phys. Rev.* **A72**, 041401 (2005).

42. K. Nagaya, Y. Teranishi, and H. Nakamura, *ACS Symposium Series 821*, American Chemical Society, Washington, DC, 2002, Chap. 7.

43. Y. Teranishi and H. Nakamura, *J. Chem. Phys.* **111**, 1415 (1999).

44. S. Nanbu, H. Nakamura, and F. O. Goodman, *J. Chem. Phys.* **107**, 5445 (1997).

45. K. Nagaya, Y. Teranishi, and H. Nakamura, *J. Chem. Phys.* **117**, 9588 (2002).

46. H. Nakamura, *J. Chem. Phys.* **110**, 10253 (1999).

47. S. Nanbu, T. Ishida, and H. Nakamura, *Chem. Phys.* **324**, 721 (2006).

48. H. Tamura, S. Nanbu, T. Ishida, and H. Nakamura, *J. Chem. Phys.* **124**, 0843131 (2006).

49. H. Tamura, S. Nanbu, T. Ishida, and H. Nakamura, *J. Chem. Phys.* **125**, 034307 (2006).

50. C. Zhu, H. Kamisaka, and H. Nakamura, *J. Chem. Phys.* **115**, 3031 (2001); *ibid*, **116**, 3234 (2002).

51. P. Oloyede, G. V. Mil'nikov, and H. Nakamura, *J. Chem. Phys.* **124**, 144110 (2006).

52. W. H. Miller, in *Dynamics of Molecules and Chemical Reactions*, R. E. Wyatt and J. Z. H. Zhang (eds.), Marcel Dekker, New York, 1996, Chap. 10.

53. W. H. Miller, Y. Zhao, and M. Ceotto, *J. Chem. Phys.* **119**, 1329 (2003).

54. C. Shin and S. Shin, *J. Chem. Phys.* **113**, 6528 (2000).

55. J. Jortner and B. Pullman, *Tunneling*, Reidel, Dordrecht, Holland, 1986.

56. A. Auerbach and S. Kivelson, *Nucl. Phys.* **B257**, 799 (1985).

57. T. Banka, C. M. Bender, and T. T. Wu, *Phys. Rev.* **D8**, 3346 (1973) ;**D8**, 3366 (1973).

58. W. H. Miller, *J. Chem. Phys.* **55**, 3146 (1971); **62**, 1899 (1975).

59. V. A. Benderskii, V. I. Goldanskii, and D. E. Makarov, *Phys. Rep.* **233**, 195 (1993); V. A. Benderskii, D. E. Makarov, and C. A. Wight, *Chemical Dynamics at Low Temperatures*, Adv. Chem. Phys. LXXXVIII, John Wiley & Sons, Inc., New York, 1994.

60. F. Cesi, G. C. Rossi, and M. Testa, *Ann. Phys.* (Leipzig) **206**, 318 (1991).

61. M. Willkinson, *Physica* **D21**, 341 (1986).

62. S. Takada and H. Nakamura, *J. Chem. Phys.* **100**, 98 (1994).

63. Z. H. Huang, T. E. Feuchtwang, P. H. Cutler, and E. Kazes, *Phys. Rev.* **A41**, 32 (1990).

64. P. Bowcock and R. Gregory, *Phys. Rev.* **D44**, 1774 (1991).

65. G. V. Mil'nikov, S. Yu. Grebenshikov, and V. A. Benderskii, *Izv. Akad. Nauk(Russ. Chem. Bull.)* **N12**, 2098 (1994); V. A. Benderskii, S. Yu. Grebenshikov, and G. V. Mil'nikov, *Chem. Phys.* **194**, 1 (1995).

66. A. Schmid, *Ann. Phys.* **170**, 333 (1986).

67. D. G. Truhlar, A. D. Isaacson, and B. C. Garrett, in *Theory of Chemical Reaction Dynamics*, M. Byer (eds.), CRC, Boca Raton, FL, 1985.

68. N. Makri and W. H. Miller, *J. Chem. Phys.* **91**, 4026 (1989).

69. K. Takatsuka, H. Ushiyama, and A. Inoue-Ushiyama, *Phys. Rep.* **322**, 347 (1999).

70. S. Takada and H. Nakamura, *J. Chem. Phys.* **102**, 3997 (1995).

71. A. I. Vainshtein, V. I. Zakharov, V. A. Novikov, and M. A. Shifman, *Sov. Phys. Usp.* **5**, 195(1982).

72. C. G. Callan and S. Coleman, *Phys. Rev.* **D16**, 1762 (1977).

73. J. L. Gervais and B. Sakita, *Phys. Rev.* **D16**, 3507 (1977).

74. V. V. Avilov and S. V. Iordanskii, *Sov. Phys. JETP* **42**, 683 (1975).

75. V. A. Benderskii, E. V. Vetoshkin, L. vonLaue, and H. P. Trommsdorff, *Chem. Phys.* **219**, 143 (1997).

76. V. A. Benderskii, E. V. Vetoshkin, and H. P. Trommsdorff, *Chem. Phys.* **234**, 153 (1998); **244**, 273 (18999); **244**, 299 (2001).

77. S. Takada, *J. Chem. Phys.* **104**, 3742 (1996).

78. G. V. Mil'nikov and A. J. Varandas, *J. Chem. Phys.* **111**, 8302 (1999).

79. W. F. Rowe, R. W. Duerst, and E. B. Wilson, *J. Am. Chem. Soc.* **98**, 4021 (1976).

80. D. W. Firth, K. Beyer, M. A. Dvorak, S. W. Reeve, A. Grushov, and K. R. Leopold, *J. Chem. Phys.* **94**, 1812 (1991).

81. T. Baba, T. Tanaka, I. Morino, and K. Tanaka, *J. Chem. Phys.* **110**, 4131 (1999).

82. S. L. Baughcum, R. W. Duerst, W. F. Rowe, Z. Smith, and E. B. Wilson, *J. Amer. Chem. Soc.* **103**, 6296 (1981).

83. F. Z. Smedarchina, W. Siebrand, and M. Z. Zgierski, *J. Chem. Phys.* **103**, 5326 (1995).

84. T. Carrington, Jr., and W. H. Miller, *J. Chem. Phys.* **81**, 3942 (1984); **84**, 4364 (1986).

85. N. Shida, P. F. Barbara, and J. Almlof, *J. Chem. Phys.* **91**, 4061 (1989).

86. T. D. Sewell, Y. Guo, and D. L. Thompson, *J. Chem. Phys.* **103**, 8557 (1995).

87. C. S. Tautermann, A. F. Voegele, T. Loerting, and K. R. Liedl, *J. Chem. Phys.* **117**, 1962 (2002).

88. K. Yagi, T. Taketsugu, and K. Hirao, *J. Chem. Phys.* **115**, 10647 (2001).

89. K. Yagi, T. Taketsugu, K. Hirao, and M. S. Gordon, *J. Chem. Phys.* **113**, 1005 (2000).

90. M. A. Collins, in *New Method in Computational Quantum Mechanics*, I. Prigogine and S. A. Rice (eds.), John Wiley & Sons, Inc., New York, 1996.

91. T. Takata, T. Taketsugu, K. Hirao, and S. Gordon, *J. Chem. Phys.* **109**, 4281 (1998).

92. K. C. Thompson, M. J. T. Jordan, and M. A. Collins, *J. Chem. Phys.* **108**, 564 (1998).

93. C. Moller and M. S. Plesset, *Phys. Rev.* **46**, 618 (1934).

94. G. V. Mil'nikov, K. Yagi, T. Taketsugu, H. Nakamura, and K. Hirao, *J. Chem. Phys.* **119**, 10 (2003).

95. G. V. Mil'nikov, K. Yagi, T. Taketsugu, H. Nakamura, and K. Hirao, *J. Chem. Phys.* **120**, 5036 (2004).

96. J. A. Pople, M. Head-Gordon, and K. Raghavachari, *J. Chem. Phys.* **87**, 5968 (1987).

97. R. J. Bartlett, *Ann. Rev. Phys. Chem.* **32**, 359 (1981).

98. T. H. Dunning, Jr., *J. Chem. Phys.* **90**, 1007 (1989).

99. R. A. Kendall, T. H. Dunning, Jr., and R. J. Harrison, *J. Chem. Phys.* **96**, 6796 (1992).

100. M. J. Frisch et al., Coputer code GAUSSIAN 98, revision A.5, Gaussian, Inc. Pittsburgh, PA, 1998.

101. MOLPRO, a package of *ab initio* programms designed by H-J. Werner and P. Knowles, Version 2002.1, R. D. Amos, A. Bernhardsson, and A. Berning, et al.

102. K. Yagi, G. V. Mil'nikov, T. Taketsugu, K. Hirao, and H. Nakamura, *Chem. Phys. Lett.* **397**, 435 (2004).

103. K. Tanaka, M. Toshimitsu, K. Harada, and T. Tanaka, *J. Chem. Phys.* **120**, 3604 (2004).

104. G. V. Mil'nikov, T. Ishida, and H. Nakamura, *J. Phys. Chem.* **110**, 5430 (2006).

105. T. J. Stuchi and R. Vieira-Martins, *Phys. Lett.* **A201**, 179 (1995).

106. H. Ushiyama, Y. Arasaki, and K Takatsuka, *Chem. Phys. Lett.* **346**, 169 (2001).

107. R. A. Marcus and N. Sutin, *Biochim. Biophys. Acta* **811**, 265 (1985).

108. A. V. G. Barzykin, P. A. Frantsuzov, K. Seki, and M. Tachiya, *Adv. Chem. Phys.* **123**, 511 (2002).

109. M. Bixon and J. Jortner, *Adv. Chem. Phys.* **106**, 35 (1999).

110. I. Rips and E. Polla, *J. Chem. Phys.* **103**, 7912 (1995); I. Rips, *J. Chem. Phys.* **104**, 9795 (1996); ibid, **121**, 5356 (2004).

111. R. R. Dognadze and Z. D. Urushadze, *J. Electranal. Chem.* **32**, 235 (1971).

112. S. F. Nelsen, M. N. Weaver, A. E. Konradsson, J. P. Telo, and T. Clark, *J. Amer. Chem. Soc.* **126**, 15431 (2004).

113. Y. Zhao and H. Nakamura, *J. Theo. Comp. Chem.* **5**, 209 (2004).

114. Y. Zhao, M. Han, W. Liang, and H. Nakamura, *J. Phys. Chem.* **A111**, 2047 (2007).

115. P. Brumer and M. Shapiro, *Coherent Control of Molecular Dynamics*, John Wiley & Sons, Inc., New York, 2003.

116. S. I. Chu, *Advances in Multiphoton Processes and Spectroscopy*, World Scientific, Singapore, 2986, vol. 2.

117. H. Partridge and S. R. Langhoff, *J. Chem. Phys.* **74**, 2361 (1981).

118. S. Magnier, M. Aubert-Frecon, and Ph. Millie, *J. Mol. Spect.* **200**, 86 (2000).

119. D. J. Tannor and S. A. Rice, *J. Chem. Phys.* **83**, 5013 (1985).

120. S. Meyer and V. Engel, *J. Phys. Chem.* **A101**, 7749 (1997).

121. S. Shi, A. Woody, and H. Rabitz, *J. Chem. Phys.* **88**, 6870 (1988).

122. C. D. Schwieters and H. Rabitz, *Phys. Rev.* **A48**, 2549 (1993).

123. R. Koslof et al., *Chem. Phys.* **139**, 201 (1989).

124. T. van Mourik et al., *J. Chem. Phys.* **115**, 3706 (2001).

125. H. Nakamura, *J. Chem. Phys.* **87**, 4031 (1987).

126. L. Pichl, H. Nakamura, and J. Horacek, *J. Chem. Phys.* **113**, 906 (2000).

127. H. Fujisaki, Y. Teranishi, and H. Nakamura, *J. Theo. Comp. Chem.* **2**, 245 (2002).

128. A. B. Alekseyev, H. P. Liebermann, D. B. Kokh, and R. J. Buenker, *J. Chem. Phys.* **113**, 6174 (2000).

129. N. Balakrishnan, A. B. Alekseyev, and R. J. Buenker, *Chem. Phys. Lett.* **341**, 594 (2001).

130. M. Irie, *Chem. Rev.* **100**, 1685 (2000).

131. M. Irie, S. Kobatake, and M. Horichi, *Science* **291**, 1769 (2001).

132. M. Murakami, H. Miyasaka, T. Okada, S. Kobatake, and M. Irie, *J. Amer. Chem. Soc.* **126**, 14764 (2004).

133. K. L. Kompa and R. D. Levine, *Proc. Natl. Acad. U. S.A.* **16**, 410 (2001).

134. A. H. Zewail, *Angew. Chem. Int. Ed.* (Engl.) **39**, 2586 (2000).

135. W. Fuss, W. E. Schmid, and S. A. Trushin, *J. Chem. Phys.* **112**, 8347 (2000).

136. P. Celani, S. Ottani, M. Olivucci, F. Bernardi, and M. A. Robb, *J. Amer. Chem. Soc.* **116**, 10141 (1994); P. Celani, F. Bernardi, M. A. Robb, and M. Olivucci, *J. Phys. Chem.* **100**, 19364 (1996); M. Garavelli, P. Celani, M. Fato, M. J. Bearpark, B. R. Smith, M. Olivucci, and M. A. Robb, *J. Phys. Chem.* **101**, 2023 (1997).

137. M. Garavelli, C. S. Page, P. Celani, M. Olivucci, W. E. Schmid, S. A. Trushin, and W. Fuss, *J. Phys. Chem.* **105**, 4458 (2001).

138. A. Hofmann and R. de Vivie-Riedle, *J. Chem. Phys.* **112**, 5054 (2001); *Chem. Phys. Lett.* **346**, 299 (2001).

139. D. Geppaet, L. Seyfarth, and R. de Vivie-Riedle, *App. Phys. B* **79**, 987 (2004).

140. H. Tamura, S. Nanbu, H. Nakamura, and T. Ishida, *Chem. Phys.* **401**, 487 (2005).

141. M. W. Schmidt, et al., *J. Com. Chem.* **14**, 1347 (1993); G. D. Fletcher, M. W. Schnmidt, and M. S. Gordon, *Adv. Chem. Phys.* **110**, 267 (1999); M. W. Schmidt, G. D. Fletcher, B. M. Bode, and M. S. Gordon, *Comp. Phys. Comm.* **128**, 190 (2000).

142. H. J. C. Jacobs and E. Havinga, in *Photochemistry of Vitamin D and Its Isomers and of Simple Triends*, Advances in Photochemistry, J. N. Pitts, S. G. Hammaond, and K. Gollnick (eds.), John Wiley & Sons, Inc., New York, 1979, pp. 305–373, vol. 11.

143. H. W. Kroto, J. R. Heath, S. C. O'Brien, R. F. Curl, R. E. Smalley, *Nature (London)* **318**, 162 (1985); J. R. Heath, S. C. O'Brien, Q. Zhang, Y. Liu, R. F. Curl, H. W. Kroto, F. K. Tittle, and R. E. Smalley, *J. Am. Chem. Soc.* **107**, 7779 (1985).

144. T. Guo, C. Jin, and R. E. Smalley, *J. Phys. Chem.* **95**, 4948 (1991).

145. K. Komatsu, M. Murata, and Y. Murata, *Science* **307**, 2291 (2001).

146. F. Hirata, *Molecular Theory of Solvation*, Kluwer Academic Press, Boston, 2003.

147. Y. Teranishi and H. Nakamura, *J. Chem. Phys.* **107**, 1904 (1997).

EXPLORING MULTIPLE REACTION PATHS TO A SINGLE PRODUCT CHANNEL

DAVID L. OSBORN

*Combustion Research Facility, Sandia National Laboratories,
Livermore, CA 94551-0969, USA*

CONTENTS

Advances in Chemical Physics, Volume 138, edited by Stuart A. Rice

I. INTRODUCTION

It is with a certain amount of hubris that chemists seek to uncover the mechanism of a chemical reaction. In truth, in an ensemble of reactant molecules the variation in initial positions and momenta of the reactants launches each occurrence of the reaction in a unique path on the potential energy surface (PES). The veracity of a mechanism we assign to a particular reaction will therefore depend on the extent to which an ensemble of such trajectories deviates from the canonical path described by the mechanism. In cases where deviations from a proposed canonical mechanism are extreme, the very concept of a mechanism may be misleading, imposing our desire for regularity on a chaotic process. In most cases, however, simplifying the true details of an ensemble of reacting molecules to a reaction mechanism adds insight that outweighs the errors we make in this abstraction.

We can construct and constrain a reaction mechanism based on kinetics experiments that measure thermally averaged rate coefficients, equilibrium constants, and product branching ratios as a function of temperature and pressure [1]. Chemical dynamics experiments conducted under single-collision conditions provide measurements of reaction cross-sections, product branching ratios, product state and angular distributions [2], and vector correlations [3] that help define mechanisms. These approaches generally probe the reactants or products directly and infer the behavior of reaction intermediates whose details are encoded in the data. In complement, a number of time- [4] and frequency-domain [5] experiments can directly probe reaction intermediates and provide important clues to the reaction mechanism. Finally, electronic structure calculations [6] of the potential energy surfaces on which reactions occur, coupled with statistical or dynamical calculations predicting the progress of reactions on these surfaces [7], play a role of ever-increasing importance in determining reaction mechanisms. These theoretical methods are especially attractive because of their flexibility to predict observables of the reaction that may be very difficult to actually determine in the laboratory.

Of all these observables, perhaps the most fundamental are product branching ratios that define which products are formed and in what quantities. Normally, each product branching fraction is assumed to be proportional to the flux though a particular path on the PES leading from reactants to products. However, it is possible that the products emerging from a single exit channel have been generated from the reactants via more than one distinct pathway on the potential energy surface. While the concept of multiple pathways leading to a single product channel is not new, the difficulty of observing this phenomenon in the laboratory has, until recently, limited the discovery of such systems. Recent advances in both experiment and theory have shown that multiple pathways leading to a single product channel are more common than previously thought. Studying the participation of multiple pathways in chemical reactions provides a stringent test of a proposed reaction mechanism, and has in some cases resolved long-standing quandaries relating experimental observations to reaction mechanisms [8].

A. Product Branching versus Pathway Branching

The product branching ratio f_n is defined as the quantity of each chemically distinguishable set of reaction products n compared to the quantity of all products. As an example, consider the hypothetical reaction

$$A + ABA \rightarrow AA + BA \qquad f_1 = 0.3 \tag{1}$$

$$\rightarrow AAB + A \qquad f_2 = 0.7 \tag{2}$$

in which we postulate product channel (1) represents 30% of the total products. In this example, if the three identical atoms "A" are distinguished with subscripts, a pathway branching ratio g_{nm} can be defined for each product channel

$$A_a + A_b BA_c \rightarrow A_a A_b + BA_c \qquad g_{1a} = 0.9 \tag{1a}$$

$$\rightarrow A_a B + A_b A_c \qquad g_{1b} = 0.1 \tag{1b}$$

Pathways (1a) and (1b) are both part of product channel (1), but their formation mechanisms are different. In this example, path (1a) represents 90% of channel (1) and may be formed through abstraction by atom A_a of atom A_b from the triatomic reactant, whereas in path (1b) A_a could add to B followed by elimination to form $A_a B + A_b A_c$. This example demonstrates why experimental detection of pathway branching is usually difficult; pathways (1a) and (1b) lead to chemically identical products, even though they represent dramatically different chemical mechanisms.

In most cases, the observables measured in the study of a chemical reaction are interpreted under the following (often valid) assumptions: (1) each product channel observed corresponds to one path on the PES, (2) reactions follow the minimum energy path (MEP) to each product channel, and (3) the reactive flux passes over a single, well-defined transition state. In all of the reactions discussed in this chapter, at least one, and sometimes all of these assumptions, are invalid.

The following sections discuss classes of reactions in which multiple pathways appear, outline methods of probing and predicting their behavior, provide representative examples of such reactions, and examine the ramifications of multiple pathways on our understanding of reaction mechanisms.

II. CLASSES OF REACTIONS EXHIBITING MULTIPLE PATHWAYS

While the classes listed below may not be truly comprehensive, they represent useful distinctions to categorize pathway competition. Furthermore, these classes are not mutually exclusive; some reactions share attributes of more than one class.

A. Abstraction versus Addition–Elimination

Many chemical reactions can be classified by either abstraction or addition–elimination mechanisms. Abstraction mechanisms are common in the reaction of radicals with closed-shell species, such as the reaction

$$OH + H_2 \rightarrow H + H_2O \tag{3}$$

in which the hydroxyl radical abstracts an H atom from molecular hydrogen. These reactions often have a significant potential energy barrier that must be overcome by translational or internal energy in order for reaction to succeed. In the case of radical–radical reactions, abstraction mechanisms usually have no entrance barrier, as in the reaction

$$H + CH_2CH \rightarrow H_2 + HCCH \tag{4}$$

In either case, abstraction mechanisms are direct (no long-lived collision complex is formed), have small entropy costs ("loose" transition states), and typically deposit large amounts of vibrational energy in the newly formed bond while the other bonds in the system act largely as spectators.

By contrast, addition–elimination mechanisms in their simplest form begin with formation of an addition complex resulting from a well on the PES, followed by dissociation of the complex, yielding products. Both the entrance to and exit from the well may be hindered by barriers on the PES. Addition mechanisms are uncommon in radical + saturated closed-shell reactions due to the difficulty of bond formation with the saturated species (ion–molecule reactions are exceptions). By contrast, additions are more common in radical + unsaturated closed-shell species, where the double or triple bond allows a low barrier or barrierless pathway for addition of the radical into the π-bond of the stable species, such as the reaction

$$CH + HCCH \rightarrow cyclic\text{-}C_3H_3 \tag{5}$$

Addition is even more facile in radical + radical reactions due to the usually barrierless formation (on the singlet-coupled surface) of a chemical bond between the two unpaired electrons, as in the reaction

$$H + HCO \rightarrow H_2CO \tag{6}$$

The collision complex is, by necessity, always formed with sufficient energy to dissociate back to reactants. If the barrier(s) to product formation are at or below the reactant energy, the complex may eliminate a fragment to form products. During the lifetime of the collision complex, the energy liberated by formation of the adduct may partially or completely randomize within the internal degrees of freedom of the complex. This process is known as intramolecular vibrational redistribution (IVR). Generally, addition–elimination mechanisms have higher entropy costs ("tight" transition states) compared to abstraction, and typically produce products with a more statistical distribution of energy among their vibrational modes.

Pathways through abstraction and addition–elimination mechanisms may lead to the same product channel. However, they will only compete significantly when the energy and entropy barriers (or free energy barriers $\Delta G = \Delta H - T\Delta S$) for the two mechanisms are comparable. In these cases, they make excellent candidates for multiple pathway studies because several experimental approaches discussed in Section III are suited to detect the competition.

B. Participation of Identical Atoms

Multiple pathways leading to the same product channel can also be observed in a reaction when there are a sufficient number of identical atoms, thereby allowing different intermediate structures to yield the same products. In these cases, the mechanisms in the two pathways are often quite similar, but involve differing positions of identical atoms on the reactants. The different pathways often involve formation of ring intermediates in which the rings have different sizes. A simple example of this class is the photodissociation of vinyl chloride [9]

$$H_2CCH_aCl + 193\,nm \rightarrow H_2CC + H_aCl \rightarrow HCCH + H_aCl \tag{7a}$$
$$\rightarrow HCCH_a + HCl \tag{7b}$$

in which atom H_a has been labeled for clarity. After internal conversion to the ground electronic state, vinyl chloride can decompose via either a three-center (α, α) or a four-center (α, β) elimination. The first pathway formally yields vinylidene (H_2CC), which will rapidly isomerize to acetylene [10], resulting in the same products as the four-center elimination.

The possibility of multiple pathways arising from identical atoms becomes greater as the total number of atoms increases because the possibility to form rings with varying numbers of atoms increases. However, ring-forming intermediates are not the only mechanisms that make use of identical atoms. The reactions $CD_3^+ + C_2H_6$ and $HO_2 + O_3$ discussed in Section V also owe their multiple pathways to the participation of identical atoms, despite having no ring intermediates.

C.   Avoiding the Minimum Energy Path

A reaction may have one and often many wells on its potential energy surface. Although the deepest well could lie in a remote part of configuration space, separate from the seemingly logical path from reactants to products, generally it lies on an intuitively reasonable path to the exit channel. The forces acting on the reactants in the entrance channel usually pull the reactants along the minimum energy path toward the first well. Due to the large density of states in a deep well on the PES, a trajectory may spend significant time trapped as a collision complex over this well until sufficient energy migrates to a coordinate that leads out of the well.

The definition of the MEP is generally identical to the intrinsic reaction coordinate (IRC) [11, 12]. The IRC is the steepest descent path from reactants or products into wells, or from transition states into the wells they separate. An ambiguity arises when a deep well on the potential energy surface does not lie on the steepest descent pathway described above, but is separated by a barrier (however small) from the IRC. It may make sense, especially in cases where the kinetic energy in this region is larger than the barrier height blocking access to this well, to include this well in the definition of the MEP.

However, deep potential wells, including those on the MEP, may be avoided in the reaction mechanism. Forces exerted on the downhill slope of saddle points or ridges on the PES can impart sufficient velocity (both magnitude and direction) to steer the trajectory past a well, just as a skilled kayaker can avoid a whirlpool in the middle of a river by choosing an appropriate velocity well before the whirlpool is encountered. Multiple pathways can arise when the initial conditions upon surmounting the barrier either facilitate or hinder the avoidance of the well. The reaction $OH^- + CH_3F$ discussed in Section V is an example of this phenomenon.

Reactions without wells can also exhibit multiple pathways due to deviation from the MEP. While many trajectories may follow the MEP over a saddle point, alternative pathways arise when forces on the PES steer away from the saddle point, typically into relatively flat regions of the PES, before finding an additional path to the same exit channel. The "roaming" mechanisms recently elucidated in the photodissociation of formaldehyde and acetaldehyde, and the reaction of $CH_3 + O$, are examples of this phenomenon, and are discussed in Section V.

D. Contribution of Multiple Electronic States

Reactions occurring on two (or more) electronic states can lead to the same product asymptote. These pathways may occur if more than one electronic state correlates adiabatically to the same asymptote (e.g., single or triplet coupling of two approaching species), or if nonadiabatic transition(s) move population from one state to another. Here, I make the distinction that products of the same structural formula do not represent the same exit channel if they are produced in different electronic states. For example, in the reaction

$$C_2H(X^2\Sigma^+) + O(^3P) \rightarrow CH(A^2\Delta) + CO(X^1\Sigma^+) \tag{8a}$$
$$\rightarrow CH(X^2\Pi) + CO(X^1\Sigma^+) \tag{8b}$$

the two different electronic states of CH formed constitute two different product channels for the purpose of this chapter. In a similar way, reactions that begin on two different electronic surfaces and terminate at a single asymptote, such as

$$C_2(X^1\Sigma_g^+) + HCCH(X^1\Sigma_g^+) \rightarrow C_4H(X^2\Sigma^+) + H(^2S) \tag{9a}$$
$$C_2(a^3\Pi_u) + HCCH(X^1\Sigma_g^+) \rightarrow C_4H(X^2\Sigma^+) + H(^2S) \tag{9b}$$

are not classified as having multiple pathways to a single product channel. Although this distinction is somewhat arbitrary, molecules in different electronic states exhibit such widely varying properties and reactivity that it seems justified to treat them as different reactive species. We will therefore only consider pathway competitions in which reaction begins on a single electronic state, and ends on a single (possibly different) electronic state, with branching to more than one electronic state during the course of the reaction.

Multiple pathways to the same product channel therefore occur via nonadiabatic transitions that lead from the initial electronic state to at least one other electronic state before converging on the product asymptote. Two examples are presented in this chapter: the photodissociations $CH_2O \rightarrow H + HCO$ and $H_2O \rightarrow H + OH$. There is evidence of similar effects in the photodissociation $HNCO \rightarrow H + NCO$ [13].

III. METHODS AND OBSERVABLES

It is challenging experimentally to study two pathways leading to a single product channel for the simple reason that the products in either case are structurally identical. Nevertheless, there are several methods, each applicable to certain classes of reactions, that can distinguish the presence of multiple pathways.

A. Isotopic Labeling

The most straightforward experimental approach is isotopic labeling of certain atoms in the reactants. The detection method must distinguish between the possible isotopologs of the products. For example, in the reaction

$$H + H_2O \rightarrow H_2 + OH \qquad \text{(direct H abstraction)} \qquad (10a)$$
$$\rightarrow [H_3O]^* \rightarrow H_2 + OH \qquad \text{(addition—elimination)} \qquad (10b)$$

the two mechanistic pathways could be distinguished if the reaction were labeled.

$$H + D_2O \rightarrow HD + OD \qquad \text{(direct D abstraction)} \qquad (11a)$$
$$\rightarrow [HD_2O]^* \rightarrow D_2 + OH \qquad \text{(addition—elimination)} \qquad (11b)$$

Detection of products by mass spectrometry can separate the different product isotopologs according to their different mass-to-charge (m/z) ratios. Optical spectroscopy can separate the isotopologs due to the mass-dependent shift of rovibrational energy levels, and hence radiative transitions, upon isotopic substitution. Both techniques can provide quantitative measurement of the pathway branching fractions g_{nm}. The isotopic labeling method typically provides the highest accuracy in measuring pathway branching fractions. These techniques have a long history in examination of reaction pathways, as demonstrated by the investigations of thermal and photochemical decomposition of acetaldehyde beginning in the 1940s [14].

In reactions (10) and (11), the two mechanistic pathways will also likely yield distinguishable product state distributions in the H_2 and OH products, offering another method for separation of the pathway contributions. However, when competing pathways have qualitatively similar mechanisms, for example, when both pathways are dominated by deep wells on the potential energy surface, isotopic labeling may be the only method to distinguish the contributions. In such cases, intramolecular vibrational redistribution of energy among the internal modes of the complex will usually be efficient, and hence the state distributions of the final products may be quite similar.

A disadvantage of this technique is that isotopic labeling can cause unwanted perturbations to the competition between pathways through kinetic isotope effects. Whereas the Born–Oppenheimer potential energy surfaces are not affected by isotopic substitution, rotational and vibrational levels become more closely spaced with substitution of heavier isotopes. Consequently, the rate of reaction in competing pathways will be modified somewhat compared to the unlabeled reaction. This effect scales approximately as the square root of the ratio of the isotopic masses, and will be most pronounced for deuterium or tritium substitution for hydrogen. Near channel thresholds, the differences in zero-point energy between isotopologs can have significant effects [15]. For heavier atoms the isotope effects are typically quite small, and in general the information gained from isotopic labeling outweighs the potential perturbations caused by isotope effects.

A limitation of isotopic labeling is that it is only useful for distinguishing competing pathways in which identical atoms are distributed among *different* product molecules. For example, in photodissociation of H_2CO, the products $H_2 + CO$ may form via a three-center elimination transition state, or via a recently discovered "roaming" mechanism[8] that does not involve this transition state. Because the identical atoms in the reactant are present in only one of the bimolecular products (H_2), isotopic labeling is not helpful in distinguishing between the pathways.

B.   Final State Distributions

In cases where two pathways to a single product channel arise from qualitatively different mechanisms, the final state distributions of the products may be sufficiently different to delineate and even quantify competing pathways. Depending on the mechanisms in play, useful information may come from many product distributions including translational energy, vibrational, rotational, and fine-structure (e.g., lambda-doubling) distributions, angular distributions, and correlations of vector quantities (e.g., velocity with angular momentum) among the products. Many experimental techniques are capable of state-specific detection of products, some of which will be discussed in Section V. Typically, these experiments are conducted at sufficiently low pressures, or with sufficiently fast time resolution, that the products may be probed before collisions modify the final state distribution.

For a useful separation of pathways, the variation in final state distributions within each pathway must be at least somewhat smaller than the variation between pathways. The aforementioned dissociation of H_2CO provides a perfect example of this technique, in which the H_2 produced through the three-center elimination leads to extensive rotational excitation of CO, with only moderate vibrational excitation of H_2. By contrast, the competing pathway involving "roaming" of one H atom leaves much less energy in CO rotation, with very significant vibrational excitation of H_2 [8].

Several general trends can be established for the interpretation of final-state distributions. Significant translational energy (E_T) deposited into the products, especially if the translational energy distribution $P(E_T)$ peaks well away from $E_T = 0$, is indicative of a pathway surmounting an exit barrier that is a significant fraction of the total available energy. Despite the fact that the reaction coordinate is only one of many degrees of freedom on the PES, the downhill gradient established by the barrier along the reaction coordinate provides a "kick" to the departing fragments, effectively channeling potential energy into translation. By contrast, a $P(E_T)$ distribution in which zero translational energy is the most probable is indicative of an exit channel where a well on the PES is connected with little or no barrier to the product asymptote.

Product vibrational excitation greater than expected by equipartition of available energy among product vibrational modes is a common signature of "direct" processes proceeding over potential barriers (e.g., abstraction pathways) in which there is no long-lived complex immediately preceding the exit channel. The time spent following such pathways is generally too short for equilibration of energy among the degrees of freedom. In these cases, the vibrational distribution may be "inverted", that is, the most probable vibrational quantum number n in a given mode is not $n = 0$. In an equilibrium harmonic vibrational distribution (i.e., described by a temperature) the population of vibrational levels $P(n)$ (with energies measured from the zero-point level) is given by [16]

$$P(n) = \exp(-h\nu n/k_B T)/\{1 - \exp(-h\nu/k_B T)\} \qquad (12)$$

This function decreases monotonically with increasing vibrational quantum number n, and hence an inverted vibrational distribution can never be described with a temperature (except for degenerate vibrations). A $P(n)$ distribution that is "thermal", or at least not inverted, is indicative of a well on the PES that is connected with little or no barrier to the product asymptote.

The degree of vibrational excitation in a newly formed bond (or vibrational mode) of the products may also increase with increasing difference in bond length (or normal coordinate displacement) between the transition state and the separated products. For example, in the photodissociation of vinyl chloride [9] (reaction 7), the H—Cl bond length at the transition state for four-center elimination is 1.80 Å, whereas in the three-center elimination, it is 1.40 Å. A Franck–Condon projection of these bond lengths onto that of an HCl molecule at equilibrium (1.275 Å) will result in greater product vibrational excitation from the four-center transition state pathway, and provides a metric to distinguish between the two pathways.

Product rotational excitation is enhanced when two fragments separate from a reaction intermediate with their centers of mass not aligned with the breaking bond. A torque may then be exerted on one fragment or both fragments that evolves into asymptotic product rotation. This kinematic effect accounts for the large rotational excitation of HCl in the three-center elimination from vinyl chloride. Product rotation can also result from angular anisotropy in the exit channel of the PES. For example, in the dissociation of the linear molecule $ABC \rightarrow A + BC$, if the minimum energy path for breaking the A—B bond also involves a decrease in the angle ABC, rotational excitation will be imparted to fragment BC even if there is not a significant impulse along the A—B bond. Similar effects have been observed in bimolecular reactions, for example, in the reactions of chlorine atoms with amines, in which exit channel interactions play a significant role in HCl rotational distributions [17].

The angular distribution of product fragments in a crossed-molecular-beams experiment, or a unimolecular photodissociation experiment, can provide information on the lifetime of the reaction intermediates. Roughly speaking, an isotropic angular distribution of fragments implies that the reaction pathway involves an intermediate that lives longer than its rotational period, while an anisotropic distribution is indicative of a short-lived, "direct" process. Abstraction or rebound pathways are typically described by the latter case, while dissociation from a well on the potential surface allows sufficient time for rotation of the complex that the angular distribution is isotropic.

Finally, fine-structure distributions and vector correlations between departing fragments can provide important distinctions between competing reaction pathways. Fine-structure distributions, such as the lambda doublet population ratio of OH radical products, can be probed by optical spectroscopy. In this example, transitions accessing different lambda-doublet states probe whether the unpaired electron is located in the plane or perpendicular to the plane of OH rotation. Vector correlations between departing products can probe the alignment and orientation of fragments with respect to each other [3]. For example, in the hypothetical photodissociation ABCD $\rightarrow$ AB + CD, if the velocity vector of the two separating fragments is parallel to the angular momentum vector of AB, we can conclude that this fragment's rotational motion is aligned to the dissociation axis in the same way a propeller is aligned to the fuselage of an airplane. Conversely, a perpendicular correlation means that AB departs with the alignment of a discus with respect to the relative velocity vector. Polarized lasers utilized in ion imaging or optical spectroscopy experiments can effectively probe these final state signatures of pathway details [18].

C. Transient Probing of Intermediates

A qualitatively different approach to probing multiple pathways is to interrogate the reaction intermediates directly, while they are following different pathways on the PES, using femtosecond time-resolved pump–probe spectroscopy [19]. In this case, the pump laser initiates the reaction, while the probe laser measures absorption, excites fluorescence, induces ionization, or creates some other observable that selectively probes each reaction pathway. For example, the ion states produced upon photoionization of a neutral species depend on the Franck–Condon overlap between the nuclear configuration of the neutral and the various ion states available. Photoelectron spectroscopy is a sensitive probe of the structural differences between neutrals and cations. If the structure and energetics of the ion states are well determined and sufficiently diverse in structure, then the time-resolved photoelectron spectra [20, 21] could reveal signatures of two different intermediate structures, representing two different pathways on the PES. Transient absorption spectroscopy and other femtosecond time-resolved techniques may also be applicable to this problem.

In principle, these approaches are very attractive because they probe multiple pathways in the critical regions where the pathways are separated, but in practice these are extremely challenging experiments to conduct, and the interpretation of results is often quite difficult. Furthermore, these experiments are difficult to apply to bimolecular collisions because of the difficulty of initiating the reaction with sufficient time resolution and control over initial conditions.

D. Kinetics Methods

A final technique that deserves mention is the measurement of reaction rate coefficients $k(T, p)$ as a function of temperature and pressure. Here the rate of formation of products (or, more commonly decay of reactants) is monitored under thermal equilibrium conditions as a function of temperature, pressure, and composition of the bath gas. Again, several general principles that help distinguish between mechanistic pathways can be observed. For example, reactions with rates decreasing with increasing temperature generally have no barrier to association of reactants, while an increasing rate with increasing temperature implies that a barrier to reaction must be surmounted. Similarly, the rate of reactions without significant wells on the potential energy surface (and hence without long-lived complexes) typically display no dependence on pressure or composition of the bath gas, because the time between collisions with the surrounding gas is longer than the reaction time.

While there is no argument that the behavior of the reaction rate as a function of temperature and pressure is a measure of the reaction mechanism, by their nature these measurements are thermal averages over all initial and final states. Because of this averaging, it is quite difficult to determine details of two independent pathways leading to the same product channel from measurements of $k(T, p)$ alone. Instead, one measures the average behavior of two competing pathways. However, if, over the range of temperatures and pressures studied, the mechanism changes from one dominant pathway to another, with a competition region between, the influence of competing pathways may be discerned [22].

IV. THEORETICAL METHODS

The value of theoretical investigation for determining competing reaction pathways leading to a single product channel cannot be overestimated. Theoretical approaches have a distinct advantage over experiment in that the labeling of atoms in the reaction is straightforward. Different pathways to a product channel are now easily distinguished. Indeed, it is almost universally true that pathway branching studies are either a combination of experiment and theory, or consist purely of theoretical evidence.

A.   Electronic Structure: Global Potential Energy Surfaces

The first step in theoretical predictions of pathway branching are electronic structure (*ab initio*) calculations to define at least the lowest Born–Oppenheimer electronic potential energy surface for a system. For a system of N atoms, the PES has $(3N - 6)$ dimensions, and is denoted $V(R_1, R_2, \cdots, R_{3N-6})$. At a minimum, the energy, geometry, and vibrational frequencies of stationary points (i.e., asymptotes, wells, and saddle points where $\partial V / \partial R_i = 0$) of the potential surface must be calculated. For the statistical methods described in Section IV.B, information on other areas of the potential are generally not needed. However, it must be stressed that failure to locate relevant stationary points may lead to omission of valid pathways. For this reason, as wide a search as practicable must be made through configuration space to ensure that the PES is sufficiently complete. Furthermore, a search only of stationary points will not treat pathways that avoid transition states.

One way to address this issue is to calculate energies of not only stationary points, but of a grid of points spanning the configuration space. Given a sufficiently extensive grid of points in the $(3N - 6)$ dimensions, one can be assured of global coverage of the potential surface. However, such an exhaustive calculation is only possible for the smallest systems. In these cases, the grid of potential energy points must be fit to an analytic function for use in classical, quasiclassical, or quantum mechanical calculations. For larger systems, where it is not easy to tell if the grid is sufficiently dense, some other measure must be used to decide whether the calculated PES spans all the important regions of the true PES. Sometimes this measure derives from the skill and intuition of an experienced theoretician. Alternatively, there are approaches that attempt to overcome the human bias from this search, one of which is discussed in Section IV.C.

B.   Statistical Methods

The development of transition state theory (TST) by Eyring [23] and by Evans and Polanyi [24] in 1935 allowed calculation of the absolute rates of chemical reactions using a classical approach built on statistical mechanics. The dividing surface separating reactants from products defines the critical molecular configuration known as the transition state. The bold assumption of TST is that the rate of the chemical reaction depends only on the properties of the transition state and the isolated reactants; other details of the potential energy surface are unimportant. Together with the related Rice Ramsberger Kassel Marcus (RRKM) [25] theory for unimolecular reactions, these statistical theories have had much success in predicting reaction rates without the burden of calculating the entire PES. Further refinements [26], such as variational TST [27], the Statistical Adiabatic Channel Model [28], and Variable Reaction Coordinate TST [29], have allowed more accurate treatment of reactions with and without barriers. Statistical methods, such as the Prior Distribution [2], Phase Space Theory [30], and the Separate Statistical Ensembles method [31], predict product state distributions for comparison with experiment. Finally, some hybrid methods, such as the Statistical Adiabatic Impulse model [32], convolve the product state predictions of purely statistical approaches with a simple kinetic impulsive model to account for dissociation over exit barriers with large amounts of excess energy.

The ability to probe specific pathways using these statistical approaches relies completely on the geometric structures determined from the calculated stationary points of the PES. In other words, the knowledge that, for example, two particular stationary points represent a three- and four-member ring intermediate, respectively, allows one to calculate rates, and therefore pathway branching ratios through these channels.

One disadvantage of statistical approaches is that they rely on two of the assumptions stated in the introduction, namely, that reactions follow the minimum energy path to each product channel, and that the reactive flux passes through a transition state. Several examples in Section V violate one or both of these assumptions, and hence statistical methods generally cannot treat these instances of competing pathways [33].

C. Trajectory Calculations

Classical and quasiclassical trajectory calculations are among the best suited theoretical approaches for calculating and quantifying the importance of competing pathways to a single product channel. These methods calculate the progress of a chemical reaction by choosing initial conditions of reactants (position and momentum) and propagating trajectories of the nuclei along the potential energy surface by numerical integration of Newton's equations of motion. Each trajectory calculated represents one "run" of the reaction. An ensemble of trajectories can be averaged to obtain product state distributions and rate coefficients. However, each trajectory can also be analyzed individually and assigned to a particular canonical pathway on the PES. The ability to observe the time-dependent position of all the atoms in a trajectory is extremely valuable in obtaining insight into the mechanistic pathways involved in a reaction. Several examples of such calculations are given in Section V.

Traditionally, trajectory calculations were only performed on previously calculated (or empirically estimated) potential energy surfaces. With the increased computational speed of modern computers, it has also become possible to employ "direct dynamics" trajectory calculations [34, 35]. In this method, a global potential energy surface is not needed. Instead, from some initial starting point of the positions and momenta of the atoms in a system, a new trajectory position is calculated from the integration of Newton's equations. An electronic structure calculation is performed at this point to find the energy and gradients along all $(3N - 6)$ coordinates. These gradients define the local forces at this point on the potential, which are used to predict the next step in the trajectory. This approach has several advantages and disadvantages compared to running trajectories on an analytic representation of the PES.

First, because the trajectories themselves define which parts of the PES are important, no time is wasted calculating regions of the potential that are never accessed by trajectories. This advantage is especially important as the number of atoms increases, making calculation of the global $(3N - 6)$ coordinate surface prohibitively expensive. Indeed, for systems larger than seven or eight atoms, only direct dynamics calculations are reasonable to attempt with current computational resources. Second, in the computation of a global PES, compromises must often be made in how many points are chosen to span configuration space. It is possible that important, but nonintuitive regions of configuration space may have few enough points that they are poorly characterized. Such poorly characterized areas could result in omission of a valid pathway. Finally, the grid of electronic structure points are typically fit to a multidimensional analytic function for traditional trajectory calculations. While an analytic representation of the potential energy optimizes the speed of the trajectory calculation, the fitting can induce artifacts into the surface that over- or underestimate the depth of wells and the height of barriers. These artifacts can be avoided in direct dynamics calculations.

However, the direct dynamics calculations are computationally expensive, and cannot employ particularly high levels of electron correlation or large basis sets. If certain regions of the potential cannot be treated to within the required accuracy using a computationally affordable level of theory, the results may have unacceptably large errors. Nevertheless, direct dynamics calculations have played and will play a critical role in the discovery and analysis of competing pathways in chemical reactions.

One hybrid approach that combines the efficiency of running trajectories on an analytical PES with the advantages of computing only the important regions of the surface is the method of "growing" potential energy surfaces of Collins and co-workers [36]. In this method, the energy, gradients $(\partial V/\partial R_i)$, and second derivatives $(\partial^2 V/\partial R_i^2)$ of some chosen initial set of molecular configurations are calculated. A trial PES is then constructed by interpolation of these data. Classical trajectory calculations are used to "explore" this PES, and the paths of the trajectories are used to determine the most important locations for further energy, gradient, and second derivative calculations. These new points are incorporated into a newly interpolated potential energy surface, and the process is repeated until convergence is achieved. Convergence is defined as the insensitivity of the calculated observables to the addition of further data points to the PES. In this method, little time is wasted calculating the potential energy in regions of configuration space that the dynamics never sample. More effort can then be spent on the regions of the surface on which the reaction results most critically depend. Furthermore, the "growth" of the surface, largely independent of human bias, should reduce the chance that an important region of configuration space goes unexplored.

Finally, rigorous quantum mechanical scattering calculations, for both unimolecular and bimolecular reactions, are of course appropriate for studying pathway competition. However, quantum methods suffer from one technical, but substantial problem. There is no "direct quantum" equivalent of direct dynamics trajectories. Instead, quantum calculations require a complete potential energy surface (although it may be in reduced dimensionality, and with absorbing boundary conditions), which is generally not feasible for systems with more than four atoms. For these reasons, quantum mechanical calculations are likely to play a more limited role in pathway branching studies than trajectory calculations, although they are employed in the final example given in Section V.

V. EXAMPLES OF MULTIPLE PATHWAY REACTION SYSTEMS

This section reviews 10 chemical systems in which multiple pathways to a single product channel play a role. The systems are chosen to give representative examples of the classes of reaction described in Section II, rather than an exhaustive catalog of all reactions in which multiple pathways make a contribution. The systems are grouped by class, although some reactions fit in more than one class. The examples begin with systems interrogated by isotopic labeling, and in which multiple identical atoms enable the pathway competition. Next, systems amenable to product state distribution studies with competition between abstraction, addition–elimination, and roaming pathways are reviewed. Two systems that avoid the minimum energy path, violating an assumption of TST, are then discussed. Finally, examples are discussed in which multiple electronic states and quantum interference are active in the dynamics.

A. $CD_3^+ + C_2H_6$: Hydrogen Atom Migration Mechanisms

The exothermic ion molecule reaction

$$CH_3^+ + C_2H_6 \rightarrow CH_4 + C_2H_5^+ \qquad \Delta H \approx -41 \, kcal \, mol^{-1} \qquad (13)$$

has been described as a classic example of a hydride-transfer reaction [37], in which a proton from ethane, along with the two electrons in its σ bond, are transferred to the methyl cation. In a pair of experiments in 1993, Herman, Bondybey, and co-workers studied both the kinetics [38] and dynamics [39] of this reaction. They utilized the isotopolog CD_3^+ and observed the following products:

$$CD_3^+ + C_2H_6 \rightarrow CD_3H + CH_2CH_3^+ \qquad \text{(hydride transfer)} \qquad (13a)$$
$$\rightarrow CD_3CH_2^+ + CH_4 \qquad \text{(methylene transfer)} \qquad (13b)$$

leading to the initial conclusion that the reaction can follow two pathways: either the expected hydride transfer, or a methylene (CH_2) transfer that in either case forms methane and an ethyl cation.

The kinetics study [38] utilized a Fourier transform–ion cyclotron resonance (FT–ICR) mass spectrometer to measure the pathway branching ratios. The ability to eject selected masses and the extremely high mass resolution of this technique ensured that the observed $CD_3CH_2^+$ was in fact a primary product of the reaction. Temporal profiles from this reaction are shown in Fig. 1. Noticeably absent from the mass spectrum are the cations $C_2D_2H_3^+$ and

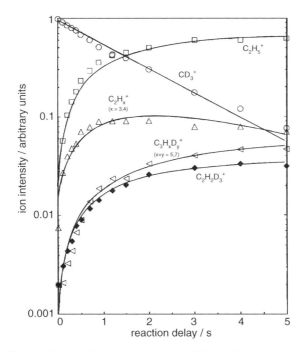

Figure 1. Time-resolved profiles of cations from the $CD_3^+ + C_2H_6$ reaction at 2.0-eV collision energy. The decay of CD_3^+ and the formation of $C_2H_5^+$ and $CD_3CH_2^+$ cations follow pseudo-first-order kinetics. Reprinted from [38] with permission from Elsevier.

$C_2DH_4^+$, which would be present if a long-lived intermediate allowed isotopic scrambling of the hydrogen atoms. In the FT–ICR study at collision energies $< 1\,eV$, the methylene-transfer channel represented $\sim 7\%$ of the pathway branching forming ethyl cations. This fraction decreased at higher collision energies.

Having established the presence of two pathways, further evidence of the pathway mechanisms can be gleaned from the crossed-beam scattering study [39]. In this experiment, CH_3^+ or CD_3^+ ions crossed a modulated, thermal beam of C_2H_6, and the translational energy and angular distributions of the positive ions were detected. At a collision energy of $1.7\,eV$, the crossed-beams study found a pathway branching of 36% for the methylene-transfer channel that is, production of $CD_3CH_2^+$. Probability density contour plots of the ethyl cation fragments are shown in Fig. 2. The most probable translational energy of the products is 10–12% of the available energy.

The contour plots show predominately forward–backward scattering, implying that a direct abstraction process is unlikely. However, plots from the isotopically labeled reaction show a slight propensity for backward scattering of $CH_2CH_3^+$ [Fig. 2(b)], and forward scattering of $CD_3CH_2^+$ [Fig. 2 (c)]. The authors interpreted this behavior as evidence of an osculating complex, that is, an intermediate with an average lifetime of one or a few rotations of the complex. In this case, the product's angular distribution retains a signature of the pathway that formed the product. From analysis of the data the authors conclude the complex lifetime is ~ 1 ps, or ~ 1–3 rotational periods of the complex.

These results, combined with the lack of isotopic scrambling in the products, led the authors to propose mechanisms for the two pathways, as shown in Fig. 3. The methyl cation attacks any of the H atoms in ethane, leading to a protonated propane intermediate ($C_3H_9^+$). The global minimum of $C_3H_9^+$ is the cyclic proponium ion, lying $\sim 6.4\,kcal\,mol^{-1}$ below the products [40]. Evidently, isomerizations within the $C_3H_9^+$ ladder shown in Fig. 3 occur without migrating D atoms. The more direct route in this mechanism, to $CH_3CH_2^+ + HCD_3$ products, is consistent with the experimental finding that this pathway is dominant. Furthermore, the two pathways predict ejection of the ethyl cation from opposite ends of the $C_3H_9^+$ intermediate, in agreement with the slight forward–backward asymmetry observed in the scattering contour diagrams of Fig. 2. The ~ 1-ps lifetime deduced for the complex appears consistent with the competition between the longer isomerization pathway and the short pathway to $CH_3CH_2^+ + HCD_3$ products. If the $C_3H_9^+$ lifetime were longer, isotopic scrambling would surely occur, while if the $C_3H_9^+$ lifetime were shorter, little of the $CH_4 + CH_2CD_3^+$ product would be produced. The branching to this pathway decreases at high collision energies, where the osculating complex may be expected to have a shorter lifetime.

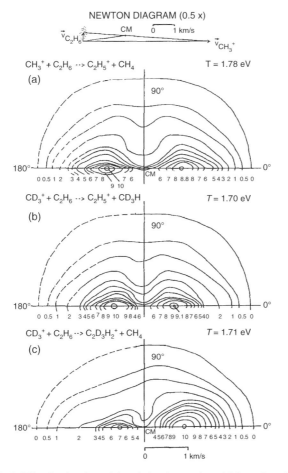

NEWTON DIAGRAM (0.5 x)

$CH_3^+ + C_2H_6 \dashrightarrow C_2H_5^+ + CH_4$ T = 1.78 eV

(a)

$CD_3^+ + C_2H_6 \dashrightarrow C_2H_5^+ + CD_3H$ T = 1.70 eV

(b)

$CD_3^+ + C_2H_6 \dashrightarrow C_2D_3H_2^+ + CH_4$ T = 1.71 eV

(c)

Figure 2. Probability density plots of the ethyl cation product. (a) from the unlabeled reaction, (b) $CH_2CH_3^+$ from the labeled reaction, and (c) $CD_3CH_2^+$ from the labeled reaction. The backward scattered ethyl cation is more probable in (b), while the forward scattered ethyl cation is more probable in (c). Reprinted from [39] with permission from Elsevier.

Finally, it is worthwhile to note that the "methylene-transfer" pathway does not involve direct transfer of a CH_2 group. An H atom migration pathway is a more accurate description of this pathway, given the experimental and theoretical evidence.

B. $HO_2 + O_3$: Dual Abstraction Pathways

Like the previous reaction, multiple pathways in the $HO_2 + O_3$ reaction arise because of the participation of identical atoms. However, in this case the two

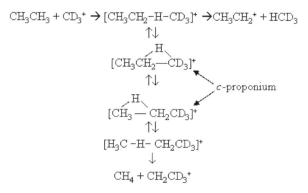

Figure 3. Proposed mechanisms of the $CD_3^+ + C_2H_6$ reaction.

competing pathways are both abstraction mechanisms, commencing on opposite ends of the HO_2 radical.

The $HO_2 + O_3$ reaction is an important contributor to the catalytic loss of ozone in the lower stratosphere through the cycle

$$HO_2 + O_3 \rightarrow HO + O_2 + O_2 \tag{14}$$

$$HO + O_3 \rightarrow HO_2 + O_2 \tag{15}$$

resulting in a net coversion of two ozone molecules to three O_2 molecules per cycle. The title reaction is also a key step in determining the balance in the HO_x family between highly reactive OH and the much less reactive HO_2 radical. Reaction kinetics studies find a strongly curved Arrhenius plot between 230 and 300 K. Nelson and Zahniser [41] speculated that this curvature might be due to a mechanism shift between O atom abstraction by ozone at low temperatures, to H atom abstraction at room temperature. The isotopic labeling scheme they utilized to distinguish between these two pathways was

$$H^{18}O^{18}O + O_3 \rightarrow H^{18}O + {}^{18}OO + O_2 \quad \text{(oxygen abstraction: } 5-12\%\text{) (16a)}$$

$$O_3 + H^{18}O^{18}O \rightarrow O_2 + OH + {}^{18}O_2 \quad \text{(hydrogen abstraction: } 95-88\%\text{)(16b)}$$

in which the unlabeled ozone is of natural isotopic abundance (99.3% $^{16}O_3$), and their pathway branching results between 226 and 355 K are given in parentheses.

Nelson and Zahniser used a moveable injector discharge flow apparatus in these studies, generating the HO_2 from the $H + O_2 + M$ reaction. They detected both ^{16}OH and ^{18}OH by laser-induced fluorescence, correlating their ratio with the pathway branching ratio. One possible complication in this scheme is the participation of isotope scrambling reactions, such as

$$OH + {}^{18}O_2 \rightarrow {}^{18}OH + O^{18}O \tag{17}$$

$$H^{18}O_2 + O_2 \rightarrow HO_2 + {}^{18}O_2 \tag{18}$$

$$H^{18}O_2 + O_3 \rightarrow HO_2 + {}^{18}O^{18}OO \tag{19}$$

The rates of these three reactions have been measured by Sinha et al. [42], and are too slow to affect the Nelson and Zahniser experiment.

The measured $[{}^{16}OH]/[{}^{18}OH]$ branching ratio versus inverse temperature is plotted in Fig. 4. If the two species are produced by two parallel pathways, the total reaction rate is a simple sum of the two pathway-resolved rates. In this case, the data points in an Arrhenius plot should fall on a straight line with a slope proportional to the difference in activation energies for the two competing pathways. A fit to the data in Fig. 4 yields the result that the barrier to O atom abstraction is $1.0 \pm 0.4 \, kcal \, mol^{-1}$ larger than for H atom abstraction. Although

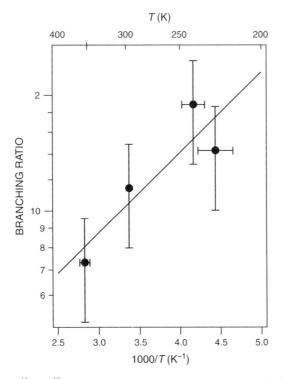

Figure 4. The $[{}^{16}OH]/[{}^{18}OH]$ branching ratios versus inverse temperature for the $H^{18}O_2 + {}^{16}O_3$ reaction. Reprinted with permission from [41]. Copyright 1994 American Chemical Society.

there is a slight temperature dependence to the pathway branching, its magnitude is insufficient to account for the non-Arrhenius behavior in the total rate coefficient of the $HO_2 + O_3$ reaction at low temperatures [43].

These measurements show the sensitivity of the pathway competition to very small differences in the abstraction barrier heights. Such pathway branching measurements are quite sensitive to the relative energetics of the potential energy surface, and can serve as a stringent test to electronic structure determinations of barrier heights. By the same token, this system demonstrates that two abstraction pathways will only have a significant competition when the barrier heights are quite similar.

C. $HCCO + O_2$: Pathways through Ring Intermediates

The final example of a reaction in which multiple pathways arise from the participation of identical atoms is the reaction $HCCO + O_2$. This reaction demonstrates another general feature of pathway competition when unsaturated species are involved, namely, the possibilities for formation of ring intermediates of different sizes.

The ketenyl radical (HCCO) is a key intermediate in the oxidation of acetylene in flames. It is mainly formed from the $O + C_2H_2 \rightarrow HCCO + H$ reaction. In lean flames, the $HCCO + O_2$ reaction is the main pathway for decay of HCCO, and this reaction has recently been shown to be the source of prompt CO_2 [44, 45].

The reaction has three plausible product channels:

$$HCCO + O_2 \rightarrow H + CO + CO_2 \qquad \Delta H_0 = -110.4\,\text{kcal mol}^{-1} \qquad (20a)$$
$$\rightarrow OCHCO + O \qquad \Delta H_0 = -1.3\,\text{kcal mol}^{-1} \qquad (20b)$$
$$\rightarrow OH + CO + CO \qquad \Delta H_0 = -86.0\,\text{kcal mol}^{-1}[46] \qquad (20c)$$

In a theoretical study, Klippenstein et al. [44] predicted that the $H + CO + CO_2$ channel is the dominant product channel in the temperature range 300–2500 K and pressures up to 100 atm. An experimental study by Osborn [45] concluded that this channel represents at least 90% of the product branching at 293 K, in agreement with theory.

The potential energy surface [47] for this reaction (Fig. 5) shows many potentially competitive pathways, labeled A–F, leading to the two most exothermic product channels. Many of these pathways can be isotopically separated by reaction of $^{18}O_2$ with HCCO in normal abundance, as diagramed in Fig. 5. Zou and Osborn used time-resolved Fourier transform emission spectroscopy to detect the CO and CO_2 products of this reaction [47]. Rotationally resolved infrared (IR) spectroscopy can easily identify all the possible isotopologs. For example, Fig. 6 shows a single

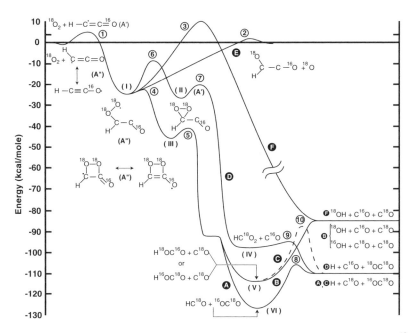

Figure 5. Potential-energy diagram including zero-point energy for the HCCO + $^{18}O_2$ reaction. Energies of reactants and products ignore differences between ^{18}O and ^{16}O. Intermediates species are denoted by Roman numerals, saddle points by Arabic numerals, and reactions paths are labeled A–F. Reproduced from [47] by permission of the PCCP Owner Societies.

spectrum 300 μs after reaction initiation in the CO_2 asymmetric stretching region for the HCCO + $^{18}O_2$ reaction. The only CO_2 product observed is $^{16}OC^{18}O$. The CO product can also be examined, and with higher signal-to-noise than CO_2. Only $C^{18}O$ is observed in the labeled reaction. Kinetic isotope effects in this system are expected to be negligible due to the small difference in mass of the two oxygen isotopes and the large amounts of available energy in this reaction.

Referring again to Fig. 5, the reaction proceeds over a small entrance barrier (1) to the initial adduct (**I**). Pathways E and F have much higher barriers than the other pathways available for decomposition of (**I**), and are therefore assumed to be negligible. There are three reaction paths (A, C, and D) leading from the initial adduct to H + CO + CO_2. These paths involve either a four-membered OCCO ring (**III**) or a three-membered OCO ring (**II**) as reaction intermediates. There is one remaining path (B), leading to OH + CO + CO.

The key pathway branching competition in this reaction is the isomerization of the initial adduct (**I**) to either the resonantly stabilized four-member ring intermediate (**III**), or the three-member ring intermediate (**II**). Formation of the four-member ring has a lower energy barrier (saddle point 4), but is more entropically constrained, as all the internal rotors of the system are eliminated.

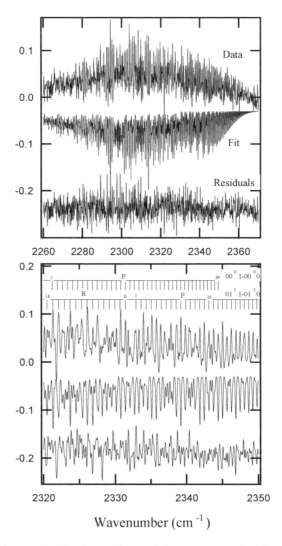

Figure 6. Time-resolved Fourier transform emission spectrum in the CO_2 asymmetric stretch region from the $HCCO + {}^{18}O_2$ reaction. Only signal from ${}^{16}OC{}^{18}O$ is observed. The fit to the data is shown inverted for clarity along with residuals. The lower panel is magnified along the wavenumber axis to emphasize the rotational resolution required. Reproduced from [47] by permission of the PCCP Owner Societies.

Formation of the three-member ring has a higher barrier ⑥, but its entropy is higher due to the remaining CO rotor. Decomposition of the three-member ring via pathway D leads to $HC{}^{18}O_2 + C{}^{16}O$, followed by dissociation of $HC{}^{18}O_2$ leading to the final products $H + C{}^{18}O_2 + C{}^{16}O$.

Decomposition of the four-member ring intermediate occurs by breaking the O—O bond over ⑤, followed by an essentially barrierless breaking of the C—C bond, yielding $HC^{18}O + {}^{16}OC^{18}O$ via path A. The weak C—H bond in HCO will certainly break given the immense exothermicity, leading to the final path A products $H + C^{18}O + {}^{16}OC^{18}O$. However, trajectory calculations [44] on this system show another possibility in the decay of the four-member ring. After the O—O bond scission over ⑤, the H atom can migrate to the unpaired electron on the ^{16}O atom. Furthermore, if rotation of the CO_2 moiety by 180° occurs, H atom migration to the ^{18}O atom of the CO_2 moiety is also possible. These options lead to the HOCO + CO well (intermediate **V**). With so much internal energy, the HOCO intermediate is unlikely to decompose to $H + CO_2$ over a tight transition state (path C), and therefore intermediate (**V**), if it is formed, will decay primarily via pathway B (over a loose transition state) to the OH + CO + CO product channel. This product channel is known to be minor ($<10\%$). Nevertheless, path B is included in the analysis of the isotopic spectra for completeness.

In summary, pathways A, B, and D are the only pathways at room temperature that contribute to products. Pathways A and D both lead to the main product channel $(H + CO + CO_2)$, and are isotopically distinguishable, whereas pathway B leads to a different product channel, but must be included in the analysis because it affects the isotope ratios. The observed spectra from the labeled reaction contain only $C^{18}O$ and $^{16}OC^{18}O$. Therefore, only upper limits can be placed on the isotope ratios. The three values needed to constrain the three pathways under consideration are the ratio limits $[C^{16}O]/[C^{18}O] < 0.16$ and $[^{18}OC^{18}O]/[^{16}OC^{18}O] < 0.30$, and the limit that the total OH + CO + CO production is <0.10 [45].

The results from these measurements show that the reaction occurs primarily through the four-member ring intermediate. The IR spectra yield the limits $\Phi_{3\text{-mem}} < 15\%$ and $\Phi_{4\text{-mem}} > 85\%$ for the total reactive flux through the two possible ring intermediates. Furthermore, the results show that hydrogen transfer during decomposition of the four-membered ring intermediate is a minor process at most. These results are in agreement with the limited number of trajectories calculated for this reaction [44].

In highly exothermic reactions such as this, that proceed over deep wells on the potential energy surface, sorting pathways by product state distributions is unlikely to be successful because there are too many opportunities for intramolecular vibrational redistribution to reshuffle energy among the fragments. A similar conclusion is likely as the total number of atoms increases. Therefore, isotopic substitution is a well-suited method for exploration of different pathways in such systems.

D. H_2CO Photodissociation: The Roaming Atom Mechanism

In a recent study, combining experimental and theoretical investigation, Suits and co-workers [8] discovered two active pathways in the photodissociation

238   DAVID L. OSBORN

CH$_2$O $\rightarrow$ H$_2$ + CO. In contrast to the previous three systems, this pathway competition cannot be studied by isotopic substitution because the identical atoms are found in only one product (H$_2$). Fortunately, the two contributing pathways can be separated because their product state distributions are markedly different.

Formaldehyde is a prototypical molecule for the study of unimolecular dissociation dynamics and has had a significant impact upon our understanding of statistical reaction models (e.g., TST, RRKM, phase-space theory) [48]. A potential energy diagram of CH$_2$O is shown in Fig. 7. Highly excited formaldehyde in its ground electronic state S_0 can be prepared with well-defined energies by internal conversion from laser-excited rovibrational levels in the S_1 state. There is a transition state (TS) for elimination to H$_2$ + CO that has a skewed structure, with both hydrogen atoms on the same side of the C—O bond. The large exit barrier leads to significant translational energy release (65% of the available energy), whereas the structure of the TS creates much rotational, but little vibrational excitation of CO. The cofragment H$_2$ is moderately vibrationally excited, but rotationally quite cold.

By contrast, the dissociation channel leading to H + HCO (the radical channel) has no barrier and a dissociation threshold [49] of 30,328.5 cm^{-1}. In 1993, van Zee et al. [50] found that excitation above this threshold led to CO rotational distributions that were bimodal. In addition to the high-J CO ($J_{max} \sim 45$) that is also produced when excitation is slightly < 30,328.5 cm^{-1}, a significant population of low J CO was observed. As one explanation for this bimodal distribution, they proposed the opening of a second pathway producing H$_2$ + CO that was in some way related to the energetic opening of the H + HCO

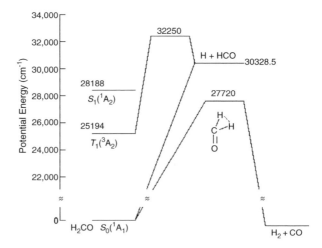

Figure 7. Potential energy diagram of CH$_2$O. After excitation to specific rovibrational levels of S_1, internal conversion leads to highly excited molecules in the ground electronic state S_0, whereas intersystem crossing populates the lowest triplet state T_1.

channel. Valachovic and co-workers [51] suggested the low-J CO might arise from dynamics on the S_0 surface that would produce an H atom loosely bound to an HCO core that has substantial rotation about its a axis.

Recently, Suits and co-workers [8, 52] used direct current (DC)-sliced ion imaging measurements of CH_2O to reinvestigate this bimodal distribution. In this technique, a single rovibrational level of CO is ionized, and its recoil velocity from the H_2 fragment is imaged. The great advantage of this technique is the ability to record full $H_2(v, J)$ distributions that are correlated with individual rovibrational levels of CO. Three translational energy distributions from this experiment with an excitation energy of 30,340.1 cm^{-1} are shown in Fig. 8. Panel A represents the $H_2(v, J)$ distribution correlated with high rotational excitation of CO, with population of $H_2(v = 0-4)$. This distribution is very similar to that seen when the photodissociation energy is just under the radical threshold of 30,328.5 cm^{-1}, and is indicative of the traditional pathway through the skewed transition state to $H_2 + CO$ products. Panel B, probing $J_{CO} = 28$, shows a bimodal distribution of $H_2(v, J)$ states, with significantly more vibrational excitation. Finally, in panel C, probing $J_{CO} = 15$, the only significant contribution is $H_2(v = 6-7)$. It is plausible that the $H_2(v = 6-7)$ population signifies the opening of a new pathway producing $H_2 + CO$.

The conclusion that highly vibrationally excited H_2 correlated with low-J CO represents a new mechanistic pathway, and the elucidation of that pathway, is greatly facilitated by comparison with quasiclassical trajectory calculations of Bowman and co-workers [8, 53] performed on a PES fit to high level electronic structure calculations [54]. The correlated H_2 / CO state distributions from these trajectories, shown as the dashed lines in Fig. 8, show reasonably good agreement with the data. Analysis of the trajectories confirms that the $H_2(v = 0-4)$ population represents dissociation over the skewed transition state, as expected.

The trajectories that produce $H_2(v = 6-7)$ show one H atom very loosely bound, moving slowly around the HCO moiety in a rather flat part of the PES, until this H atom reaches a nearly linear CHH configuration. This configuration is near the (barrierless) TS for the related abstraction reaction $H + HCO \rightarrow H_2 + CO$. The roaming H atom then abstracts the H atom that was tightly bound to the CO moiety, and departs with little rotation, but significant vibrational excitation. This mechanism is aptly named the "roaming atom mechanism". It is important to remember that in the data presented in Fig. 8, CH_2O has $\sim$12 cm^{-1} more energy than needed to form the $H + HCO$ radical channel; the system is unbound with respect to H atom loss. However, if $>$12 cm^{-1} of energy is sequestered in the HCO moiety, the H atom cannot leave until that energy is transferred into the reaction coordinate. (Hiding 12 cm^{-1} of energy in isolated HCO is not difficult: Any vibrational excitation above the zero-point level, and all rotational levels except 0_{00}, 1_{01}, and 2_{02}, represent more than this energy.) This roaming H atom must find the other H atom before IVR

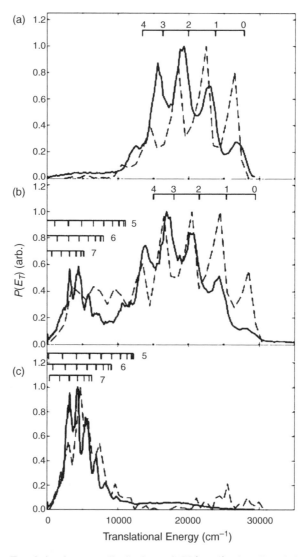

Figure 8. Translational energy distributions of CO($v = 0$) after dissociation of H_2CO at $h\nu = 30{,}340.1\,\text{cm}^{-1}$ for the CO product rotational levels (a) $J_{CO} = 40$, (b) $J_{CO} = 28$, and (c) $J_{CO} = 15$. The internal energy of the correlated H_2 fragment increases from right to left. Dashed lines are translational energy distributions obtained from the trajectory calculations. Markers indicate H_2 vibrational thresholds up to $v = 4$, and in addition odd rotational levels for $v = 5-7$. Reprinted from [8] with permission from the American Association for the Advancement of science.

succeeds in transferring enough energy to the reaction coordinate to drive the H + HCO product channel.

The overall pathway branching for the competing pathways at $hv = 30,340.1 \, cm^{-1}$ is determined to be

$$H_2CO + hv \rightarrow H_2 + CO \qquad \text{(82\% traditional TS pathway)} \qquad (21a)$$
$$\rightarrow H_2 + CO \qquad \text{(18\% roaming pathway)} \qquad (21b)$$

As the photon energy is increased $>30,340.1 \, cm^{-1}$, the product branching fraction to the H + HCO channel increases, and the molecular channel decreases. Surprisingly, the pathway branching to the roaming channel appears to increase in importance relative to the traditional TS pathway as the photon energy is increased [52].

An important point to emphasize is that the roaming atom channel forms $H_2 + CO$ without going over the minimum energy path (the skewed transition state) shown in Fig. 7. This behavior is contrary to the assumptions of RRKM and other transition state theories, which therefore cannot be used to treat this product channel without significant modification. This case highlights the value of classical and quasiclassical trajectory calculations in elucidating alternate pathways in general, and roaming atom pathways in particular.

One general question regarding roaming mechanisms is whether the roaming moiety is restricted to the light H (or D) atom, which can quite easily exhibit large amplitude motion due to its low mass. In a quote from Ref. [8], the authors speculate on this point: "A key question remaining is whether diatomic products besides H_2, or even polyatomic products, may be formed by such a [roaming] mechanism. It may be more common for the roaming species to be a hydrogen atom that can rapidly explore the accessible regions of the surface. Perhaps, however, an atom or group besides H could be the abstraction target". As discussed in Section V.E, there is already some evidence to support roaming by fragments other than H atoms.

E. CH₃CHO Photodissociation: The Roaming Mechanism Again?

Acetaldehyde is methyl-substituted formaldehyde, and has a number of similarities with its smaller cousin. In particular, when photodissociated at 308 nm, internal conversion to S_0 is rapid, and acetaldehyde can decompose via analogous radical and molecular channels

$$CH_3CHO + hv \rightarrow CH_3 + HCO \qquad \text{(threshold} = 83.6 \, kcal \, mol^{-1}) \qquad (22a)$$
$$\rightarrow CH_4 + CO \qquad \text{(threshold} = 82.3 \, kcal \, mol^{-1}) \qquad (22b)$$

where the threshold represents the asymptotic product energy for the radical channel and the TS barrier height for the molecular channel. However, additional

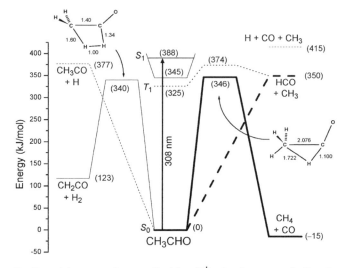

Figure 9. Potential energy diagram (in kJ mol^{-1}) showing energetically allowed photo-dissociation channels for CH$_3$CHO. Note the similar threshold energies for the four energetically allowed channels. Reprinted with permission from Dr. Scott Kable, University of Sydney.

radical and molecular channels are possible, although they have not been experimentally observed

$$CH_3CHO + h\nu \rightarrow CH_3CO + H \quad \text{(threshold} = 90.1 \, \text{kcal mol}^{-1}) \quad (22c)$$
$$\rightarrow CH_2CO + H_2 \quad \text{(threshold} = 81.3 \, \text{kcal mol}^{-1}) \quad (22d)$$

The potential energy diagram for these channels is quite similar to those in CH$_2$O, and is shown in Fig. 9.

Houston and Kable [55] have measured CO($v = 0-2, J$) product state distributions from 308-nm photodissociation of CH$_3$CHO using laser-induced fluorescence. They found bimodal rotational distributions for CO($v = 0, 1$) that can be adequately represented as the sum of two Gaussian distributions peaking at $J_{CO} \sim 30$ and $J_{CO} \sim 10$. This signature is analogous to the bimodal CO distributions from formaldehyde observed by van Zee and co-workers [50]. Could the low-J CO in this case also correspond to a roaming mechanism for the production of CH$_4$ + CO? (Note that fragmentation to H + CO + CH$_3$ is not energetically possible at 308 nm.)

In addition to CO($v = 0-2, J$) populations, Houston and Kable recorded CO Doppler profiles to measure the translational energy release, and the vector correlation between the recoil velocity vector and the angular momentum vector of CO. Together, these data paint a compelling picture that two pathways to CH$_4$ + CO are operative. The rotationally hot CO population (85% of total CO) correlates with significant translational energy release ($\langle E_T \rangle = 0.25 E_{\text{avail}}$) and substantial $\nu \perp J$ correlation of 0.7 ± 0.3. These state distributions agree with the predictions of direct dynamics trajectory calculations [56] for dissociation over the three-center TS leading to $CH_4 + CO$. By contrast, the rotationally colder CO population (15% of total CO) correlates with a small translational energy release ($\langle E_T \rangle = 0.11 E_{\text{avail}}$), no ν, J correlation, and by conservation of energy, with very highly vibrationally excited methane ($\langle E_{\text{vib}} \rangle (CH_4) = 0.87 E_{\text{avail}}$).

These product state distributions for low J CO mirror the state distributions for the roaming H atom channel in formaldehyde. Assuming, therefore, that this second pathway does represent a roaming channel, Houston and Kable propose two possibilities that are related to the two possible radical product channels (22a and c). If the roaming fragment is the aldehyde hydrogen, then the mechanism correlates with the onset of the $CH_3CO + H$ radical channel and is related to the bimolecular reaction $H + CH_3CO \rightarrow CH_4 + CO$. The second possibility involves a roaming CH_3 moiety that eventually finds the aldehyde hydrogen on route to $CH_4 + CO$. This mechanism corresponds to the onset of the $CH_3 + HCO$ radical channel, and hence to the $CH_3 + HCO \rightarrow CH_4 + CO$ bimolecular reaction. The authors argue that this pathway seems more plausible because the roaming methyl fragment is more reactive due to its carbon sp^2 hybridization, compared to the sp^3 hybridization of the CH_3 group in the CH_3CO moiety.

Differentiating between these two mechanisms may be fairly straightforward by reducing the photon energy until it is below the onset of the $CH_3CO + H$ radical channel, but still above the $CH_3 + HCO$ channel. When the total energy is insufficient for H atom loss, it should also be less likely for the H atom to move sufficiently far away from the CH_3CO core to begin roaming. If the bimodal CO distribution persists, the evidence for a roaming methyl group will be strengthened. But perhaps the most pressing need in the examination of these competing pathways is further trajectory calculations. The direct dynamics trajectories of Kurosaki and Yokohama were initiated at the TS for $CH_4 + CO$ formation, which surely biased these calculations against observation of any roaming trajectories. It may be that roaming mechanisms are more general than expected. The coordinated effort of theory and experiment will be required to clarify these mechanistic pathways.

F. $C_2H_3 + H$: Abstraction versus Addition–Elimination

The vinyl + hydrogen atom reaction is a rich example of competition between abstraction and addition–elimination pathways

$$C_2H_3 + H \rightarrow C_2H_2 + H_2 \quad \text{(abstraction)} \tag{23a}$$
$$\rightarrow [C_2H_4]^* \rightarrow C_2H_2 + H_2 \quad \text{(addition} - \text{elimination)} \tag{23b}$$

This reaction was investigated by Klippenstein and Harding [57] using multireference configuration interaction quantum chemistry (CAS + 1 + 2) to define the PES, variable reaction coordinate TST to determine microcanonical rate coefficients, and a one-dimensional (1D) master equation to evaluate the temperature and pressure dependence of the reaction kinetics. There are no experimental investigations of pathway branching in this reaction.

For the addition pathway, a three-dimensional (3D) PES was calculated by fixing the vinyl radical at its equilibrium geometry and positioning the H atom in a grid of points in and above the vinyl plane. Six 2D contour plots of this surface are shown in Fig. 10 that provide a useful visualization of the 3D surface. These plots show two barrierless addition pathways. Addition from the upper right in any panel has a larger angle of acceptance both in and out of plane, and is termed the frontside addition. Addition from the lower right is denoted backside addition.

Abstraction routes for the H atom attacking each hydrogen in the vinyl radical were also calculated. Referring to the atom positions in Fig. 10, a barrierless abstraction pathway was found for attack on the H atom in the lower left (i.e., cis to the CH hydrogen). Attack on the hydrogen in the upper left (i.e., trans to the CH hydrogen) has a barrier of $1.2 \, \text{kcal} \, \text{mol}^{-1}$. No path for abstraction could be found for attack on the CH hydrogen.

The lowest pathway for decomposition of the C_2H_4 adduct is a 1,1 elimination forming $H_2 + H_2CC$ (vinylidene) [58]. Vinylidene is then expected to rapidly isomerize to acetylene. The 1,1 elimination barrier is a tight TS, but can compete with the barrierless dissociation $C_2H_4 \rightarrow C_2H_3 + H$ because the energy of the elimination barrier lies $15 \pm 2 \, \text{kcal} \, \text{mol}^{-1}$ below the vinyl + H reactants [57].

In total, there are three barrierless entrance pathways: frontside and backside addition of hydrogen atom to vinyl, and abstraction of the cis hydrogen. All the pathways discussed so far are on the singlet PES. There are also addition pathways on the triplet surface, but these are impeded by small barriers and are not expected to be important compared to the singlet surface.

Because the three entrance pathways considered are barrierless, variational variable reaction coordinate TST [29] was employed to provide a more optimal treatment of the TS dividing surface as a function of energy. These calculations lead to the estimate that the abstraction pathway accounts for 9–10% of the total branching to $C_2H_2 + H_2$, with the addition–elimination pathway accounting for the remaining 90%. Furthermore, the frontside addition pathway is favored by a factor of ~ 2 compared to the backside addition pathway. At low pressure, stabilization by collisions into the C_2H_4 well is negligible, and all addition complexes will form $C_2H_2 + H_2$.

While there have been no experimental studies of the pathway branching to $C_2H_2 + H_2$, it is likely that these pathways could be effectively probed by both

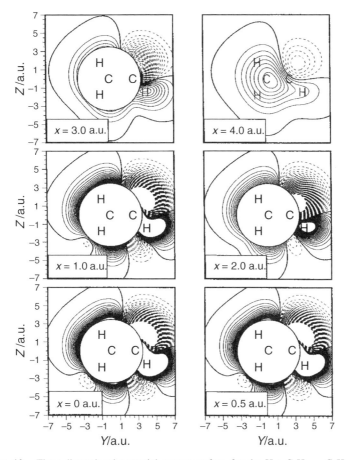

Figure 10. Three-dimensional potential-energy surface for the $H + C_2H_3 \rightarrow C_2H_4$ addition reaction. The lower left plot is taken in the symmetry plane of the vinyl radical. The other plots are taken in parallel planes at distances of 0.5–4.0 a.u. from the symmetry plane (1 a.u. = 0.52918 Å). Solid contours are positive, dashed contours are negative, and the zero-energy contour (defined to be the energy of the reactant asymptote) is shown with a heavy solid line. The contour increment is $1\,kcal\,mol^{-1}$. Reproduced from [57] by permission of the PCCP Owner Societies.

isotopic substitution and product state distribution measurements. Simultaneous application of the two approaches could be the most informative. In the case of the $C_2D_3 + H$ reaction, abstraction should produce only HD products. For the addition–elimination channel, if we assume a long-lived CD_2CDH complex, the products should be a 50:50 mixture of $D_2 + HCCD$ and $HD + DCCD$ (in the absence of a kinetic isotope effect). Observation of D_2 would then be strong evidence for addition–elimination.

In a crossed-molecular-beams experiment, the angular distributions could shed more light on the pathway branching. The direct abstraction process should be sufficiently rapid to produce backward scattered HD (with respect to the incoming H atom), while D_2 produced from addition–elimination through a long-lived complex should be forward–backward symmetric. Finally, the HD product formed by abstraction should contain significantly more vibrational excitation than the HD and D_2 formed by addition–elimination. The pathway for barrierless abstraction is essentially the limiting case of an early barrier exothermic reaction, in which translational energy in the entrance channel should be efficiently converted into HD vibration in the exit channel [59]. By contrast, the HD (or D_2) bond length at the transition state for 1,1 elimination from CD_2CDH is 0.832 Å, which is not drastically different than the equilibrium bond length $r_e = 0.742$ Å for isolated H_2. Therefore, measuring the translational energy distribution of the products, or directly probing the HD/D_2 vibrational distribution are other methods for separating the two pathways.

One further complication such experiments may face is the presence of roaming atom pathways in the decay of the highly excited CD_2CDH complex. If we assume that roaming can happen, a roaming H atom will always result in an HD product. If we assume that the D atoms are equally likely to roam and randomly abstract another hydrogen, then each roaming D atom will result in HD products one-third of the time. Therefore the roaming mechanism would produce a 50:50 mixture of HD/D_2. Assuming the dynamics will be similar to the $C_2D_3 + H$ abstraction channel, these HD and D_2 atoms would be highly vibrationally excited. While this possible pathway could complicate the interpretation of experimental results, it seems clear that the $C_2H_3 + H$ reaction is a promising candidate for detailed experimental exploration of multiple competing pathways.

G. $OH^- + CH_3F$: Avoiding the Minimum Energy Path

The reaction

$$OH^- + CH_3F \rightarrow CH_3OH + F^- \tag{24}$$

is an example of S_N2 nucleophilic substitution reactions of the general form

$$X^- + CH_3Y \rightarrow XCH_3 + Y^- \tag{25}$$

When X and Y are identical halogen atoms, the potential surface for this reaction is represented by a well for the frontside ion–dipole complex $[X \cdots CH_3Y]^-$, a TS for the $[X \cdots CH_3 \cdots Y]^-$ structure in which the methyl moiety is planar, and a backside ion–dipole complex $[XCH_3 \cdots Y]^-$ that leads to products. These

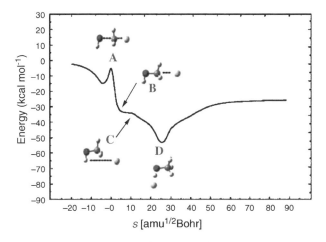

Figure 11. The minimum energy path of the OH⁻ + CH₃F reaction, not including zero-point energy. The four labeled structures are (A), the central barrier TS; (B), the nearly collinear backside well complex [HOCH₃ ··· F]⁻; (C) the transition of the F atom toward the OH moiety; (D) the hydrogen-bonded [CH₃OH ··· F⁻] structure. Reprinted from [63] with permission from the American Association for the Advancement of Science.

reactions are poorly described by statistical models, such as TS theory [60], and it is known that vibrational excitation of the reactants promotes the reaction more than translational excitation [61].

The OH⁻ + CH₃F potential energy surface [62] also shows a double-well structure, but in this case the well on the product side of the barrier represents not the traditional ion–dipole backside complex, but rather the hydrogen-bonded complex [CH₃OH ··· F⁻]. This well is the global minimum on the PES and lies on the minimum energy path toward products. Figure 11 shows the MEP for this reaction from an intrinsic reaction coordinate calculation by Sun et al. [63]. The normal backside well representing the [HOCH₃ ··· F]⁻ complex (B in Fig. 11) has been "subducted" by the deeper well of the hydrogen-bonded [CH₃OH ··· F⁻] structure (D in Fig. 11).

Direct dynamics trajectory calculations at the MP2/6-31+G* level of theory were then used to explore the reaction dynamics of this system [63]. Sixty-four trajectories were started from the central barrier shown at "A" in Fig. 11, with initial conditions sampled from a 300 K Boltzmann distribution. Of the 31 trajectories that moved in the direction of products, four trajectories followed the MEP and became trapped in the hydrogen-bonded [CH₃OH ··· F⁻] complex, with one trajectory forming products within the imposed 3-ps time limit. The remaining 27 product-bound trajectories followed a different pathway, with the F⁻ atom departing approximately collinearly along the OCF backbone at the TS. These trajectories avoided the MEP, dissociating directly to products without ever entering the hydrogen-bonded [CH₃OH ··· F⁻]

well. Approximately 90% of the products arise from this direct, non-MEP pathway.

To better understand the cause of this branching, Fig. 12 shows a 2D cut through the potential surface along the C—F bond and the FCO angle. The position of the central barrier TS is shown, with an FCO bond angle of 177.9° (i.e., nearly linear) at the saddle point. Moving down from the saddle point, the potential is initially steeply repulsive along the C—F bond, and relatively flat in the FCO bending coordinate. The flatness along the bending coordinate means there is little force pulling the trajectory toward the deep [CH$_3$OH $\cdots$ F$^-$] well, centered at $R_{C-F} \approx 3.0$ Å and $\angle$FCO $\approx 77°$. Therefore, trajectories from the TS tend to proceed directly toward products in a collinear fashion along path 1 in Fig. 12, quite oblivious to the MEP that is represented by path 2.

The surface in Fig. 12 demonstrates that there is little coupling between the C—F translation coordinate and the bending coordinate of the complex. Stated another way, the time scale for intramolecular vibrational redistribution between these coordinates is slow compared to the time scale for breaking the C—F bond. These conclusions are not obvious upon examination of the minimum energy path shown in Fig. 11, and indeed such diagrams, while generally instructive, can lead to improper conclusions because they hide the multidimensional nature of the true PES. A central assumption of statistical product distribution theories

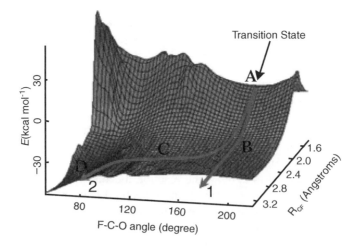

Figure 12. Two-dimensional cut through the potential surface for fragmentation of the transition state [OH $\cdots$ CH$_3$ $\cdots$ F]$^-$ complex as a function of the C—F bond length and the FCO angle. All other coordinates are optimized at each point of this PES. Pathway 1 is the direct dissociation, while pathway 2 leads to the hydrogen-bonded [CH$_3$OH $\cdots$ F$^-$] structure. The letter symbols correspond to configurations shown in Fig. 11. Reprinted from [63] with permission from the American Association for the Advancement of Science. (See color insert.)

is that IVR is fast compared to reaction. In light of the result that 90% of the products are formed by a non-MEP pathway, treatment of this reaction by statistical methods (which assume equilibration in the deep wells of the surface) may lead to incorrect conclusions regarding the nature of the reaction mechanism.

H. $CH_3 + O$: Two Pathways Around the Transition State

The reaction of a methyl radical with an oxygen atom has two main product channels

$$CH_3 + O \rightarrow H_3CO^* \rightarrow H + CH_2O \qquad\qquad (f_1 = 0.82 \pm 0.04) \quad (26a)$$
$$\rightarrow H_2 + HCO^* \rightarrow H_2 + H + CO \qquad (f_2 = 0.18 \pm 0.04) \quad (26b)$$

The second channel, producing CO, was first observed by Seakins and Leone [64], who estimated 40% branching to this channel. Later measurements by Fockenberg et al. [65] and Preses et al. [66] concluded the branching to CO is $\sim$18%. Note that decomposition of formaldehyde formed in reaction (26a) is not a possible source of CO due to the large barrier for formaldehyde decomposition. Marcy et al. [67] recently combined time-resolved Fourier spectroscopy experiments with direct dynamics classical trajectory calculations to examine the mechanism of the CO product channel. They observed two pathways for CO formation, neither of which involve crossing a TS.

The mechanism begins with formation of the methoxy intermediate (CH_3O) in a deep well on the PES, as shown in Fig. 13. Surprisingly, no TS could be located for the direct elimination of H_2 from methoxy. Although a path was found to $H + H_2 + CO$ through $CH_3O \rightarrow H_2COH \rightarrow H_2 + HOC \rightarrow H_2 + H + CO$, the highest barrier on this pathway is drastically higher than that for $H_2COH \rightarrow H_2CO + H$. It is unlikely that this pathway is significant, and no trajectories following this pathway were observed. The key question to explore in the formation of CO from methoxy $+ O$ is the pathway for $CH_3O \rightarrow H_2 + HCO$. Once HCO is formed with large amounts of internal energy, it simply dissociates to give $H + CO$.

As pointed out in the $OH^- + CH_3F$ reaction, 1D schematic potential energy diagrams may be misleading. To gain a better understanding of the PES topology, two 2D cuts through the 9D surface are shown in Figs. 14 and 15. In Fig. 14, two of the C—H bond lengths are varied over a grid of values and the other seven coordinates are optimized at each point. There are four wells and four TS saddle points on this diagram, which is symmetric about the diagonal. Note that the symmetric pathway along the diagonal connecting CH_3O to $H_aH_b + HCO$ goes over a maximum in the potential (that is not a first-order saddle point) at C—H distances of 1.6 Å.

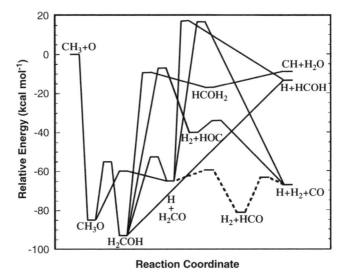

Figure 13. Potential energy diagram of stationary points of the $CH_3 + O$ reaction. The results are from CCSD(T)/aug-cc-pvtz//CCSD(T)/cc-pvdz calculations including zero point energy. Reprinted with permission from [67]. Copyright 2001 American Chemical Society.

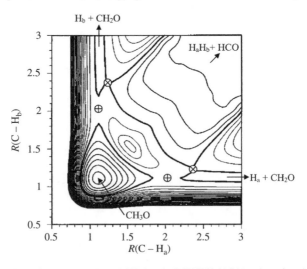

Figure 14. Potential energy surface of CH_3O at the B3LYP/6-31G* level as a function of two C–H bond lengths. All other coordinates are optimized at each point. The darkest contour lines represent the energy ($-66.0 \text{ kcal mol}^{-1}$ with respect to $CH_3 + O$) of the saddle point for $H + H_2CO \rightarrow H_2 + HCO$. The lightest contours are higher in energy and the darker contours are lower. The contour interval is 5 kcal mol^{-1}. The length scales are in Angstroms. The saddle point for $CH_3O \rightarrow H + H_2CO$ is denoted by $\oplus$ ($-63.4 \text{ kcal mol}^{-1}$). The saddle point for the abstraction $H + H_2CO \rightarrow H_2 + HCO$ is denoted by $\otimes$ ($-66.0 \text{ kcal mol}^{-1}$). Reprinted with permission from [67]. Copyright 2001 American Chemical Society.

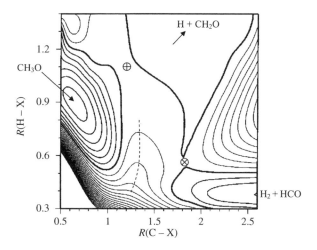

Figure 15. Potential energy surface of CH_3O at the B3LYP/6-31G* level as a function of $R(H-X)$ and $R(C-X)$, where X is the midpoint of two of the hydrogens. The saddle point for $CH_3O \rightarrow H + H_2CO$ is denoted by $\oplus$ ($-63.4\,kcal\,mol^{-1}$). The saddle point for the abstraction $H + H_2CO \rightarrow H_2 + HCO$ is denoted by $\otimes$ ($-66.0\,kcal\,mol^{-1}$). The dotted line denotes a prominent ridgeline. Other conventions are the same as in Fig. 14. Reprinted with permission from [67]. Copyright 2001 American Chemical Society.

A different 2D cut through the full PES is shown in Fig. 15, where the coordinates are described in the figure caption. The same saddle points can be seen as in Fig. 14. In Fig. 15, the minimum energy path for the critical transformation $CH_3O \rightarrow H_2 + HCO$ takes a circuitous route. It begins in the CH_3O well, moves up and right over the saddle point ($\oplus$) toward the $H + CH_2O$ well in the upper right corner. The MEP then continues downward in Fig. 15, over the saddle point ($\otimes$), and into the $H_2 + HCO$ well. The dotted line in Fig. 15 denotes a prominent ridgeline that corresponds to the maximum in Fig. 14 at C—H distances of 1.6 Å. This ridgeline blocks a direct path from CH_3O to $H_2 + HCO$.

Because no TS for $CH_3O \rightarrow H_2 + HCO$ could be located, Marcy et al. [67] carried out direct dynamics trajectory simulations to determine the mechanism for formation of CO. These calculations were done at the B3LYP/6-31G* level due to its semiquantitative description at a relatively modest computational cost. The trajectories were initiated from a dividing surface characterized by a C—O separation of 2.65 Å, which is approximately the distance between C and O at the TS for the initial addition to form CH_3O. From the 240 trajectories propagated with total angular momentum $J = 50$, the $H_2 + HCO$ product branching fraction was found to be 0.15 at room temperature, in good agreement with the experimental value of 0.18 ± 0.04 from Preses et al. [66] Marcy et al. [67] concluded that the ridgeline shown in Fig. 15 that blocks the direct pathway for $CH_3O \rightarrow H_2 + HCO$ is the main reason the product branching to the CO channel is as small as it is.

Three sample trajectories forming $H_2 + HCO$ are shown in Fig. 16. Each trajectory (a–c) is plotted twice; once on the 2D PES of Fig. 14 (right column) and

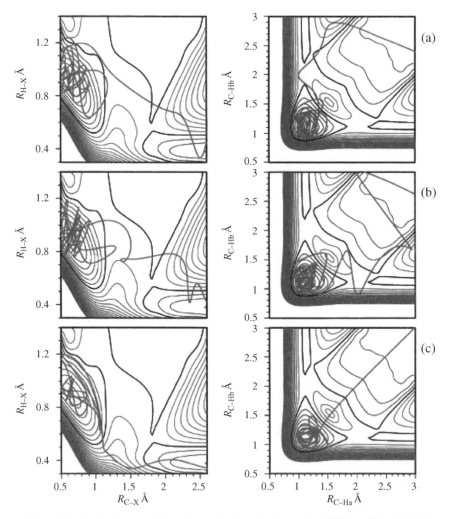

Figure 16. Projections onto 2D surfaces of trajectories (in green) of $CH_3O \rightarrow H_2 + HCO$. The left column is a projection onto the surface of Fig. 15. The right column is a projection onto the surface of Fig. 14. The black contour represents the saddle point energy for the $H + H_2CO \rightarrow H_2 + HCO$ reaction. Blue contours are lower in energy; red contours are higher. Reprinted with permission from [67]. Copyright 2001 American Chemical Society. (See color insert.)

once on the 2D PES of Fig. 15 (left column). Of the trajectories that form $H_2 + HCO$, two pathways are observed. The first two trajectories [Figs. 16(a and b)] demonstrate the first pathway. Here the trajectory orbits for some time in the methoxy well, then leaves the well passing essentially through the transition state for $CH_3O \rightarrow H + H_2CO$. However, the trajectories do not continue into the formaldehyde valley, but are steered near the TS for $H + H_2CO \rightarrow H_2 + HCO$ (as shown in the right column of Fig. 16). Note that these trajectories lead to substantial excitation of H_2 vibrational motion. These two trajectories were termed "frustrated H loss trajectories" by Marcy et al. [67] They represent, however, the same roaming atom mechanism [68] as observed in $H_2CO \rightarrow H_2 + CO$ photodissociation, characterized by an H atom on a very flat potential that finds its way to an abstraction channel yielding H_2.

The trajectory in Fig. 16(c) is an example of the second pathway toward $H_2 + HCO$. This trajectory is quite striking in that it leaves the CH_3O well by passing over the ridgeline [Fig. 16(c) left column] directly to products. The same trajectory plotted in the right column demonstrates that this trajectory completely avoids all the saddle points on this surface, and therefore also avoids the MEP. This pathway leads to much lower vibrational excitation of H_2 than observed in the pathway of Figs. 16(a and b). Although one might reasonably speculate that this especially interesting trajectory is also especially rare, in fact the majority of trajectories forming $H_2 + HCO$ follow this pathway. The reason these trajectories follow this pathway over a ridge is not well understood, but may be related to effective orbital angular momentum barriers [67].

Knyazev has developed a statistical method based on RRKM theory to treat reactions that pass over a ridge [33]. The method places the dividing surface along the PES ridge separating reactants from products. One assumption of this method is that the Hamiltonian describing motion along the ridge is separable from all other degrees of freedom. However, the exact definition of the ridge dividing surface is not unique. Using this method, he predicts the branching fraction of the CO product channel to be 4–7%, depending on the potential energy surface used.

In summary, there appears to be no transition state for the $CH_3 + O \rightarrow H_2 + H + CO$ product channel, but nevertheless this channel represents about one-fifth of the total products. There are two pathways to these products. The main pathway involves a direct passage from CH_3O to $H_2 + HCO$ over a ridgeline in the potential energy surface. This ridgeline can also be viewed as a saddle point of order >1. The minor pathway approximately follows the sequential pathway $CH_3O \rightarrow H + H_2CO \rightarrow H_2 + HCO$ passing near the transition states for these individual steps. This pathway, termed a "frustrated H loss pathway" in the original chapter, is a roaming atom channel, and represents the first reported case of this mechanism in a bimolecular reaction.

I. H_2CO Photodissociation: Participation of Multiple Electronic States

The photodissociation of formaldehyde has two product channels

$$CH_2O + h\nu \rightarrow H_2 + CO \tag{27a}$$

$$\rightarrow H + HCO \tag{27b}$$

Section V.D described the competition of two pathways in the $H_2 + CO$ molecular channel. There are also multiple pathways to the radical channel producing $H + HCO$. In all cases, highly vibrationally excited CH_2O is prepared by laser excitation via the $S_1 \leftarrow S_0$ transition. In the case of the radical channel discussed in this section, multiple pathways arise because of a competition between internal conversion $(S_1 \rightarrow S_0)$ and intersystem crossing $(S_1 \rightarrow T_1)$, followed by evolution on these electronic states to the ground-state $H + HCO$ product channel. Both electronic states S_0 and T_1 correlate adiabatically with $H + HCO$ products, as shown in Fig. 7.

Because the pathway to $H + HCO$ on S_0 is barrierless (with a loose TS), whereas the pathway on T_1 has an exit barrier (tight TS), the dissociation dynamics of the two pathways can be expected to differ markedly. Measuring the translational energy release and the product state distributions of the HCO fragment are therefore appropriate experimental techniques for exploring this competition.

The pathway branching hinges on the rates of nonradiative transitions out of S_1. The efficiency of internal conversion (IC) versus intersystem crossing (ISC) depends on the coupling between the particular $S_1(\nu, J)$ state excited and nearby states in the S_0 and T_1 manifolds, respectively. The IC process is facilitated by second-order nonadiabatic vibronic interactions that couple S_1 with S_0 [69], whereas the ISC process coupling S_1 with T_1 arises from second-order spin–orbit coupling [70]. The competition between these two pathways has been examined experimentally by Chuang et al. [71] and Dulligan et al. [72], who concluded that the flux through each pathway fluctuates substantially with excitation wavelength. In the energy range near the T_1 barrier, the density of states in $T_1(\sim 0.3/cm^{-1})$ is substantially smaller than the density of S_0 states $(\sim 100/cm^{-1})$ because T_1 lies $25,194\,cm^{-1}$ higher than S_0. The fact that ISC can compete with IC implies that the spin–orbit couplings driving the former are significantly larger than the vibronic coupling driving the latter. Note that the triplet pathway is accessible below the T_1 barrier due to quantum mechanical tunneling.

Recently Valachovic et al. [51] studied the competition between S_0 and T_1 pathways as a function of excitation energy in the range $1103-2654\,cm^{-1}$ above the $H + HCO$ threshold using photofragment ion imaging and H-atom Rydberg time-of-flight spectroscopy (HRTOF). The energy range encompasses levels well below and above the T_1 barrier. Although the energy of the barrier on the T_1 surface is only known to within $\sim 200\,cm^{-1}$, this accuracy is sufficient to ensure that excitation below the barrier energy will yield H + HCO produced *only* via the S_0 state. The product state distributions in this case serve as a reference for the S_0 role when analyzing data at energies where the triplet state may also contribute.

Because dissociation on S_0 is barrierless, the product state distributions should be well approximated by statistical theories, especially when the excess energy is small, as in the Valachovic study. Product state distributions arising from the S_0 pathway should be characterized by small translational energy release, but significant rovibrational excitation of HCO. This signature is demonstrated in the top panel of Fig. 17, which shows a HRTOF spectrum with

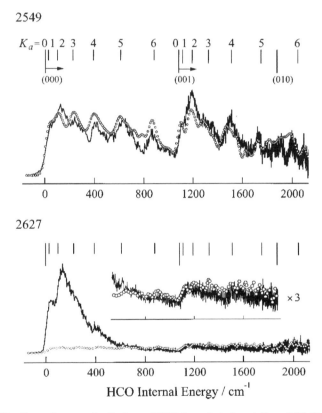

Figure 17. Internal energy distributions of HCO from photodissociation of CH_2O at 2549 cm^{-1} (upper panel) and 2627 cm^{-1} (lower panel) above the threshold for the H + HCO channel. The HCO vibrational thresholds are labeled with their quantum numbers, and combs label the K_a stack thresholds. The open circles show predictions of the SSE/PST model. The upper panel is indicative of an S_0 dominant pathway. In the lower panel, T_1 is dominant, but S_0 structure can still be observed. Reprinted with permission from [51]. Copyright 2000, American Institute of Physics.

excitation energy 2549 cm^{-1} above the H + HCO threshold. Note that the distribution is plotted as a function of HCO internal energy. This distribution shows both vibrational excitation of HCO and significant a-axis rotational excitation extending at least to $K_a = 6$. This distribution is reproduced remarkably well by a separate statistical ensembles–phase space theory (SSE/PST) [31] calculation also shown in Fig. 17. Note that this excitation energy lies ~ 629 cm^{-1} above the T_1 barrier, and yet the distribution appears to be almost perfectly represented by a pure singlet pathway.

The lower panel of Fig. 17 shows another HRTOF spectrum at slightly higher excitation energy. This spectrum is dominated by HCO molecules in their ground vibrational state with only moderate rotational excitation. This signature is assigned to the triplet pathway, because the T_1 barrier is expected to channel energy efficiently into translation, leaving little rovibrational excitation in HCO. The SSE/PST simulation (appropriate for the singlet pathway only) for the 2627-cm^{-1} spectrum agrees well with the data for energies >800 cm^{-1}. Because the triplet pathway signature is confined to low HCO internal energies, it is possible to discern a small contribution of the singlet pathway even when the triplet pathway dominates. The converse is not true because the S_0 signature populates essentially all energetically accessible internal states of HCO.

In general, the singlet mechanism dominates for energies below the T_1 barrier, although when a close match between the excited S_1 level and a T_1 resonance exists, the triplet contribution can be substantial. Excitation in the first ~ 600 cm^{-1} above the T_1 barrier is characterized by sporadic changes in the singlet–triplet pathway branching. The fluctuations arise because the triplet energy levels are sparsely and irregularly spaced, but when S_1/T_1 levels are close in energy, the stronger ISC coupling strength leads to domination by the triplet pathway. For excitation energies >600 cm^{-1} above T_1 barrier, the triplet pathway dominates, presumably due to sufficient state density of triplet levels that a strong S_1/T_1 overlap is common. However, the singlet pathway always persists at a low level because, despite the weaker S_1/S_0 coupling strength per S_0 state, the density of S_0 states remains high.

These experiments use the product state distribution technique to allow a qualitative characterization of the competition between multiple electronic states. In contrast to the pathway competition in the molecular channel of formaldehyde (Section V.D), where the correlated product state distributions delineate the two channels quite cleanly, it will likely more often be the case that the product state distribution method allows only qualitative separation, due to overlapping distributions. Nevertheless, such experiments provide critical insight into pathway competition.

Finally, note that the competition between IC and ISC is believed to be a general feature in the photochemistry of ketones RCOR′ [73], of which CH$_2$O is the simplest member. There is evidence that pathway branching to the S_0 state is enhanced when the R, R' ligands are aromatic (as in diphenyl ketone) or quasiaromatic (as in dicyclopropyl ketone) [74]. While pathway competition may be common in larger ketones, the detailed characterization of this competition, and the factors that control it, are not as yet understood.

J. H₂O Photodissociation: Chemical "Double Slits"

When two pathways on the potential energy surface lead to one exit channel, the wavelike nature of molecules should manifest itself by constructive and destructive interference between the outgoing wave functions as they recombine in the common exit channel. This phenomenon is analogous to Thomas Young's observation in 1801 that light passed through a pair of apertures creates interference fringes when the light is recombined on a screen at a particular point. This phenomenon gave the first firm proof that light could be described as a wave. For molecular interference to be observed, the initial conditions of the system must be coherently prepared, that is, with a well-defined phase of the molecular wave function. If the lengths of two reaction pathways differ as a function of some observable (e.g., final rotational state), we should observe constructive and destructive interference in this observable. For a bimolecular reaction, this level of control over initial conditions is essentially impossible. However, molecules cooled in a supersonic expansion to one or a few quantum states, and excited by coherent radiation to a dissociative state, can show this behavior. Dixon and co-workers [75] demonstrated this phenomenon in the dissociation of H₂O at 121.6 nm. In related work, Rakitzis and co-workers [76] showed that photodissociation of ICl, through excitation to multiple electronic states, leads to interference between multiple pathways, as observed in the electronic angular momentum helicity.

The combined experimental and theoretical investigation of Dixon et al. [75] applied the H-atom Rydberg time-of-flight method to measure the translational energy distribution of H atoms from the photodissociation

$$H_2O(\tilde{X}^1A_1) + h\nu \rightarrow H_2O(\tilde{B}^1A_1) \rightarrow H_2O(\tilde{X}^1A_1) \rightarrow H(^2S) + OH(X^2\Pi) \quad (28)$$

The initial excitation at 121.6 nm accesses the $\tilde{B}(^1A_1)$ electronic state of H₂O, which has a linear equilibrium geometry. Several product channels are possible from this point [77], but the only one discussed here is $H(^2S) + OH(X^2\Pi)$, which arises from internal conversion back to the ground electronic state. The internal conversion process is facilitated by two conical intersections, at the linear H—OH geometry (⊗a in Fig. 18), and the linear H—HO geometry (⊗b in Fig. 18), respectively, as shown in Fig. 18. These conical intersections "funnel" population from the $\tilde{B}$ to the $\tilde{X}$ state. This process leads to highly rotationally excited OH ($N_{max} = 45$) due to the strong force applied along the bending

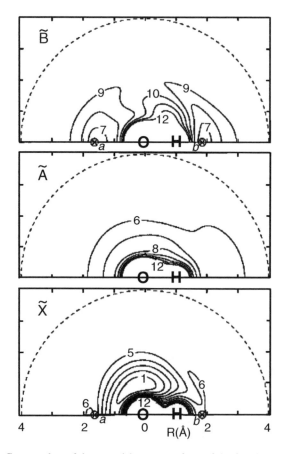

Figure 18. Contour plots of the potential energy surfaces of the first three electronic states of H_2O. The polar plots depict the movement of one H atom around OH with an OH bond length fixed at 1.07 Å. Energies are in electron volts relative to the ground electronic state. The $\tilde{X}$ and $\tilde{B}$ states are degenerate at the conical intersection (denoted by $\otimes$) in the (a) H—OH geometry and (b) H—HO geometry. Reprinted from [75] with permission from the American Association for the Advancement of Science.

coordinate as H_2O moves from its bent geometry in the Franck–Condon region of the $\tilde{B}$ state to the conical intersections at 0 and 180°.

The H_2O molecules are cooled in a supersonic expansion to a rotational temperature of $\sim 10\,K$ before photodissociation. The evidence for pathway competition is an odd–even intensity alteration in the OH product state distribution for rotational quantum numbers $N = 33-45$. This intensity alternation is attributed to quantum mechanical interference due to the N-dependent phase shifts that arise as the population passes through the two different conical intersections.

Dixon et al. [75] use a simple quantum mechanical model to predict the rotational quantum state distribution of OH. As discussed by Clary [78], the component of the molecular wave function that describes dissociation to a particular OH rotational state N is approximated as

$$\Psi_N = \psi_1 + m_N\psi_2 \qquad (29)$$

where ψ_1 and ψ_2 are the component wave functions exiting the $\tilde{B}$ state at H—OH and H—HO conical intersections, respectively. The quantum number m_N denotes the symmetry of the rotational wave function of OH with respect to the two geometries H—OH and H—HO. It takes the values ± 1 depending on whether the final OH rotational quantum number N is even or odd. The probability of producing the final OH rotational state N is proportional to

$$|\Psi_N|^2 = (\psi_1)^2 + (\psi_2)^2 + 2m_N\psi_1\psi_2 \qquad (30)$$

The third term on the right-hand side is the interference term that changes sign with even or odd N. This term may be more or less significant depending on the magnitude of the product $\psi_1\psi_2$. This model qualitatively agrees with the experimental results, predicting the odd–even alternation of intensity with N, though it overestimates the signal contrast. An overestimation of the interference contrast is expected, since the calculation assumes all H_2O molecules are in their rotationless state, while in the experiment $\sim76\%$ of the molecules are in rotational levels other than the ground state. Increasing the number of quantum states in the initial conditions will increasingly average out any quantum mechanical interferences.

Figure 19 shows the outgoing scattering waves from the simple model plotted on the same coordinate system as in Fig. 18. The top panel shows significant wave function amplitude moving from the Franck–Condon region (shown in gray) toward both conical intersections. In the lowest panel of Fig. 19, strong interference is seen between the waves emerging from the two conical intersections on the ground electronic state surface. Asymptotically, this interference determines the even–odd alternation in N. It is important to note that no even–odd alternation was observed in similar HRTOF experiments conducted with an initial H_2O rotational temperature of ~100 K [77]. Quantum interference from multiple pathways provides an exquisite test of our understanding of the PES and the dynamical calculations performed on it, but this method will probably only be feasible for very small photodissociation systems at very cold temperatures (to reduce the number of initial quantum states populated). An alternative method to avoid averaging out quantum effects is to perform true quantum state-to-state measurements, which are, unfortunately, quite difficult to accomplish.

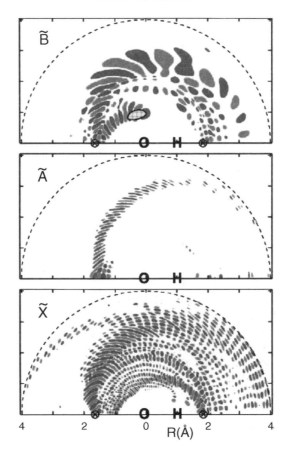

Figure 19. Outgoing waves on the first three electronic states of H_2O following excitation of the $(\tilde{B}^1A_1) \leftarrow (\tilde{X}^1A_1)$ transition. Green and red lobes have positive and negative amplitudes, respectively. The gray-filled contour on the $\tilde{B}$ surface represents the Franck–Condon region of excitation from the ground state. Reprinted from [75] with permission from the American Association for the Advancement of Science. (See color insert.)

VI. CONCLUSIONS AND FUTURE OUTLOOK

In the study of reaction mechanisms, it is almost always easier to detect final products than reaction intermediates. Although it is often the case that each detected product channel represents one pathway on the PES, this chapter demonstrates many examples where this assumption fails.

In some cases presented here, each pathway follows a "traditional" route through wells and/or over saddle points toward products. In these cases, a competition arises because the rate-limiting energy and entropy barriers of the two pathways are similar in magnitude. The effect of entropy can lead to temperature dependence of pathway branching ratios. These cases are labeled as *statistical pathway branching*, because the concepts underlying TS theory and other statistical models can describe their pathway branching with equal fidelity as single pathway processes, given accurate descriptions of the stationary points on the PES. The examples A, B, C, F, and I of Section V fall into this category.

In other cases, competition arises between a traditional pathway that follows the minimum energy path, and an alternative pathway that avoids it. These cases are labeled as *dynamical pathway branching*, because statistical methods, which assume reactions occur via the MEP and over saddle points, cannot treat these nontraditional pathways. Instead, these systems must be explored theoretically by dynamical methods that rely on a potential energy surface accurate not only at stationary points, but also at many intervening nuclear configurations. The shape of the PES really matters. The examples D, E, G, H, and perhaps also F of Section V are in this category. In the $OH^- + CH_3F$ case, forces arising from the shape of the PES encourage pathways that avoid the minimum energy path. In the roaming mechanism cases, the roaming fragment has almost enough energy to dissociate and orbits the other fragment until an abstraction channel is (presumably randomly) accessed. The key factor in common is the flatness of the PES in at least one coordinate, which leads to inefficient coupling (slow IVR) between this coordinate and the rest of the system, and hence the possibility that the non-MEP trajectory does not collapse to the MEP.

For both statistical and dynamical pathway branching, trajectory calculations are an indispensable tool, providing qualitative insight into the mechanisms and quantitative predictions of the branching ratios. For systems beyond four or five atoms, direct dynamics calculations will continue to play the leading theoretical role. In any case, predictions of reaction mechanisms based on examinations of the potential energy surface and/or statistical calculations based on stationary point properties should be viewed with caution.

The realization that roaming mechanisms play an important role in several unimolecular and bimolecular reactions makes it worthwhile to discuss when such channels may be expected to contribute. Most examples of roaming mechanisms to date involve highly vibrationally excited closed-shell singlet species (H_2CO, CH_3CHO, perhaps C_2H_4) that can decay to closed-shell products ($H_2 + CO, CH_4 + CO, H_2 + C_2H_2$). In all these cases, the critical point in the roaming pathway is access to a barrierless radical + radical abstraction channel (e.g., $H + HCO \rightarrow H_2 + CO$), in which the H atom being extracted is only weakly bonded ($D_0 < 30\,\text{kcal mol}^{-1}$) to the radical core. Furthermore, the energetic threshold to form these radical + radical products is similar to the threshold for forming the closed-shell products. The energized closed-shell molecule can be formed either by photoexcitation of the ground-state molecule, or by addition in a radical + radical reaction.

Generalizing from these observations, it seems reasonable that all $H+C_nH_{2n+1}$ reactions are candidates for roaming mechanisms. The alkyl radical can add an H atom to form an energized alkane adduct. If one of the H atoms in the adduct gains almost enough energy to dissociate, it can orbit the alkyl core at fairly long distances. The alkyl radical core has weak C—H bonds on carbon atoms adjacent to the carbon from which the roaming atom departed. If the roaming atom finds one of these weak C—H bonds, it can undergo a barrierless abstraction to form H_2 + an alkene. A similar route will also be possible for H atoms reacting with C_nH_{2n-1} radicals and other less saturated radicals. Such mechanisms could aptly be named addition-roaming–abstraction pathways.

However, as the $CH_3 + O$ reaction shows [67], the roaming atom mechanism is not limited to closed-shell adducts, nor is it limited to abstraction pathways that are barrierless (see Fig. 13). Finally, the photodissociation of CH_3CHO implies that the roaming fragment may not be limited to H atoms [55]. It remains to be seen how universal roaming mechanisms are, but they have the potential to be competing pathways in a large number of reactions.

The study of multiple pathways leading to a single product channel provides a stringent test of our understanding of the potential energy surface and the calculations that use it to predict reaction outcomes. Although there are not many examples to date of pathway competitions, the increasing prominence of such systems, coupled with advances in experiment and theory that facilitate their study, promises a rich future in this normally hidden facet of reaction mechanisms.

Acknowledgments

The author thanks Dr. Larry Harding, Dr. Stephen Klippenstein, Dr. Jim Miller, Dr. Craig Taatjes, Dr. David Chandler, Dr. Ward Thompson, and Dr. Joel Bowman for helpful discussions. This work is supported by the Division of Chemical Sciences, Geosciences, and Biosciences, the Office of Basic Energy Sciences, the U. S. Department of Energy. The work at Argonne was supported under DOE Contract Number W-31-109-ENG-38. Sandia is a multiprogram laboratory operated by Sandia Corporation, a Lockheed Martin Company, for the National Nuclear Security Administration under contract DE-AC04-94-AL85000.

References

1. J. I. Steinfeld, J. S. Francisco, and W. L. Hase, *Chemical Kinetics and Dynamics*, Prentice Hall, New Jersey, 1999.

2. R. D. Levine and R. B. Bernstein, *Molecular Reaction Dynamics and Chemical Reactivity*, Oxford University Press, 1987.

3. G. E. Hall and P. L. Houston, *Annu. Rev. Phys. Chem.* **40**, 375 (1989).

4. A. Stolow, A. E. Bragg, and D. M. Neumark, *Chemical Reviews* **104**, 1719 (2004).

5. D. M. Neumark, *Phys. Chem. Chem. Phys.* **7**, 433 (2005).

6. W. J. Hehre, L. Radom, P. V. Schleyer, and J. A. Pople, *Ab Initio Molecular Orbital Theory*, John Wiley & Sons, Inc., Hoboken, 1986.

7. T. Baer and W. L. Hase, *Unimolecular Reaction Dynamics*, Oxford University Press, New York, 1996.

8. D. Townsend, S. A. Lahankar, S. K. Lee, S. D. Chambreau, A. G. Suits, X. Zhang, J. Rheinecker, L. B. Harding, and J. M. Bowman, *Science* **306**, 1158 (2004).

9. S. R. Lin, S. C. Lin, Y. C. Lee, Y. C. Chou, I. C. Chen, and Y. P. Lee, *J. Chem. Phys.* **114**, 160 (2001).

10. K. M. Ervin, J. Ho, and W. C. Lineberger, *J. Chem. Phys.* **91**, 5974 (1989).

11. K. Fukui, *Acc. Chem. Res.* **14**, 363 (1981).

12. C. Gonzalez and H. B. Schlegel, *J. Chem. Phys.* **90**, 2154 (1989).

13. M. Zyrianov, Th. Droz-Georget, A. Sanov, and H. Reisler, *J. Chem. Phys.* **105**, 8111 (1996).

14. S. W. Benson, *The Foundations of Chemical Kinetics*, Krieger Publishing, FL, 1982.

15. R. Schinke, S. Y. Grebenshchikov, M. V. Ivanov, and P. Fleurat-Lessard, *Annu. Rev. Phys. Chem.* **57**, 625 (2006).

16. D. A. McQuarrie, *Statistical Mechanics*, Harper Collins, New York, 1976.

17. C. Murray and A. J. Orr-Ewing, *Int. Rev. Phys. Chem.* **23**, 435 (2004).

18. J. I. Cline, K. T. Lorentz, E. A. Wade, J. W. Barr, and D. W. Chandler, *J. Chem. Phys.* **115**, 6277 (2001).

19. A. H. Zewail, *J. Phys. Chem. A*, **104**, 5660 (2000).

20. D. R. Cyr and C. C. Hayden, *J. Chem. Phys.* **104**, 771 (1996).

21. I. Fischer, D. M. Villeneuve, M. J. J. Vrakking, and A. Stolow, *J. Chem. Phys.* **102**, 5566 (1995).

22. J. A. Miller and S. J. Klippenstein, S. H. Roberston, *Proc. Combust. Inst.* **28**, 1479 (2000).

23. H. Eyring, *J. Chem. Phys.* **3**, 107 (1935).

24. M. G. Evans and M. Polanyi, *Trans. Faraday Soc.* **31**, 875 (1935).

25. R. A. Marcus and O. K. Rice, *J. Phys. Colloid Chem.* **55**, 894 (1951); R. A. Marcus, *J. Chem. Phys.* **20**, 359 (1952).

26. D. G. Truhlar, B. C. Garrett, and S. J. Klippenstein, *J. Phys. Chem.* **100**, 12771 (1996).

27. D. G. Truhlar and B. C. Garrett, *Acc. Chem. Res.* **13**, 440 (1980).

28. M. Quack and J. Troe, *Ber. Bunsenges. Phys. Chem.* **78**, 240 (1974).

29. S. J. Klippenstein, *J. Chem. Phys.* **96**, 367 (1992).

30. P. Pechukas and J. C. Light, *J. Chem. Phys.* **42**.

31. C. Wittig, I. Nadler, H. Reisler, M. Noble, J. Catanzarite, and G. Radhakrishnan, *J. Chem. Phys.* **83**, 5581 (1985).

32. D. H. Mordaunt, D. L. Osborn, and D. M. Neumark, *J. Chem. Phys.* **108**, 2448 (1998).

33. V. D. Knyazev, *J. Phys. Chem. A* **106**, 8741 (2002).

34. R. Car and M. Parrinello, *Phys. Rev. Lett.* **55**, 2471 (1985).

35. K. Bolton, W. L. Hase, and G. H. Peslherbe, *Multidimensional Molecular Dynamics Methods*, D. L. Thompson, (ed.), World Scientific, London, 1998, pp. 143–189.

36. M. J. T. Jordan, K. C. Thompson, and M. A. Collins, *J. Chem. Phys.* **102**, 5647 (1995).

37. F. H. Field and F. W. Lampe, *J. Am. Chem. Soc.* **80**, 5587 (1958).

38. C. Berg, W. Wachter, T. Schindler, C. Kronseder, G. Niedner-Schatteburg, V. E. Bondybey, and Z. Herman, *Chem. Phys. Lett.* **216**, 465 (1993).

264 DAVID L. OSBORN

39. M. Farnik, Z. Dolejsek, Z. Herman, and V. E. Bondybey, *Chem. Phys. Lett.* **216**, 458 (1993).
40. P. M. Esteves, C. J. A. Mota, A. Ramirez-Solis, and R. Hernandex-Lamondeda, *J. Am. Chem. Soc.* **120**, 3213 (1998).
41. D. D. Nelson and M. S. Zahniser, *J. Phys. Chem.* **98**, 2101 (1994).
42. A. Sinha, E. R. Lovejoy, and C. J. Howard, *J. Chem. Phys.* **87**, 2122 (1987).
43. S. C. Herndon, P. W. Villalta, D. D. Nelson, J. T. Jayne, and M. S. Zahniser, *J. Phys. Chem. A* **105**, 1583 (2001).
44. S. J. Klippenstein, J. A. Miller, and L. B. Harding, *Proc. Combust. Inst.* **29**, 1209 (2002).
45. D. L. Osborn, *J. Phys. Chem.. A* **107**, 3728 (2003).
46. B. Ruscic *et al.*, *J. Phys. Chem. A*, 2002, **106**, 2727.
47. P. Zou and D. L. Osborn, *Phys. Chem. Chem. Phys.* **6**, 1697 (2004).
48. C. B. Moore and J. C. Weisshaar, *Annu. Rev. Phys. Chem.* **34**, 525 (1983).
49. A. C. Terentis and S. H. Kable, *Chem. Phys. Lett.* **258**, 626 (1996).
50. R. D. van Zee, M. F. Foltz, and C. B. Moore, *J. Chem. Phys.* **99**, 1664 (1993).
51. L. R. Valachovic, M. F. Tuchler, M. Dulligan, Th. Droz-Georget, M. Zyrianov, A. Kolessov, H. Reisler, and C. Wittig, *J. Chem. Phys.* **112**, 2752 (2000).
52. S. A. Lahankar, S. D. Chambreau, D. Townsend, F. Suits, J. Farnum, X. Zhang, J. M. Bowman, and A. G. Suits, *J. Chem. Phys.* **125**, 044303 (2006).
53. X. Zhang, J. L. Rheinecker, and J. M. Bowman, *J. Chem. Phys.* **122**, 114313 (2005).
54. X. Zhang, S. Zou, L. B. Harding, and J. M. Bowman, *J. Phys. Chem. A* **108**, 8980 (2004).
55. P. L. Houston and S. H. Kable, *Proc. Natl. Acad. Sci. USA* **103**, 16079 (2006).
56. Y. Kurosaki and K. Yokohama, *J. Phys. Chem. A* **106**, 11415 (2002).
57. S. J. Klippenstein and L. B. Harding, *Phys. Chem. Chem. Phys.* **1**, 989 (1999).
58. J. H. Jensen, K. Morokuma, and M. S. Gordon, *J. Chem. Phys.* **100**, 1981 (1994).
59. J. C. Polanyi, *Acc. of Chem. Res.* **5**, 161 (1972).
60. W. L. Hase, H. Wang, and G. H. Peslherbe, *Adv. Gas Phase Ion Chem.* B3B, 125 (1998).
61. V. F. DeTuri, P. A. Hintz, and K. M. Ervin, *J. Phys. Chem. A* **101**, 5969 (1997).
62. J. M. Gonzales, R. S. Cox, S. T. Brown, W. D. Allen, and H. F. Schaefer, *J. Phys. Chem. A* **105**, 11327 (2001).
63. L. Sun, K. Song, and W. L. Hase, *Science* **296**, 875 (2002).
64. P. W. Seakins and S. R. Leone, *J. Phys. Chem.* **96**, 4478 (1992).
65. C. Fockenberg, G. E. Hall, J. M. Preses, T. J. Sears, and J. T. Muckerman, *J. Phys. Chem. A* **103**, 5722 (1999).
66. J. M. Preses, C. Fockenberg, and G. W. Flynn, *J. Phys. Chem. A* **104**, 6758 (2000).
67. T. P. Marcy, R. R. Diaz, D. Heard, S. R. Leone, L. B. Harding, and S. J. Klippenstein, *J. Phys. Chem. A* **105**, 8361 (2001).
68. S. J. Klippenstein and L. B. Harding, private communication.
69. R. G. Miller and E. K. C. Lee, *J. Chem. Phys.* **68**, 4448 (1979).
70. D. J. Clouthier and D. A. Ramsay, *Annu. Rev. Phys. Chem.* **34**, 31 (1983).
71. M. C. Chuang, M. F. Foltz, and C. B. Moore, *J. Chem. Phys.* **87**, 3855 (1987).
72. M. J. Dulligan, M. F. Tuchler, J. Zhang, A. Kolessov, and C. Wittig, *Chem. Phys. Lett.* **276**, 84 (1997).

73. N. J. Turro, *Modern Molecular Photochemistry*, Benjamin/Cummings Publication Company, Menlo Park, NJ, 1978; J. Michel and V. Bonacic-Koutecky, *Electronic Aspects of Organic Photochemistry*, Wiley–Interscience, New York, 1990.

74. S. M. Clegg, B. F. Parsons, S. J. Klippenstein, and D. L. Osborn, *J. Chem. Phys.* **119**, 7222 (2003).

75. R. N. Dixon, D. W. Hwang, X. F. Yang, S. Harich, J. J. Lin, and X. Yang, *Science* **285**, 1249 (1999).

76. T. P. Rakitzis, S. A. Kandel, A. J. Alexander, Z. H. Kim, and R. N. Zare, *Science* **281**, 1346 (1998).

77. D. H. Mordaunt, M. N. R. Ashfold, and R. N. Dixon, *J. Chem. Phys.* **100**, 7360 (1994).

78. D. C. Clary, *Science* **285**, 1218 (1999).

PHOTOELECTRON CIRCULAR DICHROISM IN CHIRAL MOLECULES

IVAN POWIS

School of Chemistry, University of Nottingham, Nottingham NG7 2RD, UK

CONTENTS

Advances in Chemical Physics, Volume 138, edited by Stuart A. Rice
Copyright © 2008 John Wiley & Sons, Inc.

I. INTRODUCTION

Symmetry breaking associated with chiral phenomena is a theme that recurs across the sciences—from the intricacies of the electroweak interaction and nuclear decay [1–3] to the environmentally influenced dimorphic chiral structures of microscopic planktonic foraminifera [4, 5], and the genetically controlled preferential coiling direction seen in the shells of snail populations [6, 7].

At the molecular level, considerable capital, both intellectual and financial, is invested in the study of chiral phenomena. Perhaps the question of broadest scope concerns the phenomenon of homochirality—the observation that all terrestrial life displays a unique invariant handedness of its molecular building blocks, such as the amino acids. An important school of thought argues that an understanding of the origin of life is inseparable from an understanding of the homochirality that accompanies it [8, 9]. If homochirality is actually a prerequisite for life [10], the detection of a chiral imbalance, or enantiomeric excess (ee), by planetary missions probing for prebiotic molecules is considered a likely signature for some form of extraterrestrial life [11–14].

More prosaically, homochirality causes a differentiation in the responses of living organisms to opposite handed forms (enantiomers) of chiral molecules that are encountered in their environment. Over 800 odor molecules that are relevant to the food and fragrance industries have been identified that are chiral and whose enantiomers are perceived to have very different smells [15]. Similarly, responses to pharmaceuticals can be enantiomer specific. Seven of the top 10 selling drugs on the market have a chiral active ingredient and increasingly such drugs are being developed in single enantiomer form, taking a rapidly expanding market share with a predicted value of \$15–25 billion by 2008/2009 [16, 17].

The investigation of chiral molecular phenomena, and associated technique development, thus finds potentially significant practical applications to place alongside the fundamental interest of this topic. This chapter will examine the recently investigated phenomenon of photoelectron circular dichroism (PECD) that arises from a dissymmetry in the angular distribution of photoelectrons emitted from randomly oriented molecular enantiomers when these are ionized by circularly polarized radiation. The associated chiral asymmetries are found to be $\sim 10\%$, several orders of magnitude greater than those typically encountered from dilute, randomly oriented samples. It will be further argued that the study of PECD and the underlying symmetry breaking provides a far more sensitive probe of the photoelectron dynamics and of molecular structure and conformation than more traditional photoionization experiments, and that it opens up new avenues for complementary spectroscopic investigation and characterization of chiral biomolecules.

II. CHIROOPTICAL ASYMMETRIES

A. Absorption Circular Dichroism

Ever since Pasteur's work with enantiomers of sodium ammonium tartrate, the interaction of polarized light has provided a powerful, physical probe of molecular chirality [18]. What we may consider to be conventional circular dichroism (CD) arises from the different absorption of left- and right-circularly polarized light by target molecules of a specific handedness [19, 20]. However, absorption measurements made with randomly oriented samples provide a dichroism difference signal that is typically rather small. The chirally induced asymmetry or dichroism can be expressed as a Kuhn g-factor [21] defined as:

$$g = (I_{\mathrm{lcp}} - I_{\mathrm{rcp}})/\langle I \rangle \qquad (1)$$

where $I_{\mathrm{rcp(lcp)}}$ is the signal obtained with right (left) circularly polarized light (CPL) and $\langle I \rangle$ is the mean or unpolarized signal. Kuhn g-factors rarely exceed 0.01% making them a challenging quantity to measure. Consequently, the technique is nearly always applied to non-dilute samples in the liquid phase. The usable wavelengths are commonly restricted by the availability of transmission polarizing optics (i.e., $\lambda/4$-wave plates) to the region $>110\,\mathrm{nm}$, and are usually further restricted by solvent absorption to $>160\,\mathrm{nm}$.

Nevertheless, valence absorption CD has become a widely used technique to distinguish enantiomers on a purely physical basis and nowadays is also routinely used to infer conformation of biomolecules, such as proteins [22]. The fundamental weakness of valence CD signals arises because the phenomenon is due to interference between the electric and (much weaker) magnetic dipole $(E1 \cdot M1)$ interaction terms . Occasionally, when a transition is nominally electric dipole forbidden but magnetic dipole allowed, as in the $n \rightarrow \pi^*$ excitation of carbonyl groups, larger percentage asymmetries are obtained [23] because the interfering terms become commensurate with one another. Enhancement of the CD asymmetry can also be obtained in the solid state due to scaling effects in the non-electric dipole contributions. Photoabsorption by aerosols of chiral species can show asymmetries of the order of 0.05% [24] reaching an "amazingly high" 10% in highly ordered nanocrystals of tyrosine enantiomer [25]. Extending the approach to core level excitations of ordered systems in the X-ray region allows for similar enhancements of the electric quadrupole interaction term [26, 27] so that CD can be generated by electric dipole–electric quadrupole ($E1 \cdot E2$) interference with comparable asymmetry factors to those in the valence shell; but asymmetries as high as a "spectacular" 12.5% have been reported in the case of an oriented single crystal of a cobalt complex [28].

While it is clear that orientation can enhance the chiral asymmetry there are, however, issues to do with unquantified near-neighbor interactions in the condensed phases. In solution, for example, there are concerns that a significant proportion of the measured CD signal may be contributed by the induced chiral structure of the solvation shell of achiral solvent molecules around the sample molecule [29–31]. This can prove detrimental to efforts to assign stereo-chemical configuration by comparison with theory. For benchmarking theoretical models, or assigning absolute configuration, it is highly desirable to be able to perform measurements on isolated gas-phase enantiomers. This is certainly possible [31–33], but being an intrinsically dilute medium that lacks orientational order and any of the near-neighbor enhancement effects operating in the condensed phase, gas-phase measurements are always going to be restrained by the weakness of the conventional CD effect.

B.   Forward–Backward Asymmetry in Angle-Resolved Chiral Molecule Photoemission

Many readers will have some familiarity with the standard expressions for the angular distribution of photofragments ejected from a randomly oriented (gas-phase) molecule by perfectly polarized light:

$$I_{\text{lin}}(\theta) = 1 + \beta P_2(\cos\theta) \qquad (2)$$

$$I_{\text{cpl}}(\theta) = 1 - (\beta/2)P_2(\cos\theta) \qquad (3)$$

Equation 2 applies to linearly polarized light, where θ is the ejection angle measured with respect to the electric field vector, while Eq. (3) applies to circular polarization and θ is measured relative to the photon propagation direction. The function P_2 is the second Legendre polynomial. The parameter β takes a value ranging from -1 to $+2$; these limits for linear polarization correspond, respectively, to pure $\sin^2$ or $\cos^2$ distributions while $\beta = 0$ corresponds to an isotropic distribution. Therefore β characterizes the anisotropy of the distribution, and is commonly also referred to as an asymmetry parameter; because we will soon be discussing other asymmetry in the angular distribution we will avoid this latter terminology.

Less widely appreciated is the fact that the angular distribution functions Eqs. (2 and 3) are actually subcases of a more general form. It was first proposed by Ritchie [34] that, even in the pure electric-dipole approximation, another term was required for completeness, and that hence the general photoionization angular distribution function, normalized over the surface of a unit sphere, should be written as [35]:

$$I_p(\theta) = 1 + b_1^{\{p\}} P_1(\cos\theta) + b_2^{\{p\}} P_2(\cos\theta) \tag{4}$$

where P_1 is the first Legendre polynomial. The coefficients $b_1^{\{p\}}$ and $b_2^{\{p\}}$ depend on the photoionization dynamics. They also depend explicitly on the polarization as expressed by the superscript p ($p = 0$ linear polarization, $p = \pm1$ left, right circular polarization). These coefficients will be shown to have the following symmetry relationships [34, 35]:

$$b_1^{\{0\}} = 0 \tag{5}$$
$$b_1^{\{+1\}} = -b_1^{\{-1\}} \tag{6}$$
$$b_2^{\{0\}} = -(1/2)b_2^{\{\pm1\}} \tag{7}$$

From the condition Eq. (5) it is readily seen that for linearly polarized light the P_1 term vanishes from Eq. (4) and the more familiar form of Eq. (2) is recovered for linear polarization. Evidently, for linear polarization $b_2^{\{0\}}$ is just equivalent to the β anisotropy parameter. Equation 7 also introduces the additional prefactor of $-\frac{1}{2}$ that occurs for circular polarization, as seen in Eq. (3).

For achiral molecules, the restriction $b_1^{\{p\}} = 0$ is quite general, applying for *any* polarization; only for circular polarization and *chiral* molecules are nonzero values of the $b_1^{\{\pm1\}}$ parameters possible. It will be shown that the reason is the non-equivalence of the ±1 spherical tensor components of the photoionization matrix elements when the molecular potential has a preferential "handedness." In this special case, the P_1 Legendre polynomial (which is actually just a simple $\cos\theta$ term) is retained in Eq. (4). Hence, a forward–backward asymmetry in the angular distribution can be anticipated as θ swings through 90°. Furthermore, the antisymmetry of $b_1^{\{\pm1\}}$ [Eq. (6)] leads to a reversal of the forward–backward asymmetry when the helicity of the radiation is exchanged, thus giving rise to a CD (or difference between the response to left- and right-handed circular polarization in the angle-resolved photoelectron spectrum). A similar reversal is anticipated when the handedness of the enantiomeric molecular target is exchanged.

This forward–backward asymmetry of the photoelectron distribution, expected when a randomly oriented sample of molecular enantiomers is ionized by circularly polarized light, is central to our discussion. The photoelectron angular asymmetry may conveniently be defined in a manner analogous to the Kuhn g-factor as the difference in the electron fluxes through equal area elements $\sin\theta d\theta d\phi$ on the unit sphere in the $0°$ and $180°$ directions ratioed to the *mean* flux through an area element (averaged over the whole sphere):

$$\gamma = [I_p(0°)\sin\theta d\theta d\phi - I_p(180°)\sin\theta d\theta d\phi]/\sin\theta d\theta d\phi$$

$$\gamma = 2b_1^{\{p\}} \tag{8}$$

where the second line follows from Eqs. (4 and 6). This choice of definition establishes a clear link between the asymmetry and the anticipated $b_1^{\{\pm1\}}$ parameters.

1. A Simple Analogy

A detailed derivation of the photoionization differential cross-section expression, leading ultimately to the angular distribution in Eq. (4), is provided in Appendix A. This will help provide a detailed understanding of the photoelectron dynamics that determine the angular distribution parameters, as will be discussed in a subsequent section, but for now it may help develop the reader's appreciation of this phenomenon to provide a simple, if necessarily inexact, mechanical analogy.

Consider a threaded rod, representing a molecular enantiomer, that lies away from an observer. If the observer reaches out and spins a nut on the rod clockwise with his right hand, the nut will travel forward, away from the observer, and will shortly fly off the rod. Here, the angular momentum imparted to the nut (electron) by the observer's hand (photon) causes it to be ejected in a specific direction from the rod (molecular enantiomer) in the observer's reference frame. This is mediated by the interaction between the chiral thread of the rod and nut (the chiral molecular potential). If the rod is turned through $180°$ and the action repeated, the nut (electron) still departs in the same direction, away from the observer. Hence, the orientation of the rod (molecule) in the observer's frame does not alter the direction in which the nut (electron) is ejected.

Now, consider repeating these operations, but with the use of the observer's left hand to spin the nut in an anticlockwise direction as seen by the observer. It should be apparent that, in either orientation of the rod, the nut will now travel backward (toward the observer). This must suggest that an electron would be ejected in a reversed direction following excitation by a photon of opposite helicity, again mediated through the chiral interaction of the electron with the enantiomer's potential field.

Finally, the reader is invited to consider the above sequence of observations when the right-hand threaded rod is substituted by an assembly with a left-hand thread. The direction of the nut's travel will in each case be the opposite of that already described. We infer that exchanging the handedness of the molecular enantiomer will cause there to be a reversal of the observed (lab frame) direction of photoelectron ejection.

C. Photoelectron Circular Dichroism

While it is clear that a direct measurement of the angular distribution, Eq. (4), with a given helicity of light should be capable of yielding the $b_1^{\{\pm 1\}}$ angular parameter, it is often more convenient to examine the dichroism, or difference, obtained with opposite helicities of the light (or, possibly, of the enantiomer). From Eq. (4) and the antisymmetry property Eq. (6) one obtains an expression for the PECD:

$$I_{\text{lcp}}(\theta) - I_{\text{rcp}}(\theta) = (b_1^{\{+1\}} - b_1^{\{-1\}})P_1(\cos\theta) = 2b_1^{\{+1\}}\cos\theta \qquad (9)$$

The PECD measurement clearly takes the form of a cosine function with an amplitude given entirely in terms of the single chiral parameter, b_1. It therefore provides exactly the same information content as the γ asymmetry factor defined above [Eq. (8)]. Experimental advantages of examining the PECD rather than the single angular distribution $I_P(\theta)$ are likely to include some cancellation of purely instrumental asymmetries (e.g., varying detection efficiency in the forward–backward directions) and consequent improvements in sensitivity.

However, any practical, quantitative use of Eq. (9) presupposes that fully normalized distribution functions $I_{\text{lcp,rcp}}(\theta)$ are available. If normalization by the mean of the two angular intensities is included, as in the Kuhn g-factor Eq. (1), we obtain:

$$\Gamma(\theta) = (I'_{\text{lcp}}(\theta) - I'_{\text{rcp}}(\theta))/[(I'_{\text{lcp}}(\theta) + I'_{\text{rcp}}(\theta))/2]$$
$$= 2b_1^{\{+1\}}\cos\theta/[1 + b_2^{\{\pm 1\}}P_2(\cos\theta)] \qquad (10)$$

where the primes indicate non-normalized experimental measurements. It can be seen that knowledge of the second angular parameter $b_2^{\{\pm 1\}}$ is in general still required.

Figure 1 illustrates an actual experimental observation of the photoelectron CD obtained from the $h\nu = 10.3$-eV photoionization of (R)-camphor [36]. Full details of this system and the experimental technique are considered later, but for now we can note that the figure shows the intensity differences $[I_{\text{lcp}}(\theta) - I_{\text{rcp}}(\theta)]$ that have been observed by an electron imaging technique, in a plane that contains the photon beam propagation direction. It is sufficient to realize that photoelectron velocity maps onto the radial coordinate in the circular distribution patterns; thus faster electrons are found at larger radius. In this example, there are clearly two ring structures, an outer-ring structure corresponding to ionization of the highest occupied molecular orbital (HOMO)

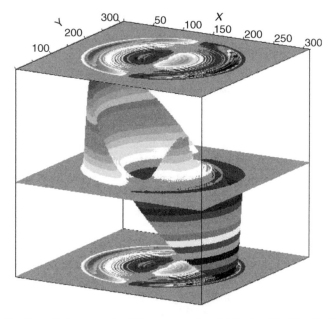

Figure 1. Photoelectron circular dichroism angular distribution $[I_{lcp}(\theta) - I_{rcp}(\theta)]$ for the $h\nu = 10.3$-eV photoionization of (R)-camphor, as imaged with the photon beam propagating along the x axis. The x, y axis scales are the physical pixel coordinates of the detector.

orbital, and an inner-ring structure corresponding to slower electrons from ionization of the HOMO-1 orbital.

It is very obvious from the peaks and troughs displayed in Fig. 1 that the anticipated dissymmetry between forward and backward electron ejection directions (relative to the photon beam direction) is borne out by experiment. Moreover, one sees that the dissymmetry lies in opposite directions for ionization of the two energetically accessible orbitals observed here.

A different view is provided in Fig. 2, which displays the chiral asymmetry parameter measured across the $h\nu = 21.2$-eV photoelectron spectrum for both the (R)- and (S)- enantiomers of glycidol. The sign and magnitude of the γ-factor for a given enantiomer vary along the photoelectron spectrum. Note that these changes correlate with different predicted orbital ionization energies [38] such that the PECD spectrum can be seen to indicate underlying bands that are not adequately resolved in the simple intensity curve of the photoelectron spectrum recorded under identical conditions. Equally significant, the forward–backward asymmetry measured by the γ-factor reverses when the enantiomer is changed, as anticipated above, such that the (R)- and (S)-enantiomer curves mirror one another across the $\gamma = 0$ axis.

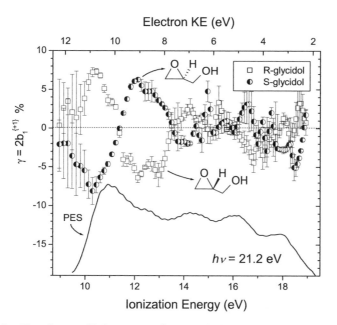

Figure 2. Photoelectron chiral asymmetry factor, γ, obtained as a function of electron kinetic energy at $h\nu = 21.2\,\text{eV}$ for the (R)- and (S)- enantiomers of glycidol. Also included is a moderate resolution photoelectron spectrum recorded under identical conditions. Data from Refs. [37, 38].

Quite generally, we will find that the magnitude and direction of these chiral dissymmetries varies greatly with both electron energy and initial orbital. The prediction or interpretation of such characteristics falls beyond the capability of the simple, intuitive analogy presented in II.B.1, so that we must now turn to consider the quantum interference effects that control the observable distributions in order to enhance our predictive abilities. A reader wishing to pass over these details at first encounter will find a summary of the deductions made at the start of the subsequent Section, IV.

III. PHOTOELECTRON DYNAMICS IN CHIRAL MOLECULES

A phenomenological description of the differential cross-section for emission of photoelectrons into solid angle Ω in the lab frame can be written, assuming random molecular orientation and an axis of cylindrical symmetry defined by the photon polarization, as

$$\frac{d\sigma}{d\Omega} = \frac{\sigma}{4\pi} \sum_j b_j^{\{p\}} P_j(\cos\theta) \tag{11}$$

If the isotropic coefficient $b_0^{\{p\}}$ is specified to be unity, σ is just the total (integrated) cross-section. In Appendix A, an alternative quantum mechanical expression for this cross-section is obtained in the electric dipole approximation. By comparing the two expressions, it can be seen that the Legendre polynomial coefficients in Eq. (11) may be obtained from the inner summation terms in Eq. (A.15). Hence, the Legendre polynomial coefficients are

$$
\begin{aligned}
B_j^{\{p\}} = 2\pi \sum_{\substack{lmv \\ l'm'v'}} & (-i)^{l-l'} e^{i(\sigma_l - \sigma_{l'})} (-1)^{p-m-v} \frac{\sqrt{(2l'+1)(2l+1)}}{(2j+1)} \\
& \times \langle 1 - p, 1p | j0 \rangle \langle 1 - v', 1v | jv - v' \rangle \\
& \times \langle l'0, l0 | j0 \rangle \langle l' - m', lm | jm - m' \rangle \\
& \times [e_{v'}^1 f_{l'm'v'}^{(-)}]^* e_v^1 f_{lmv}^{(-)} \delta_{v'-v,m'-m}
\end{aligned}
\tag{12}
$$

In order to fix $b_0^{\{p\}} = 1$, the coefficients $B_j^{\{p\}}$ in Eq. (12) have simply to be renormalized, dividing each by $B_0^{\{p\}}$, that is, $b_j^{\{p\}} = B_j^{\{p\}}/B_0^{\{p\}}$.

In Eq. (12), l, m are the photoelectron partial wave angular momentum and its projection in the molecular frame and v is the projection of the photon angular momentum on the molecular frame. The presence of an alternative primed set l', m', v' signifies interference terms between the primed and unprimed partial waves. The parameter σ_l is the Coulomb phase shift (see Appendix A). The $f_{lmv}^{(-)}$ are dipole transition amplitudes to the final-state partial wave l, m and contain dynamical information on the photoionization process. In contrast, the Clebsch–Gordan coefficients (CGC) provide geometric constraints that are consequent upon angular momentum considerations.

A.   Geometric Considerations

The role of the Clebsch–Gordan coefficients (CGC) in shaping the structure and properties of the photoelectron angular distribution function has been discussed previously [34, 35]. Triangle conditions on the first CGC in Eq. (12) restrict j to the range $0\ldots 2$ so justifying the limitation of the summation in Eq. (4) to no more than the second Legendre polynomial term. In the specific case of linearly polarized light ($p = 0$), the first CGC in Eq. (12) is actually also zero for $j = 1$; hence, $b_1^{\{0\}} = 0$ [Eq. (5)] and the familiar Yang theorem result, embodied in Eq. (2), is obtained whereby the lab-frame anisotropy depends only on the isotropic term (P_0) and the second Legendre polynomial term (P_2). The conventional electron anisotropy parameter, β, is thus $b_2^{\{0\}}$ in the present notation. If, instead, circularly polarized light is considered ($p = \pm 1$), then for $j = 2$ the value of the first GCG, and hence $b_2^{\{\pm 1\}}$, is halved compared to that for $j = 2, p = 0$. With the additional phase, $(-1)^p$, appearing in Eq. (12) another well-known result is recovered; the effective β parameter for circularly polarized light is reduced by a factor of $-\frac{1}{2}$ compared to the linearly polarized case, as indicated by Eqs. (3 and 7).

Significantly, when switching from linear to circularly polarized light the first CGC in Eq. (12) acquires nonzero values of $\mp 1/\sqrt{2}$ for $j = 1$, $p = \pm 1$. Consequently, with circularly polarized light the first-order Legendre coefficients are not automatically constrained to be zero, as anticipated by Eq. (4); the change of sign of the CGC for opposite circular polarizations provides a corresponding change of sign in the $b_1^{\{\pm 1\}}$ parameters, and hence the antisymmetry noted in Eq. (6).

The polarization state $p = \pm 1$ is necessary but not, however, a sufficient condition for a nonzero P_1 coefficient in the angular distribution. Recalling the general CGC symmetry relation

$$\langle l_1 m_1, l_2 m_2 | l_3 m_3 \rangle = (-1)^{l_1 + l_2 - l_3} \langle l_1 - m_1, l_2 - m_2 | l_3 - m_3 \rangle \qquad (13)$$

and applying it to the second CGC when $j = 1$ shows that it will vanish for $v = v' = 0$ and that it changes sign as $(v', v) \rightarrow (-v', -v)$ – implying a net cancellation of terms [34] over the full summation on v, v' *unless* $f_{lmv=1}^{(-)} \neq f_{lmv=-1}^{(-)}$ – a condition satisfied only by chiral molecules for which the projections $v = \pm 1$ are inequivalent. Achiral molecules cannot then have a nonzero $b_1^{\{p\}}$ parameter for *any* light polarization state. Only for chiral molecules lacking a plane of symmetry will this second condition allowing for the retention of a P_1 term in the angular distribution Eq. (4) be satisfied. Furthermore, the antisymmetry of the $f_{lmv}^{(-)}$ matrix elements upon the reflection required to convert one enantiomeric form to another is then such as to exchange signs on the $b_1^{\{p\}}$ coefficients in much the same way as changing the light helicity does.

Applying the symmetry condition Eq. (13) to the third CGC implies that $l' + l + j = $ even. Along with the triangle rule this requires that $l' = l$ for the isotropic term $j = 0$, $l' = l \pm 1$ for $j = 1$, and $l' = l \pm 2, 0$ for $j = 2$. These deductions determine the nature of the interference implicit in each of the angular parameters $B_j^{\{p\}}$. The fourth CGC in Eq. (12) further regulates the interference between different m projections since $|m - m'| \leq j$. Furthermore the delta function in Eq. (12) requires that just as $v' \neq v$ when $j = 1$, so too $m' \neq m$.

B.   Continuum Scattering Phase Shifts

The partial wave basis functions with which the radial dipole matrix elements $f_{lmv}^{(-)}$ are constructed (see Appendix A) are **S**-matrix normalized continuum functions obeying incoming wave boundary conditions.

$$\psi_{lm}^{(-)}(\vec{r}) \xrightarrow{r \rightarrow \infty} (\pi k)^{-\frac{1}{2}} (2ir)^{-1} \sum_{L'} [e^{i\chi_{l'}} \delta_{LL'} - S_{LL'}^* e^{-i\chi_{l'}}] Y_{L'}(\hat{r}) \qquad (14)$$

where k is the electron momentum, L is shorthand for l, m and

$$\chi_l = kr - \frac{l\pi}{2} + \ln(2kr)/k + \sigma_l \tag{15}$$

Asymptotically, these functions assume the characteristics of the single *outgoing* wave, lm, appropriate to the simple description of physical observables, such as the angular distribution, but when approaching the inner-molecular ion core region they have a mixed lm character from the various *incoming* wave components [39, 40]. The S-matrix completely specifies the asymptotic outcome of the short-range mixing of the lm waves by scattering between the photoelectron and anisotropic molecular ion core.

Choosing the continuum inhomogeneity differently yields the alternative K-matrix normalized wave functions:

$$\psi_{lm}^K(\vec{r}) \xrightarrow{r\to\infty} \frac{1}{\sqrt{\pi k}} r^{-1} \sum_{L'} [\sin\chi_l \delta_{LL'} + K_{LL'} \cos\chi_l] Y_{L'}(\hat{r}) \tag{16}$$

These functions have the real, long-range radial oscillatory behavior of Coulomb waves (argument χ_l) with an asymptotic phase shift determined by the mixing of the regular and irregular (sin and cos) forms. The K-matrix defines this mixing and represents the scattering of a specific lm into more mixed angular momenta caused by the short-range electron–molecule interaction. The S- and K-matrix normalized functions are related by a simple linear transformation.

The continuum electron-phase shifts induced by the short-range scattering off the chiral molecular potential are most conveniently introduced by a third choice of continuum function, obtained by diagonalizing the K-matrix by a transformation $\mathbf{U}$, resulting in a set of real eigenchannel functions (apart from normalization) [41]:

$$\begin{aligned}
\psi_\alpha^E(\vec{r}) &= \sum_L \psi_L^K(\vec{r}) U_{L\alpha} \\
&= \frac{1}{\sqrt{\pi k}} r^{-1} \sum_L [\sin\chi_l + \tan\xi_\alpha \cos\chi_l] Y_L(\hat{r}) U_{L\alpha}
\end{aligned} \tag{17}$$

where ξ_α is a channel phase shift characteristic of the short-range molecular ion potential. Because these various continuum bases are related by unitary transformations, the S-matrix normalized functions can be expressed in terms of the eigenchannel functions as [42]:

$$\psi_{lm}^{(-)}(\vec{r}) = \sum_\alpha U_{L\alpha} e^{-i\xi_\alpha} \psi_\alpha^E(\vec{r}) \tag{18}$$

Substitution of Eq. (18) into the partial wave expansion formula [Eq. (A.3)] gives for the continuum state

$$\Psi_{\vec{k}}^{(-)}(\vec{r}) = \sum_{lm} i^l e^{-i\sigma_l} Y_{lm}^*(\hat{k}) \sum_{\alpha} U_{L\alpha} e^{-i\xi_\alpha} \psi_\alpha^E(\vec{r}) \qquad (19)$$

from which it can be seen that net continuum phase shifts, $\eta_{lm\alpha}$, can be represented using factors $e^{-i(\sigma_l + \xi_\alpha)}$ that add the short-range molecular scattering phase to the Coulomb phase.

C. Quantum Interference and Phase in the Angular Distribution

The last two CGC in Eq. (12) evidently dictate that rather different partial wave interference contributions are made to each of the angular parameters. This will impact on the dynamical information conveyed by each one. Equally important, the phase subexpression

$$(-i)^{l-l'} e^{i(\sigma_l - \sigma_{l'})} \qquad (20)$$

that appears in Eq. (12) consequently plays subtly differing roles. By writing the complex exponential as $\cos(\sigma_l - \sigma_{l'}) + i\sin(\sigma_l - \sigma_{l'})$ it can be seen that the effect of the $(-i)^{l-l'}$ in this expression is to select out either the cosine or the sine term as the real part of Eq. (20), depending on whether $(l - l')$ is even or odd. Only the real parts of the summands in Eq. (12), and similar, can contribute toward physically observable properties, so that the contribution made by relative phase shifts alternates according to the interfering ℓ-waves.

First to be considered is the isotropic P_0 coefficient. The $B_0^{\{p\}}$ parameter is proportional to the integrated cross-section, σ [Eq. (11)]. In fact, the preceding arguments show that when $j = 0$, $l'm' = lm$, and so there are no interference cross terms in this case. Consequently, as is already widely recognized, the integrated cross-section displays no dependence on the relative phase of the final continuum channels.

The summation for the coefficient of the P_2 term likewise includes phase insensitive contributions where $l'm' = lm$, but also now one has terms for which $l' = l \pm 2$, which introduce a *partial* dependence on relative phase shifts— specifically on the cosine of the relative phase shift. Again, this conclusion has long been recognized; for example, by an explicit factor $\cos(\eta_{l-1} - \eta_{l+1})$ in one term of the Cooper–Zare formula for the photoelectron β parameter in a central potential model [43].

In the more novel case of the P_1 coefficient, *only* interference terms for which $l' = l \pm 1$ arise, so now *all* summands contributing to $B_1^{\{\pm 1\}}$ contain phase shift information. This hints that the chiral $b_1^{\{\pm 1\}}$ parameters could be more sensitive to phase differences than traditional β parameters, and one would certainly expect that both will be more sensitive to dynamical details than the phase insensitive cross-section measurement. But the full significance of the fact that unusually now the phase dependence occurs in the form of the *sine* of the phase difference $(\sigma_l - \sigma_{l'})$ also merits further exploration, as does the inclusion of the additional phase shifts contributed by the short-range molecular interaction potential.

Figure 3 shows the variation of Coulomb phase shifts σ_l. Changes between successive shifts become less as both ℓ and the energy increase. The variation of the functions $\cos(\sigma_l - \sigma_{l-2})$ and $\sin(\sigma_l - \sigma_{l-1})$, expected to feature in respectively the $b_2^{\{p\}}$ and $b_1^{\{\pm 1\}}$ evaluations, are also shown. As the angular differences get smaller the cosine function approaches 1 and is at its most slowly varying; conversely the sine function varies most rapidly in the same region

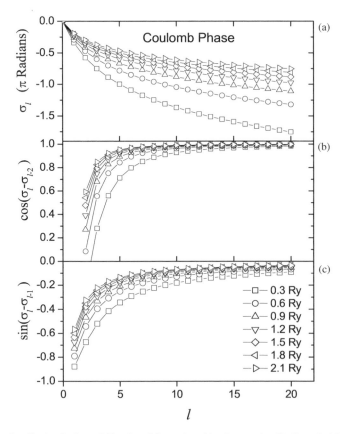

Figure 3. Coulomb phase shifts plotted for various kinetic energies (Rydbergs); (a) σ_l relative to the $\ell = 0$ wave; (b) $\cos(\sigma_l - \sigma_{l-2})$; (c) $\sin(\sigma_l - \sigma_{l-1})$.

where it is approaching zero. Thus $\cos(\sigma_l - \sigma_{l-2})$ quickly reaches a limiting value ~ 1, while $\sin(\sigma_l - \sigma_{l-1})$ continues to show much greater variation to higher ℓ. Similarly, the cosine terms converge to a limiting value with increasing energy more rapidly than do the sine terms. These differences are of course even more pronounced in percentage terms because of the different magnitudes of the sine and cosine functions.

These distinctions become more profound if the Coulomb phase difference in Eq. (20) is considered to be replaced by a net phase difference $\eta_{lm} - \eta_{l'm'}$ that includes the short-range phase shift induced by electron scattering off the chiral molecular potential (see Section III.B). Small differences in the molecular scattering phase between interfering channels will cause much greater fluctuation in the contribution made to the chiral term due to the rapidly varying sine function around zero. Likewise, small variations in the phase of a given channel, perhaps as a result of change in molecular conformation diffused across the molecule, are likely to have a seemingly disproportionate effect on the summation for the $b_1^{\{\pm 1\}}$ parameter because of the rapidly fluctuating sine term. An additional sensitization to small phase variations around zero can be attributed to the $\sin(\eta_{lm} - \eta_{l'm'})$ term since, unlike a cosine function, it will introduce a dependence on sign as well as magnitude of small phase differences.

When experimental results are later introduced, it will be seen that the significance of the final-state scattering in PECD measurements is confirmed by the observation that for C $1s$ core ionizations, which must therefore proceed from an initial orbital that is achiral by virtue of its localized spherical symmetry, there is no suggestion that the dichroism is attenuated. The sense of the chirality of the molecular frame in these cases can only come from final-state continuum electron scattering off the chiral potential. Generally then, the induced continuum phase shifts are expected to be of paramount importance in quantifying the observed dichroism.

D. PECD and CDAD

It may be worthwhile to compare briefly the PECD phenomenon discussed here, which relates to *randomly* oriented chiral molecular targets, with the likely more familiar Circular Dichroism in the Angular Distribution (CDAD) that is observed with oriented, achiral species [44–47]. Both approaches measure a photoemission circular dichroism brought about by an asymmetry in the lab frame electron angular distribution. Both phenomena arise in the electric dipole approximation and so create exceptionally large asymmetries, but these similarities are perhaps a little superficial.

In CDAD, a chiral experimental geometry is created about a fixed molecular orientation, and the asymmetry in the electron distribution can be observed in directions mutually perpendicular to the photon propagation direction and the molecular axis. The pertinent angular distribution is essentially a molecule frame fixed distribution, and the quantum interference structure of the asymmetry expression describing this is quite different from the PECD expressions given here. Hence, there are some fundamental differences between the two experiments.

At a phenomenological level too, there are differences since the CDAD effect disappears in directions parallel to the photon beam, whereas PECD asymmetry is maximized in these directions. Conversely, the PECD asymmetry disappears in directions perpendicular to the photon beam where the maximum CDAD asymmetry can be found.

Of course, it is possible to contemplate experiments that examine photoionization of oriented chiral molecules. An expression has been given for the angle integrated (total) ionization cross-section in such circumstances [48] and CDAD-type measurements have been reported on adsorbed chiral molecules [49, 50], but the interplay of natural and geometric chirality in angle-resolved dichroism measurements remains very much a topic for future investigation.

IV. THEORETICAL MODELING OF PECD

It is perhaps surprising that the key features of Ritchie's original work [34] identifying this novel asymmetry in the photoelectron angular distribution from randomly oriented chiral molecules [Eq. (4)], and his inference that as a pure electric dipole interaction it ought to give rise to much stronger asymmetries than natural absorption CD, lay practically dormant for many years. One other early, independent theoretical study [48] provided an estimate of the expected strength of the PECD effect, but it was not until much more recently that realistic quantitative calculations became feasible [35, 51]. These confirmed the expected magnitude of the asymmetry and suggested a dynamical richness of PECD measurements. Fortuitously, these developments also coincided with the wider availability of circularly polarized vacuum ultraviolet (VUV) and soft X-ray (SXR) synchrotron radiation that it is required for fully practicable experimental investigations. These advances in both theoretical and experimental capabilities underpin current interest in the topic. Experimental aspects are covered in a later Section V, but here theoretical modeling of PECD phenomena is discussed.

It was shown in the preceding section that PECD can be anticipated to have an enhanced sensitivity (compared to the cross-section or β anisotropy parameter) to any small variations in the photoelectron scattering phase shifts. This is because the chiral $b_1^{\{\pm 1\}}$ parameter is structured from electric dipole operator interference terms between adjacent ℓ-waves, each of which depends on the sine of the associated channels' relative phase shifts. In contrast, the cross-section has no phase dependence, and the β parameter has only a partial dependence on the cosine of the relative phase. The distinction between the sine and cosine dependence is significant where small phase differences are concerned; the sine function will vary most rapidly around zero, and depend on sign, as well as magnitude. It can thus be anticipated that PECD is sensitized to small phase shifts, but in a manner whose detail becomes hard to intuit. Numerical computation has therefore much to contribute to understanding specific detail of PECD.

A. Computational Methods

The numerical computation of PECD effects requires calculation of the radial dipole matrix elements, $f_{lmv}^{(-)}$ for evaluating the parameters $B_j^{\{p\}}$ from Eq. (12) . The challenge, as in all photoionization dynamics calculations, lies in evaluating the continuum functions. However, these challenges are particularly accute in application to PECD for three reasons. First, the molecules of interest are very much larger than those normally treated by such calculations—for example, the chiral terpenes that are featured in our investigations are all C_{10} molecules. Moreover, such chiral molecules, by definition, lack any elements of symmetry (although some chiral species may possess a C_2 rotational axis). Consequently, full molecule calculations are necessarily large scale and lack any scope for efficiency savings achieved by exploitation of molecular symmetry. It is also found that calculations for the $b_1^{\{\pm1\}}$ parameter tend to converge more slowly than do either cross-section or β parameter calculations, further raising the computational effort demanded [52, 53].

To date, two different methods have been applied. The author's group have extended the continuum multiple scattering (CMS-Xα) method [40, 54] to treat PECD problems [35, 36, 51, 52, 55–57]. Briefly, the continuum multiple scattering treatment is a parameterized approach that requires the construction of a self-consistent ground-state potential in which the exchange contribution to an effective one-electron potential is represented using the Xα local density approximation [58]. This proceeds by partitioning the molecule into overlapping spherical regions about each atomic center, enclosed within an outer, spherically symmetric region. Both bound and continuum electron wave functions can then be expressed in a basis of spherical harmonic functions, truncated at some value, ℓ_{max} in each spherical zone of the potential, with radial terms obtained by direct numerical integration. For the continuum problem, the self-consistent Xα potential does not have the correct asymptotic form (Coulomb attraction for ion plus electron) and so is post-modified to have this property before the continuum functions are calculated. Further relevant computational details are to be found in the cited references.

Stener and co-workers [59] used an alternative B-spline LCAO density functional theory (DFT) method in their PECD investigations [53, 57, 60–63]. In this approach a normal LCAO basis set is adapted for the continuum by the addition of B-spline radial functions. A large single center expansion of such functions is augmented in a multicenter treatment by functions placed on additional centers, allowing improved efficiency in the convergence, especially with asymmetric molecules such as these. The system is then solved at the DFT Kohn–Sham level, using an appropriate exchange-correlation functional (mainly the LB94 potential), which has the correct asymptotic Coulomb behavior.

Despite computational differences, both these approaches currently embody the same fundamental physical model: independent electron, frozen core, fixed nuclear geometry. In both cases, real **K**-matrix normalized functions [Eq. (16)] are obtained, then transformed to the **S**-matrix form [Eq. (14)] required for matrix element evaluation. While both methods rely on some degree of parameterization—the B-spline DFT calculation in the exchange-correlation functional, CMS-Xα through its Xα local potential model—CMS-Xα is undoubtedly the more semiempirical, and has the most severe approximation in its partitioning of the molecular potential into spherical regions. Indeed, CMS-Xα is sometimes disparaged for this reason. It is interesting to compare the relative performance of these two methods before relying on either to interpret experimental subtleties.

1. PECD in Oxiranes

A comprehensive theoretical investigation of PECD in a series of substituted oxiranes has been presented by Stener et al. [53] using the B-spline method. Variations in the predicted dichroism were found as both a function of initial orbital, and of the chemical substitution about the epoxy ring. These substitutions do not induce very significant geometry changes in the optimized ring structure.

Figure 4 shows results obtained in this study for four specific orbital ionizations in the molecule (S)-methyl oxirane. The HOMO orbital (16a) can be characterized as the oxygen $2p$ lone-pair electrons, while the HOMO-1 (15a) is a C—C—O ring σ bonding orbital. The deeper lying orbital (9a) is a C—C bonding π orbital, while finally we have selected the methyl group C $1s$ core orbital to present here. Included in this figure are CMS-Xα calculations for the same orbitals.[1] These latter are calculated at a geometry optimized at the MP2/6-31G(p,d) level, and utilize a very similar sized basis set expansion ($\ell_{max} = 15$) to the B-spline results.

In Fig. 5, a similar selection of results for the *trans*-difluoro-oxirane molecule is presented, again comparing B-spline and CMS-Xα calculations as before. In this case, the outer two orbitals have a reversed ordering; the O lone pair (orbital 9b) is the HOMO-1 while the HOMO (11a) has C—C—O σ-bonding characteristics. The 5b level is analogous to the 9a in methyl oxirane,

---

[1]When comparing calculations, or calculations and experiment, it is important to ensure consistent sign conventions are applied. This requires that the absolute configuration of the enantiomer and photon helicity used in a calculation be clearly specified. See Appendix B

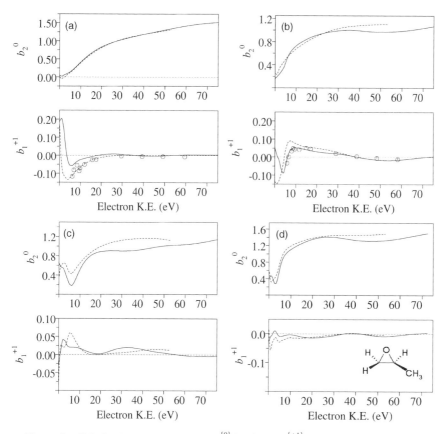

Figure 4. Calculated angular parameters $b_2^{\{0\}}(\equiv \beta)$ and $b_1^{\{+1\}}$ for (S)-methyl oxirane: solid curves CMS-Xα; broken curves B-spline (Ref. [53]). Results are given for ionization from four different orbitals: (a) 16a (HOMO); (b) 15a (HOMO-1) ; (c) 9a; (d) 4a (C 1s). Experimental data points for $b_1^{\{+1\}}$ are included where available for (a) and (b). Taken from Ref. [62].

being a C—C π-bonding orbital. However, as difluoro-oxirane has a C_2 axis, the two C atoms are symmetry equivalent in this case. Symmetric and antisymmetric combinations of the 1s orbitals can be formed, but as these would not be expected to be resolved experimentally the results are presented together in combination in the fourth panel.

Overall, it can be seen from Figs. 4 and 5 that in predicting the β parameters almost identical results are obtained in five out of the eight cases. Significantly perhaps, these include the more highly localized O- lone pair and C 1s orbitals. The worst agreement in both molecules is seen for the more delocalized ring π orbital.

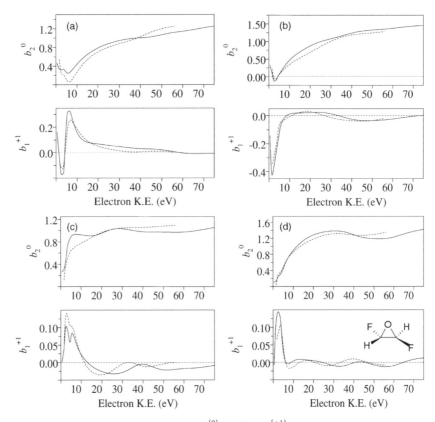

Figure 5. Calculated angular parameters $b_2^{\{0\}} (\equiv \beta)$ and $b_1^{\{+1\}}$ for *trans*-(2S,3S)-difluoro-oxirane: solid curves CMS-Xα; broken curves B-spline (Ref. [53]). Results are given for ionization from four different orbitals: (a) 11a (HOMO); (b) 9b (HOMO-1) ; (c) 5b; (d) 3a/2b C 1s.

Turning to the predictions for the chiral $b_1^{\{+1\}}$ parameters, it must first be said that the two calculations are remarkably consistent with one another in the results they provide. This may be especially true for the difluoro molecule, with again the ring π orbital proving the worst match. Nevertheless, even here the key characteristics (magnitude, shape, switch of sign, etc.) are clearly equally captured by both calculations. The methyl oxirane provides an opportunity for comparison with experimental data [62] that are available for the two outer orbitals. From Fig. 4 it would appear that the B-spline calculation performs marginally better in the HOMO ionization, the CMS-Xα in the HOMO-1 ionization; in truth the differences between either calculation and experiment are likely to be of no special significance. It is then very gratifying that these two independent calculations provide such close agreement, as clearly they capture the same underlying physics, and are in accord with experiment where this can be checked.

One can, however, learn from the small discrepancies that are apparent. First, the more delocalized and out-of-plane orbitals produce the biggest net differences in the angular parameters and here, at least, it is not unreasonable to suppose that this may be a consequence of their being diffused over several of the discrete regions, including the interstitial region, in the crudely approximated spherically partitioned $X\alpha$ potential. Second, the most significant region for differences is found within a few electronvolts of threshold. For example, in Fig. 4b where the predicted $b_1^{\{+1\}}$ parameters differ strongly in sign and magnitude. Unfortunately, this is also an experimentally difficult region, but intuitively one may expect the outgoing electron to be most sensitive to the chiral molecular potential at low energy, and so the threshold region to be most challenging for theoretical modeling. Conversely, intuition suggests that the photoelectron should be least sensitive to the chiral potential at higher energies, and so in all cases here, and indeed in all other studies to date, it is no surprise to see the dichroism decaying to zero with increasing electron kinetic energy.

Let us continue to examine the general characteristics of PECD emerging from either of these computational approaches. These pure electric dipole calculations continue to show [35, 51] that, at least within a few tens of an electronvolt of threshold, the chiral asymmetry is of the order of 0.1, with possible peaking at considerably greater values. Second, these $b_1^{\{+1\}}$ parameters and the corresponding asymmetry factors, γ, oscillate and may even change sign as a function of electron energy. As intimated previously, this variation is ultimately a consequence of the origin of the phenomenon in quantum interference effects that pass beyond simple intuitive prediction. Nor does the peak magnitude of the dichroism generally appear to scale with the degree of localization of the initial orbital, as can be seen in the subset of the data presented here. Furthermore, as noted by Stener et al. [53] in their full study, the chiral parameter displays far greater variation with chemical substitution than either the cross-section or β parameter. Evidence for this can be seen here comparing the C–C σ orbital data for the two oxiranes in Figs. 4 and 5. A similar comparison can be made for the O lone pair, and so such sensitivity to substitution seemingly applies regardless of whether or not the initial orbital is heavily localized and inherently "achiral". One infers that it is the final-state scattering diffused over the full chiral molecular potential rather than the initial orbital that lends this enhanced sensitivity. These observations are consistent with the explanation advanced in Section III.C that the quantum structure of the $b_1^{\{\pm\}}$ parameter renders it uniquely (in the present context) sensitive to even small changes in scattering phase that might be induced by any chemical substitution, even when such changes occur at relatively remote sites in the molecule.

2. PECD in Camphor

Another detailed comparative study of B-spline and CMS-Xα calculations that has been presented [57] addresses core and valence-shell PECD in camphor. For this molecule, a substantial amount of relevant experimental PECD data for the core [56] and valence-shell ionization [36, 56, 64, 65] is now available, allowing a full three-way comparison to be performed. Detailed discussion of the interpretation of the experimental results achieved with these calculations is deferred until Section VI.B, but it is helpful here to summarize the conclusions regarding the computational approaches.

For the carbonyl carbon $1s$ core level ionization, excellent quantitative agreement of the $b_1^{\{1\}}$ parameters is found, both between the alternative calculations and between either calculation and experiment (see Section VI.B.I). Given the spherical, therefore achiral, nature of the $1s$ initial orbital in these calculations, any chirality exhibited in the angular distribution must stem from the final-state photoelectron scattering off the chiral molecular ion potential. Successful prediction of any non-zero chiral $b_1^{\{1\}}$ parameter is clearly then dependent on a reliable potential model describing the final state. At this level, there is nothing significant to choose between the potential models of the two methods.

For the HOMO and HOMO-1 valence-shell ionizations, the three-way agreement CMS-Xα—B-spline—experiment is good but no longer exact [36, 57]. It may be supposed that these contrasting observations imply a common difficulty for the two theoretical methods when specifically applied to the valence shell, and one that ends up being differently approximated by them. Both the LCAO B-spline DFT and the CMS-Xα methods describe their respective potentials in a static formalism, so the response of the electron density to the external electromagnetic field is completely neglected in either case. In fact, it is already suspected from work on smaller systems that the omission of such response effects becomes a much greater deficiency in valence-shell ionization due to the greater response effects anticipated [66–68]. These response effects become very large in the presence of highly polarizable valence electrons, typically when $3p$ shells are filled as in Ar or second-row hydrides [66]. At the much higher photon frequencies necessary to ionize core orbitals, valence-shell cross-sections are quite small, and therefore any valence response effects are by then strongly reduced. Further work is required to establish the importance of inclusion of response effects in modeling PECD.

B. Convergence of the Partial Wave Expansion

A common finding of computational PECD studies is that a relatively large partial wave expansion, typically running to $\ell_{max} \geq 15$ is required. Chiral molecules necessarily are of very low, or no, symmetry, and hence are quite irregularly shaped. Intuitively, it might be rationalized that a greater spatial irregularity of the molecule would necessitate a higher harmonic content for the description of properties that must in some way sense that "shape" (e.g., the photoelectron angular distribution). Hence, it may be reasonable that chiral molecules may well need a higher ℓ_{max} for an adequate description of their photoionization dynamics than is more typically employed for more symmetric molecules.

Figure 6 shows some of the results obtained as part of a large systematic test of the impact of basis set size on convergence in CMS-Xα calculations performed for the (R)-enantiomer of carvone. [38, 52]. The angular basis used for the atomic sphere regions as well as the asymptotic region were varied; the asymptotic ℓ_{max} proved most critical and the data in the figure represent a balanced choice of atomic and asymptotic ℓ_{max} for small medium and large

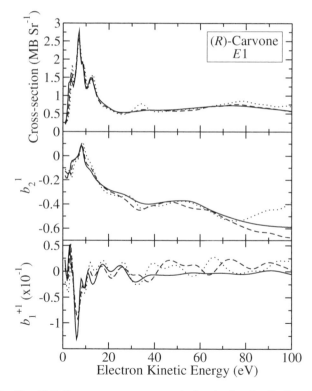

Figure 6. The CMS-Xα convergence tests made for the for the C=O carbonyl C $1s^{-1}$ ionization of the most stable (E1) conformer of (R)-Carvone. .... Small basis, $\ell_{max} = \{8, 4, 1\}$; – – – – Medium basis, $\ell_{max} = \{10, 6, 2\}$; _____ Large basis, $\ell_{max} = \{18, 10, 3\}$.

basis sets. These are designated by the truncation limits applying to the outer (asymptotic) region, the C or O atomic regions, and the H atomic regions respectively, namely $\{\ell_{out}, \ell_{C\&O}, \ell_H\}$.

It is apparent from this figure that the two calculated angular parameters are both more sensitive than the cross-section to a premature truncation, and that the the chiral $b_1^{\{1\}}$ parameter is more sensitive than the $b_2^{\{p\}}$, or β, parameter. The indicated choice of an asymptotic $\ell_{max} = 18$ represents a larger than usual (by a factor ≈ 2) outer sphere ℓ_{max} value, but is required for adequate convergence avoiding exaggerated oscillations at higher electron kinetic energies, especially in the chiral $b_1^{\{\pm1\}}$ term. This finding has been replicated in other chiral molecule PECD CMS-Xα calculations referred to in this chapter.

A similar convergence study has been reported for the B-spline calculation for *trans*-2,3 dimethyloxirane molecule. et al. [53] The authors in fact comment that the $b_1^{\{1\}}$ parameter is no more demanding on basis set size than the β parameter, but their data ([53] Fig. 1) show that when the asymptotic ℓ_{max} is increased from 10 to 15 there is a significantly greater improvement in the former angular parameter. A value of $\ell_{max} = 15$ was chosen for all subsequent B-spline calculations for oxiranes, and the same limit tends to be applied in the other reported B-spline calculations of chiral molecule PECD [60, 61].

Why should calculations of $b_1^{\{\pm1\}}$ evidently converge more slowly than those for $b_2^{\{p\}}$, and both more slowly than cross-section calculations? It seems likely that this is a consequence of the very different interference structure of these three terms and their different phase dependence—particularly the slower convergence of the $\sin(\eta_{lm} - \eta_{l'm'})$ terms in the $b_1^{\{\pm1\}}$ expression (see Section III.C). When phase differences are small, as they will be for higher ℓ waves that are partly attenuated in the molecular core region by centrifugal barriers, the rapidly varying sine term will lend them a disproportionate influence on the final value of $b_1^{\{1\}}$.

C. Molecular Conformation and Substitution Effects

In one of the very first realistic computational studies of PECD effects, performed for the amino acid alanine [51], it was noted that different results were obtained at each of three fixed geometries corresponding to low lying conformations identified in previous structure investigations [69–71]. A later combined experimental–theoretical study of PECD in 3-hydroxytetrahydrofuran [61] further looked at the influence of presumed conformation and concluded that while the predicted cross-section, σ, and β parameters were mildly affected by conformation, the chiral $b_1^{\{\pm1\}}$ parameters were much more strongly conformer dependent. The B-spline calculated PECD curves for the two conformations considered show very pronounced differences (see Fig. 7) and the distinctions are easily sufficient to distinguish and identify the dominant conformer present in the experimental sample.

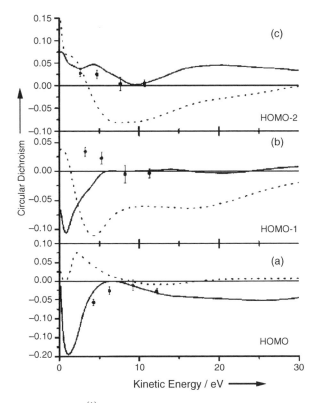

Figure 7. The PECD $b_1^{\{1\}}$ curves calculated for outer three orbital ionizations of hydroxytetrahydrofuran by the B-spline method. The solid curves refer to the conformation calculated to be most stable, the broken curves to a more energetic conformer. Experimental points with error bars are included for comparison. Reproduced, with permission, from Ref. [61].

Two wider ranging, more systematic investigations of conformational dependence have since been performed to establish whether the conformational sensitivity noted in the above PECD studies may generally provide a means for identifying and distinguishing gas-phase structure of suitable chiral species. The B-spline method has been applied to the model system (1R,2R)-1,2-dibromo-1,2-dichloro-1,2-difluoroethane [60]. Rotation around the C–C bond creates three stable conformational possibilities for this molecule to adopt. The results for both core and valence shell ionizations reaffirm an earlier conclusion; σ and β are almost unaffected by the rotational conformation adopted, whereas the PECD varies significantly. For the C 1s ionization to show any sensitivity at all to the relative disposition of the halogen atoms further reinforces the point made previously in connection with the core level PECD phenomenon,

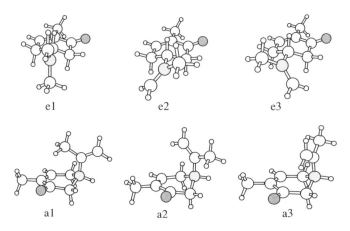

Figure 8. Equatorial **e1, e2, e3** and axial **a1, a2, a3** conformations of the carvone molecule. The asymmetric (chiral) carbon is shaded light gray, the $=CH_2$ carbon in the isopropenyl tail is shaded mid-gray and the carbonyl oxygen atom is shaded dark gray. Taken from Ref. [38].

that in such circumstances it is the delocalized continuum state that has the opportunity to sense the molecular chirality, and must therefore predominantly determine the PECD.

A study on conformational dependence of the PECD in carvone [38, 52] provides further corroboration of the picture emerging from these investigations. The carvone molecule has a six-membered ring with an isopropenyl tail grouping, and it is this tail that can move to create six different conformational structures, pictured in Fig. 8. These divide into two groupings; those with the tail axial to the ring, and those with it equatorial. In each case, three rotational conformers are obtained from rotation of the tail group. The axial and equatorial groups interconvert via a ring inversion, but otherwise the ring structure can be assumed quite rigid.

CMS-Xα calculations were performed to examine the $C=O$ carbon $1s$ PECD following an experimental investigation of this system [55]. Results of the calculation for $b_1^{\{1\}}$ for the conformers of (R)-carvone are shown in Fig. 9. Once again it is clear that there are major differences in the predicted $b_1^{\{1\}}$ curves, principally seen between the axial and equatorial groupings, but also significant variations among the rotational conformers in each category. This finding is all the more remarkable when it is recognized that the initial C $1s$ orbital is localized at some distance from the rotating tail group. Yet again, this highlights the role of multiple scattering, apparently effective over a range extending well beyond near neighbor, in the continuum state.

There were indications from other experimental work [72–74] that only the lower energy equatorial conformers are significantly populated at modest

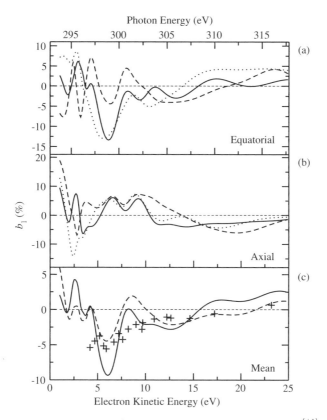

Figure 9. The $C=O$ carbon $1s^{-1}$ photoelectron dichroism parameter, $b_1^{\{\pm 1\}}$ for various conformational forms of (R)-Carvone. Equatorial conformations (a): **e1** ___; **e2** – – – –; **e3** . . . . . . Axial conformations (b): **a1** ___; **a2** – – – –; **a3** . . . . . . Panel c shows these CMS-Xα results averaged using relative populations derived from either B3LYP/6-31G** calculations (___) or MP2/6-31G** calculations (– – – –). Mean experimental values (Ref. [55]) are included in this panel for comparison.

temperatures, and an average PECD from these three is compared against experiment in Fig. 9. However, the agreement is improved by taking a Boltzmann weighted average over all six, suggesting the conclusion that the more energetic equatorial forms are also thermally populated. Perhaps the extent of agreement does not provide definitive evidence on this point, but nevertheless there are generally some clear indications that PECD may provide useful structural information.

In the same CMS-Xα study [52], a series of (R)-carvone derivatives, all in their single most stable conformation, were examined with a view to determining the influence of chemical substitution on the PECD. Both the

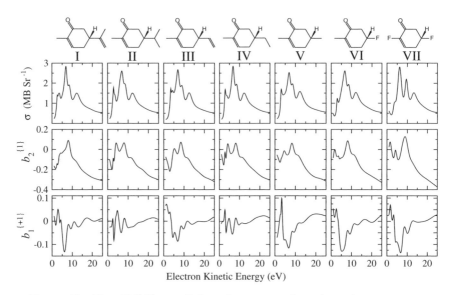

Figure 10. The CMS-Xα predictions for cross-section, the anisotropy parameter $b_2^{\{\pm 1\}}(\equiv -\beta/2)$, and the chiral $b_1^{\{+1\}}$ parameter in the carbonyl C $1s$ photoionization of (R)-carvone (**I**) and its indicated derivatives.

C $1s^{-1}$ ionization from the carbonyl group, and the HOMO^{-1} ionization of the carbonyl oxygen lone pair were included, with similar conclusions. In Fig. 10 the results for the **C**=**O** core ionization are summarized, with the structure of each derivative being indicated at the top of the figure.

The integrated cross-sections, σ, calculated for this family of molecules are very similar to one another; all have a number of features < 15 eV (notably at 3, 7, 10, and 12 eV) that can be tentatively attributed to shape resonances. Some supporting evidence can be gleaned from Hartree–Fock (HF) calculations. Certain molecular shape resonances are known to correlate with valence-like antibonding virtual orbitals that provide quasibound states embedded in the ionization continuum. Minimal basis set HF calculations identify at least two potential candidates—low lying π^* virtual orbitals with energies ∼0.2 and ∼0.4 hartrees. These virtual orbitals are located about the ring, are independent of the tail group, and are common to all the molecules **I–VII**.

The $b_2^{\{\pm 1\}}$ angular-distribution parameters show somewhat more molecule-to-molecule variation, but are essentially still quite similar to one another. They also display low energy structures that readily correlate with the noted features in the cross-section curves, and the variations are mainly confined to changes in the relative intensity of these. Much bigger differences are, however, found in the calculated chiral $b_1^{\{+1\}}$ parameters, though here as well there is a common characteristic (strong minimum at $\sim 6\,\text{eV}$) and more detailed structures that seemingly correlate with the resonant features in the cross-sections. It was similarly noted in B-spline studies on four oxirane derivatives discussed previously [53] that the chiral $b_1^{\{1\}}$ parameter changed far more on substitution than either σ or β.

In an effort to better understand the differences observed upon substitution in carvone possible changes in valence electron density produced by inductive effects, and so on, were investigated [38, 52]. A particularly pertinent way to probe for this in the case of core ionizations is by examining shifts in the core electron-binding energies (CEBEs). These respond directly to increase or decrease in valence electron density at the relevant site. The CEBEs were therefore calculated for the $C{=}O$ C $1s$ orbital, and also the asymmetric carbon atom, using Chong's ΔE_{KS} method [75–77] with a relativistic correction [78].

Results for these CEBEs are presented in Table I. As can be seen, for the carvone variants I–V the various substitutions have absolutely no effect at the carbonyl $C{=}O$ core, and are barely significant at the chiral center that lies between the carbonyl and substituent groups in these molecules. Only upon fluorine substitution at the tail (molecule VI) does the $C{=}O$ CEBE shift by one-half of an electronvolt; the second F atom substitution adjacent to the $C{=}O$ in the difluoro derivative, VII contributes a further 0.6-eV shift. This effect can be rationalized due to the electron-withdrawing power of an F atom. Paradoxically, it is these fluorine-substituted derivatives, VI, VII, that arguably produce $b_1^{\{1\}}$ curves most similar to the original carvone conformer, I, yet they are the only ones to produce a perturbation of the ground-state electron density at the C $1s$ core. This contributes further evidence to suggest that, at least for the C $1s$

TABLE I
Table of Relativistically Corrected Core Electron Binding Energies Calculated Using the $\Delta E_{KS}(\text{PW86} - \text{PW91}) + C_{\text{rel}}$ Method.

Molecule[b]	$\Delta E_{KS} + C_{\text{rel}}$ (eV)	
	C=O	C*[c]
I	292.477	290.920
II	292.502	290.699
III	292.571	291.037
IV	292.540	290.792
V	292.584	290.945
VI	293.040	293.274
VII	293.619	293.651

[a]Data taken from Ref. [38].
[b]Molecule structures as defined in Fig. 10.
[c]The asymmetric carbon at the chiral center.

ionization, PECD originates very much as a final-state scattering effect that senses changes in the molecular potential (electron density) at some distance from the initial localization site.

Of course, these studies raise the general question as to why the $b_1^{\{1\}}$ parameters are so much more sensitive to conformation than either σ or β, and how this sensitivity can extend to small changes in configuration or substitution occurring at sites relatively far removed from the initial orbital scattering center. The most plausible explanation, again, would seem to be the unique phase dependence of this term, discussed in Section III.C. The appearance of relative phase in a sine term seems likely to amplify the impact of even quite small phase differences, such as may be expected when weak multiple scattering remote from the initial site localization is subtly modified by conformational changes.

D. PECD and Continuum Resonances

It is well known that the value of the β parameter, more than the cross-section σ, often shows a strong response to resonant structure embedded in the continuum. Given the sensitivity exhibited by the $b_1^{\{1\}}$ parameter in the foregoing there must be an *a priori* expectation that it would also show a strong response to resonant behavior. Computational methods do not yet exist to deal with autoionization phenomena in the systems of interest here, but one electron shape resonances can, in principle, be examined.

A possible role for shape resonances has been postulated in a number of the photoionization studies mentioned above [52, 53, 57, 60], although it has to be noted that except for camphor [57], the evidence for the existence of the shape resonance is not definitive. (It also then remains an open question how any such resonances, inferred from fixed geometry calculations, would manifest themselves in practice in large, and sometimes floppy, molecules, such as these chiral species.)

In general terms, it has been seen here that the $b_1^{\{1\}}$ parameter curves are almost always more structured than β parameter curves. The latter are known from years of study to broadly conform to a pattern (in the absence of resonances) that starts from a small value at threshold and over a span of a few tens of electronvolts approaches the positive limit ($\beta = +2$), essentially monotonically. Empirically, small distinctions between σ and π orbital ionizations can be discussed, and of course there are many significant exceptions to such broad expectations. In contrast, there is clearly far more variability, and much less intuitive predictability in the detail of the $b_1^{\{1\}}$ curves we have seen. That being the case, while suggested shape resonant features in σ and β parameter curves can sometimes apparently map onto features in the $b_1^{\{1\}}$ curves [55, 57, 60] these are no more prominent than other structure and seem unlikely, by themselves, to provide visual clues to the presence of a resonance.

From the available evidence Stener and co-workers [53, 60] conclude that the chiral $b_1^{\{1\}}$ parameter is more sensitive to small asymmetries in the molecular potential than to continuum collapse effects at resonance. At present, such conclusions must be provisional as there is little direct evidence. There is also no evidence regarding likely behavior at autoionization resonances, and this too deserves attention.

E. Future Developments

From the discussion provided here, it should be apparent that the current level of theory can provide a reliable description for available experimental data that in favorable cases approaches near exact agreement. Certainly, the prediction of shape, magnitude, and importantly sign of the $b_1^{\{1\}}$ curves is already sufficiently good to indicate that PECD could now serve to establish absolute configuration of an enantiomer (R- or S-) where this is initially unknown.

The two computational methods, CMS-Xα and LCAO B-spline DFT, for now provide consistent, comparable results [57] with little to choose between them in comparison with experiment in those cases presented here (Sections I.D.1.a and I. D.a.2). The B-spline method holds the upper hand aesthetically by its avoidance of a model potential semiempirically partitioned into spherical atomic regions. More importantly it offers greater scope for future development, particularly as the inevitable increases in available computing power open new doors.

A common problem for both methods lies in the use of potentials that do not possess the correct net attractiveness. This can have the consequence that continuum features appear shifted in energy. In particular, there is evidence that the LB94 exchange-correlation potential currently used for the B-spline calculations, although possessing the correct asymptotic behavior for ion plus electron, is too attractive, and near threshold features can then disappear below the ionization threshold. An empirical correction can be made, offsetting the energy scale, but this can mean that dynamics within a few electronvolts of threshold get an inadequate description or are lost. There is limited scope to "tune" the Xα potential, principally by adjustment of the assumed α parameter, but for the B-spline method a preferable alternative for the future may well be use of the SAOP functional that also has correct asymptotic behavior, but appears to be better calibrated for such problems [79].

The threshold region is in fact particularly challenging for theoretical modeling. Intuitively, one anticipates that a slow outgoing photoelectron is most sensitive to the molecular potential, thus requiring a more accurate potential. It is also likely that electron correlation effects play an important role in this region [57] and one can look to any further developments in exchange correlation potentials being readily introduced into the B-spline method as they become available.

Both current methods share the limitation of being conducted for fixed nuclear geometry—typically the ground-state equilibrium geometry. However, as made clear in Section I.D.3, the chiral $b_1^{\{1\}}$ parameter appears to be strongly dependent on the assumed structure. This is then a concern particularly for nonrigid molecules and those where low frequency or large amplitude vibrations are possible. The consequences of the methyl group free rotation in methyloxirane have been explored by repeating B-spline calculations for a range of rotation angles [60]. Predictably, the $b_1^{\{1\}}$ parameter curves obtained differ from one another (see Fig. 11), but the sheer magnitude of their variability

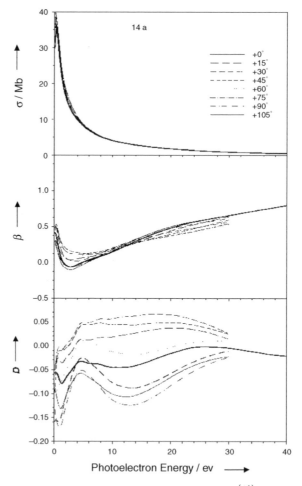

Figure 11. Cross-section, σ, β parameter, and chiral $D(\equiv b_1^{\{+1\}})$ for the 14a orbital of (S)-methyloxirane for different rotational positions of the methyl group. Reproduced, with permission, from Ref. [60].

is still unexpected. Perhaps fortuitously, when a conformationally averaged result is formed there is a significant mutual cancellation of individual deviations from the equilibrium geometry result, so that the average is very close to the original calculation at a fixed equilibrium geometry. It is not known whether such cancellations to the equilibrium result can always be relied upon. Dramatic geometry dependent variations, such as encountered in this study [60], promote some consideration of potential vibrational coupling effects and development of methods for the realistic inclusion of vibrational modes constitute a goal for photoionization dynamics calculations generally.

Another general goal, but one already identified (Section IV.A.2) to be of specific interest for PECD investigations, is the inclusion of the electron density's response to the external field. Applying a time-dependent (TD) DFT formalism in the B-spline method provides a means to accomplish this. The time-dependent perturbation creates an induced potential, as a result of Coulomb and exchange-correlation responses, that provides a descriptor for interchannel coupling effects. Continuum–continuum coupling mechanisms can be expected to permit intensity borrowing between channels, whereas the ability to model reliably discrete–continuum coupling, means that autoionization phenomena can be naturally incorporated in a TD–DFT calculation.

While TD–DFT continuum calculations for molecules, such as camphor, are not yet quite practicable, efforts to create highly parallel computer codes capable of tackling this scale of problem are expected to be fruitful soon. In the meantime TD–DFT studies for computationally less demanding small molecules [66–68] or highly symmetric molecules, such as SF_6 [79], have provided indications of the general value of the inclusion of electron response effects.

V. EXPERIMENTAL TECHNIQUES FOR PECD MEASUREMENT

Despite the first prediction [34] of a measurable PECD effect being a few decades old, it is only in the last few years that experimental investigations have commenced. Practical experiments have needed to await advances in experimental technology, and improvements in suitable sources of circularly polarized radiation in the vacuum ultraviolet (VUV) and soft X-ray (SXR) regions needed for single-photon ionization have been been key here. In the meantime, developments in other areas, principally detectors, also contribute to what can now be accomplished.

Synchrotron radiation provides a convenient source of tunable VUV and SXR radiation. Natural synchrotron radiation, emitted by relativistic electrons, is linearly polarized in the plane of their orbit, which is traditionally the configuration used to collect the radiation. However, it is well known that the polarization becomes elliptical if observed above or below the plane of the orbit.

In practice, this is difficult to exploit fully; one is essentially required to misalign the optical beamline path to accept out-of-plane radiation. The degree of ellipticity (misalignment) has to be traded against intensity and pure circular polarization states are not achieved. Moreover, it is difficult to obtain symmetric degrees of left and right polarization by moving above and below the orbit plane, and very difficult to achieve reproducibility in beamline settings and in optical alignment into the spectrometer when doing this. For this reason, the first investigation of PECD [65, 80] made using this generation technology at the BESSY I synchrotron (Berlin) chose rapid switching of the gas-phase enantiomer in preference to that of the light polarization in order to demonstrate the asymmetry in photoemission with minimal contribution from purely instrumental asymmetries and long-term drifts.

The advent of helical magnetic insertion devices at newer generation synchrotron sources has greatly improved this position, providing highly pure, symmetric left- and right-circular polarization that can be rapidly switched.

The following section first summarizes the capabilities of these more recent, insertion device equipped synchrotron sources that have been used for PECD measurements by the author and co-workers, then the experimental techniques employed with them to obtain PECD measurements.

A. Polarized Light Sources

An undulator, briefly, is a periodic array of magnetic poles inserted into the straight sections of a synchrotron storage ring [81, 82]. The stored relativistic electrons passing through this array are forced into oscillating transverse excursions, and radiate as they do so. Interference effects at each oscillation result in intense light emission in the forward direction, relatively narrowly peaked about a central wavelength that is determined by the magnetic field strength. In a planar undulator, the light is linearly polarized parallel to the plane of the forced electron oscillations.

Helical undulators build on this principle by using two orthogonal magnetic field arrays [82, 83]. These permit transverse excursions in perpendicular x and y directions. If the arrays have a relative longitudinal shift, this introduces a phase to the induced perpendicular excursions and when the phase is $\pm 90°$ the electron trajectory can follow left- or right-handed corkscrew paths The emitted radiation is correspondingly right- or left-handed CPL.

1. Polarimetry

Accurate experimental determinations require that not only the handedness of the produced light, but its exact degree of polarization, are known. The theoretical performance of an undulator may be in practice be degraded by magnetic defects, and the optical beam can be further depolarized by reflections along the beamline. Again, the dephasing on optical elements can in principle be predicted, but for some combinations of material and wavelength, the Fresnel coefficients will be too rapidly varying and/or poorly known for this to be relied upon. Long-term changes may also arise due to contamination build-up on reflecting surfaces. Where possible, this all mandates the *in situ* checking of polarization at the experimental chamber with a suitable polarimeter.

A complete description of the polarization state can be given by the set of Stokes parameters

$$S_0 = \sqrt{S_1^2 + S_2^2 + S_3^2} + S_4 \tag{21}$$

where S_0 is the total flux, S_1 corresponds to the difference in flux between horizontal and vertical linear polarizations, S_2 to the difference in flux linear polarizations tilted $45°$ and $135°$, S_3 is the circular polarization state or the difference between rcp and lcp components of the light, and S_4 is the unpolarized component. The Stokes parameters are normally quoted in normalized form $s_3 = S_3/S_0$, and so on.

On an undulator beamline, s_4 would normally only be non-zero as a consequence of some defect or optical pollution from some source. The dichroism and asymmetry factor, γ, will scale with s_3 so that the ideal for PECD measurements would be to have $|s_3| \to 1$ and $s_4 \to 0$.

2.   *UE56/2 Beamline, BESSY II*

The UE56/2 SXR beamline at the BESSY II synchrotron (Berlin, Germany) consists of two APPLE-II type undulators, producing linear or elliptical polarizations [84], in series with an intervening chicane [85, 86]. This chicane can be set to produce an angular displacement of 400 μrad between the optical beams from the two undulators that are then propagated effectively in parallel along the beamline, passing through a plane grating monochromator before being brought back to a common focus in the spectrometer by the final refocusing mirror. The lateral displacement of the two beams prior to this mirror permits a mechanical selector or chopper to be inserted and so enables rapid selection of either beam. For CD-type measurements, the two undulators can be set to produce light of opposite circular polarization state. Spectra corresponding to the two opposite photon helicities may thus be recorded in an interleaved fashion. Switching at 0.1 Hz is sufficiently fast that signal variations due to long-term drifts in gas pressure, decaying beam intensity, and so on, are common to both spectra, and ideally will cancel when a dichroism signal is generated by subtraction of one spectrum from another.

In situ polarimetry could be performed using an eight axis SXR polarimeter [87] with multilayer optics mounted immediately behind the experimental chamber. Full determinations of the four Stokes parameters are time consuming,

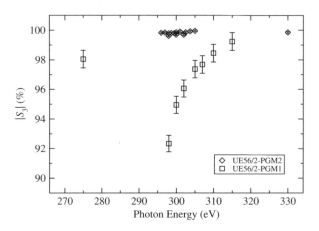

Figure 12. The degree of circular polarization $|s_3|$ measured behind the experimental chamber on two optically similar branches fed by the same undulator pair of the UE56/2 beamline at BESSY II. Data taken from Ref. [55].

but an abbreviated method [88] acceptable for ongoing checks permits the magnitude[2] of s_3 to be obtained more rapidly, assuming $s_4 \approx 0$.

It is revealing to look at an example of such data in Fig. 12. The UE56/2 beamline has two nominally identical branches with replicated optical paths, but that have been used for different experiments. The measurements show a high degree of circular polarization, $|s_3| > 98\%$, except near carbon K edge where the polarization reduces, but only on one branch. It is postulated that this is a consequence of carbon contamination on a beamline optic in that branch. These results demonstrate the necessity to be alert to such possible causes of degradation and to perform polarization checks where possible rather than rely on theoretical predictions.

3.   SU5 Beamline, LURE

The SU5 beamline [89] at the Super-ACO synchrotron (LURE, Paris[3]) employed an electromagnetic undulator to produce fully variable polarization in the VUV region [83, 90, 91]. This beamline was equipped with a gas filter for the suppression of unwanted higher order radiation [92] and had a VUV polarimeter [93] permanently installed just before the experimental chamber that could be rapidly lowered into the beam for polarization determinations. Full polarization analyses had been performed in commissioning, with s_3 values ranging from 0.9 to 0.96 for rcp and from 0.9 to 0.99 for lcp [93]. The remainder was determined

---

[2] The sign of s_3 can be reliably assumed from the undulator setting.

[3] LURE recently closed, but the concept of this beamline is to be reimplemented in the DESIRS beamline under construction at the new Soleil synchrotron outside Paris.

to be the s_4 unpolarized component, with negligible linear components s_1, s_2. Just as on the UE56/2 beamline the *in situ* polarimeter could be used in the rapid scan mode for quicker verification of the s_3 magnitude during experimental runs, if so desired.

In principle, the electromagnetic undulator design permits a rapid polarization switch by reversal of the current in one of the magnetic arrays. Although a rapid switching mode was not offered on SU5 (polarization change typically required a few minutes to complete), it is planned that such a mode will be available on the upgraded DESIRS beamline once it has been commissioned. This would then be expected to offer similar advantages to those noted for the rapid switching on UE56/2.

4. CIPO Beamline, Elettra

The circular polarization (CIPO) beamline at the Elettra synchrotron (Trieste, Italy) operates in the VUV–SXR range with radiation from a combination permanent magnet–electromagnetic elliptical wiggler [94, 95]. This does not achieve full circular polarization in the VUV region, but rather an elliptical output with principal axis lying in the horizontal plane ($s_1 > 0, s_2 = 0, s_3 < 1$). Unfortunately, in the VUV region no polarimetry data are available, but calculations indicate the degree of circular polarization achieved by the wiggler may be $\sim 80\%$, estimated to be no worse than 70% delivered at the experimental chamber [95, 96]. In PECD experiments, we have calibrated the polarization state by deduction from cross-comparison of results at a few fixed energies previously studied on the SU5 beamline where accurate polarimetry data was available [36]. Because the horizontal magnetic field array in the insertion device is electromagnetic, fast current reversal to switch left- and right-handed elliptical polarizations is possible, with the usual potential benefit for dichroism measurements.

B. Photoelectron Imaging for PECD

An electron imaging technique has been developed by the author and his co-workers for recording PECD data. In this approach the three-dimensional (3D) photoelectron angular distribution is projected onto a two-dimensional (2D) position sensitive detector. [97, 98] Assuming a cylindrical axis of symmetry (in this case the photon beam propagation direction) the 3D velocity distribution can in principal be reconstructed by one of several variants based on the inverse Abel transformation [98]. The potential advantages for measurement of the chiral asymmetry in the photoelectron angular distribution are evident; the full photoelectron energy and angular distribution [Eq. (4)] is captured, and can hence be recovered, in a single measurement that uses the full 4π solid angle of the photoemission, guaranteeing maximum efficiency and count rate.

Many algorithms for implementing the data inversion of a projected image have been proposed, but a limitation common to many is that they operate along chords perpendicular to the symmetry axis in the image, from the outer edge, toward the center. A consequence is that noise tends to accumulate in the center along the symmetry axis. For measurement of the forward–backward electron asymmetry this is particularly unfortunate since the directions $(0°, 180°)$ in which the asymmetry is expected to be peaked are the most obscured by such noise.

This has motivated development of a new fitting procedure [99] that builds upon an earlier idea to fit a 2D projection image with a set of basis functions having known inverse Abel integrals—the basis set expansion or Basex method [100]. In our development, the image is first transformed from the Cartesian coordinates of the detector to polar coordinates [99]. This transformed image is then fitted with $\{r, \theta\}$ basis functions where the angular terms are restricted to just those Legendre polynomial functions expected from the underlying physics—in the present case P_0, P_1, P_2. Operating on the image in $\{r, \theta\}$ space is better adapted to the polar symmetry of the photoionization process (or indeed many photofragmentation experiments) and brings immediate efficiency benefits through a much reduced basis set size and an improved quality of fit. Indeed, this polar modified (pBasex) inversion procedure seems to be quite generally beneficial [99]. The accumulation of noise toward the center now moves to the central mid-point area of the original image, where the information content is already minimal. The fitted radial distribution directly maps the electron speed distribution (effectively the angle integrated photoelectron spectrum) while the angular distribution is directly and fully expressed in the fitted Legendre coefficients obtained at each resolved radius–energy.

Our imaging spectrometer is also a special development of the now ubiquitous velocity map imaging (VMI) device [101]. The velocity mapping conditions use gridless electrodes for charged-particle extraction and acceleration toward the detector to create an immersion lens whose transfer function ideally maps all image trajectories of given velocity to the same point on the detector plane, regardless of their initial position in the source. This can substantially eliminate the smearing of an image by trajectories that start from different locations in a source volume of finite dimensions.

For a synchrotron radiation experiment, this feature is particularly advantageous; although modern beamlines produce a small spot size at the experiment ($\leq 250\,\mu m$) ionization volumes may nevertheless be quite extended along the photon beam direction due to the gentle focusing ($f \sim 1$ m). Unlike many tightly focused laser experiments there is thus no pronounced beam waist. A second aspect motivating our development is that frequently a greater range of kinetic energies—perhaps several tens of electronvolts—can be accessed and is of interest in a typical synchrotron experiment. The ability to image high energy electrons fully onto a detector is set by limits on physical apparatus dimensions and achievable extraction fields. To overcome these limitations, we implemented a coupled double einzel lens in the drift space of the VMI

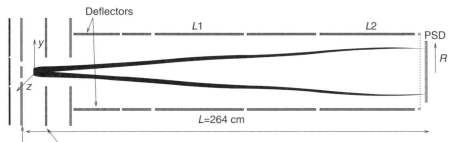

Figure 13. Modified Velocity Map Imaging spectrometer showing the double einzel lens, L_1, L_2, and 5-eV kinetic energy initially transverse trajectories from an extended source volume with $V_{rep} = 3000$ V, $V_{ext} = 0.695 \times V_{rep}$, and $V_{L1} = V_{L2} = 1000$ V. Taken with permission from Ref. [102]. Copyright (c) 2005, American Institute of Physics.

spectrometer [102] with the intention of zooming the image size. The lens design was optimized with a genetic algorithm, and the final choice increases the maximum energy that can be imaged by a factor of up to 2.5 while maintaining good focusing properties for a source of several millimeters extension. A schematic showing the lens structure and the velocity mapping by means of ray-traced trajectories is shown in Fig. 13. An added bonus from this new lens is an improvement in the radial focusing performance when the lens is operated below full strength compared to operation of the spectrometer without.

The spectrometer is fitted with a skimmed c.w. supersonic molecular beam source. Many chiral species of interest are of low volatility, so a heated nozzle–reservoir assembly is used to generate, in a small chamber behind a 70-μm pinhole, a sample vapor pressure that is then seeded in a He carrier gas as it expands through the nozzle [103]. Further details of this apparatus are given elsewhere [36, 102, 104].

1.   Data Acquisition and Analysis

A single image recorded for a fixed enantiomer and fixed circular polarization state in principle carries the full information sought consisting, after inversion, of the parameters $b_1^{\{\pm 1\}}, b_2^{\{\pm 1\}}$ and the radial distribution function $n(r)$. After mapping to energy space $n(r)$ gives $n(E)$, the photoelectron spectrum (PES). The full cylindrically symmetric 3D electron distribution before projection onto the 2D detector, and which we seek to recover by the image inversion, can be written

$$i(r, \theta) = n(r)I^{\{p\}}(\theta; r) \tag{22}$$

where $I^{\{p\}}(\theta; r)$ is the angular distribution Eq. (4) but with possible energy dependence of the $b_j^{\{p\}}$ parameters explicitly acknowledged through a parametric dependence on r, the radial coordinate. However, there are advantages to obtaining and working from both the lcp and rcp images. This is discussed fully elsewhere [36] and summarized here.

Ideally, if rapid polarization switching is available recordings with the two polarizations are interleaved, so minimizing any differences in the accumulated data that could be attributable to long-term drifts. Failing that, a pair of images with the alternative polarizations have been recorded in sequence, taking care to accumulate the same integrated electron count in each. Typically, this total count may be of the order of 10^7–10^8 electrons per image. In practice, because of the decay of the stored electron current and corresponding decrease in light intensity, this requirement may mean longer acquisition times for the second image.

The two images may be subtracted to create a difference image that reveals the dichroism directly. The 3D distribution recovered upon inversion, $D(r, \theta)$, is thus

$$D(r, \theta) = i_{\text{lcp}}(r, \theta) - i_{\text{rcp}}(r, \theta)$$

$$= n_{\text{lcp}}(r) I^{\{+1\}}(\theta; r) - n_{\text{rcp}}(r) I^{\{-1\}}(\theta; r) \tag{23}$$

$$= n(r) 2 b_1^{\{+1\}}(r) \cos(\theta) \tag{24}$$

where the last step follows from the properties of Eq. (4) and an assumption that the radial distribution, or angle integrated PES, is independent of polarization so that $n_{\text{lcp}}(r) = n_{\text{rcp}}(r) \equiv n(r)$. A pBasex inversion can thus be performed for a single r-dependent angular term. Compared to an image inversion of a single image the benefit is twofold. First, for the data inversion and associated statistics, two data sets are combined and then fitted with fewer free parameters providing better performance. More importantly, this approach helps treat the consequences of a less than perfect degree of circular polarization, $|s_3| < 1$. In most circumstances, lcp and rcp settings are likely to provide the same actual degree of circular polarization. If so, any contributions made by the s_1, s_2, and/or s_4 components of the light source should cancel out on taking the differences; the dichroism image will only be sensitive to the s_3 components.

Extracting quantitative values for $b_1^{\{+1\}}$ from Eq. (24) requires dividing the fitted $\cos(\theta)$ angular coefficient by $2n(r)$. This can be separately estimated by adding the pair of 2D images, and further summing the combined image to obtain an angle integrated pure radial distribution. This is inverted by fitting purely radial forward Abel functions to yield an estimate $n'(r)$, which operation can be performed relatively noise free [36]. Now, however, $n'(r)$ may contain a contribution from residual s_1 and/or s_4 components of the light source. A simple correction procedure scales $n'(r)$ by the polarization rate factor s_3/s_0 to remove the non-s_3 contribution to the total electron yield. The corrected $n(r)$ can then be applied in Eq. (24) to obtain $b_1^{\{+1\}}$.

The above determination of $b_1^{\{+1\}}(r)$ requires no knowledge of the $b_2^{\{\pm1\}}(r)$ parameter. This may, however, be extracted from a summed image that retains the angular information of interest by an appropriate pBasex inversion fitting for a P_2 coefficient. This determination will not be adversely affected by any residual s_4 component of the light source. The P_2 term's dependence on a lcp, rcp, and unpolarized light beams is identical: $b_2^{\{\pm1\}} = b_2^{\{unpol\}} = -\frac{1}{2}\beta$. Any residual linear component is more problematic. First, the cylindrical symmetry about the light beam propagation is broken by the linear polarization; strictly this invalidates the assumptions and hence the application of Abel-based inversion procedures. Even if this is circumvented an attempt to still fit the experimental data with a Legendre polynomial around the propagation axis will lead to erroneous estimates of the $b_2^{\{\pm1\}}$ coefficient, creating some uncertainty in this coefficient when the light source is elliptically polarized.

C. Non-imaging PECD Measurement

A limitation of the photoelectron imaging technique in the SXR region appears to be the presence in such circumstances of a background of very much higher energy electrons (from the valence shell, Auger processes, . . .). Since there is no energy filtering as such, a proportion of these electrons are also detected, degrading the signal-to-noise ratio. Worse, these very fast electrons are only partially focused onto the detector under focusing conditions that are optimized to capture e.g., K-edge electrons having up to a few tens of electronvolts. Without having captured the full 3D distribution of this background count in the 2D image, the image inversion algorithms cannot properly recognize the background contribution and so cannot be relied upon.

For this reason in the core ionization region explored by the author and co-workers at the BESSY II synchrotron, we have relied upon a different approach, recording the dichroism at a fixed observation angle using a dispersive electrostatic analyzer that filters out electrons other than those in the energy range of interest [55, 56]. The same approach using fixed angle detection with one or more analyzers has been used by other groups in the VUV excitation region [62, 65].

An observation of the dichroism at a fixed angle is capable of yielding the chiral $b_1^{\{1\}}$ parameter via the asymmetry factor Γ defined in Eq. (10). A potential difficulty for this approach is that extraction of $b_1^{\{1\}}$ from a normalized asymmetry factor $\Gamma(\theta)$ requires a knowledge also of the second parameter $b_2^{\{\pm1\}} \equiv -\frac{1}{2}\beta$ as can be seen in Eq. (10). To determine this requires measurements to be made for at least one other detector angle. This requirement can, however, be circumvented by choosing the magic angle, 54.7° to the light propagation axis, for performing the single dichroism measurement; at this angle the value of the second Legendre polynomial term is zero and so the exact value of $b_2^{\{\pm1\}}$ becomes irrelevant.

In comparison with the PECD imaging approach this method has some disadvantages. Measurement at 54.7° reduces the observable asymmetry by a factor $\cos(54.7°) = 0.58$. Unlike the imaging, which accepts a full 4π steradian solid angle, the angular acceptance by the dispersive analyzer has to be severely restricted to achieve angular resolution and only a single electron kinetic energy monitored at a time. This all constitutes a large reduction in accepted count rate, and hence sensitivity, although at BESSY the latter effect can be ameliorated somewhat by the use of a hemispherical analyzer equipped with a multichannel detector to provide some energy multiplexing. The bandwidth has proved sufficient to monitor simultaneously a number of photoelectron peaks [55]. On the plus side, the dispersive analyzer offers potentially greater energy resolution, especially for higher mean energies, than the imaging achieves, and has the better rejection of background electrons as noted.

Those fixed-angle measurements reported to date have all used either a heated effusive inlet, or heated gas cell for sample admission [55, 56, 61, 62, 65]. Probably the higher sample number densities these sources generate, compared to a supersonic beam source, provides some compensation for the reduced collection efficiency in the fixed-angle measurement.

1. Data Acquisition and Treatment

A data taking methodology has been developed, optimized for the particular characteristics of the UE56/2 BESSY beamline [55]. As discussed in Section V.A.2, the beamline provides rapid polarization switching allowing interleaved recording of spectra with lcp and rcp. However, the source consists of two undulators in sequence that follow different beam paths up to and including the final beamline refocusing mirror. Despite careful alignment, the two delivered beams cannot be guaranteed to be identical in intensity and final focus, and may thus contribute some instrumental asymmetry. To overcome this four data sets A–D are recorded one after the other. Each data set consists of a pair of spectra for opposite helicities that are designated

$$\{a^{u+}, a^{d-}\}, \{b^{u-}, b^{d+}\}, \{c^{u-}, c^{d+}\}, \{d^{u+}, d^{d-}\} \tag{25}$$

where a^{u+} denotes a binned spectrum recorded on the $u$pstream undulator set to positive helicity, a^{d-} that recorded simultaneously on the $d$ownstream undulator set to negative helicity, and so on.

The first and last data sets, A and D, are thus recorded using an opposite undulator configuration (u^+, d^-) to the second and third, B, C (u^-, d^+). While this sequence is recorded the stored beam current will decay, and with it the light intensity, but this sequencing is intended to balance out the data acquisition time spent in either undulator configuration. At the end, equal counts are obtained in each data set, and for each undulator configuration.

The data are then combined in various ways [55]. First, to aid identifying possible instrumental contributions to the measured asymmetry two spectra are formed, each combining measurements with both helicities, but on a given undulator. Differences between these two combination spectra can then be attributed to undulator characteristics rather than any dichroism. In this manner, a small energy discrepancy (~ 20 meV) was detected, affecting photolines, but not Auger lines. It was inferred that this suggested small shifts in photon energy, perhaps due to small angular differences in the two optical beam paths at the monochromator [55]. The energy scale calibration of spectra on the two undulators was subsequently corrected before any further use.

An alternative combination pair of spectra is then formed, taking the geometric mean of two undulator spectra of positive helicity, and of the two recorded with negative helicity. These two spectra of given light helicity each contain corresponding corrected contributions from both undulator sources, and it can be shown [55] that the instrumental asymmetries are effectively canceled by this procedure.

VI. PECD EXPERIMENTAL CASE STUDIES

The list of molecules whose PECD has been experimentally studied is quickly expanding, and in the VUV valence shell region now includes the prototypical chiral species camphor [36, 64, 65], bromocamphor [65, 80], fenchone and carvone [38], methyl oxirane [62, 63], glycidol [37, 38], and 3-hydroxytetrahydrofuran [61]. Studies of camphor [56], fenchone [38], and carvone [55] have all been extended to cover the SXR C $1s$ core region.

Passing reference has been made to some of these experiments, and some of their results briefly presented, in preceding sections. What follows here is a personal selection of case studies presented in greater detail. These are chosen to highlight various additional aspects of current investigation of the PECD phenomenon.

A. Fenchone

Figure 14 shows a carbon $1s$ core region of the X-ray photoelectron spectrum (XPS) of the fenchone molecule. The structure of fenchone is indicated as an inset in Fig. 14 and can be seen to be a bicylic ring structure, very similar to camphor. Fenchone has two stereogenic centers, or asymmetric carbons, where the rings fuse, but is not a true diasteroisomer as the rigid structure constrains one center to always receive the opposite designation of the other as defined by the Cahn–Ingold–Prelog sequence rules [105]. The assignment, guided by a $\Delta E_{KS}(\text{PW86} - \text{PW91}) + C_{\text{rel}}/\text{TZP}//\text{MP2}/6\text{-}31\text{G(d,p)}$ [75–78] calculation of the core-binding energies for this molecule, is very straightforward; the C=O

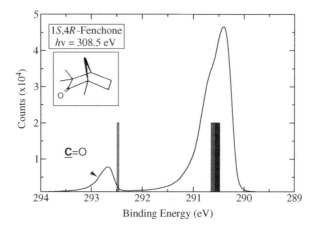

Figure 14. The C 1s core region XPS of fenchone recorded with a photon energy $h\nu = 308.5$ eV. Included in the figure are bars indicating calculated $\Delta E_{KS}(\text{PW86} - \text{PW91}) + C_{\text{rel}}$ core-binding energies. Data taken from Ref. [38]. The inset shows the structure of the (1S,4R)-enantiomer.

carbonyl 1s CEBE shows a significant shift and produces a well-resolved peak corresponding to this single orbital at a BE of 292.7 eV. All the other C 1s orbitals have very similar CEBEs and so all contribute to the single main peak in the spectrum without being individually resolved.

The well-resolved **C**=O 1s peak in the fenchone XPS provides an excellent opportunity to examine PECD from a single, well-characterized initial orbital. As has been previously mentioned, it might be thought that such a localized, spherically symmetric initial orbital would not be sensitive to the molecular enantiomer's handedness, but as can be seen in Fig. 15 (a) the dichroism in the electron yield recorded at the magic angle is sufficiently large to be easily visible by eye as a difference in the intensity of the lcp and rcp spectra.

The second row in Fig. 15 shows examples at the three selected photon energies of the **C**=O 1s difference spectra obtained for both enantiomers. After normalization by the mean spectrum the asymmetry factor $\Gamma(54.7°)$ is plotted along the bottom row. After correction for the $\cos(54.7°)$ term arising from the specific experimental geometry the net forward–backward asymmetry, γ, can be estimated to reach a peak $\sim 15\%$ in the $h\nu = 298.7-$ eV photoionization.

In accordance with the predictions that can be made on the basis of just the electric dipole approximation (see Section III.A) the observed dichroism is equal, but of opposite sign for the two enantiomers. This could be seen also in the valence shell ionization results for glycidol presented in Fig. 2. The added significance here is that a contribution to the angular distribution by higher order

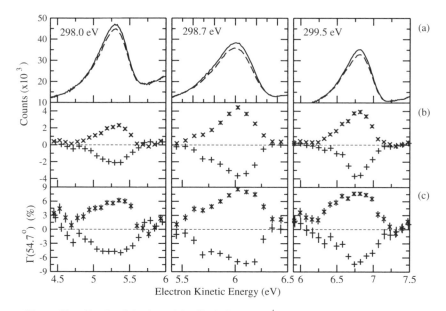

Figure 15. Circular dichroism of the C=O C 1s peak (BE = 292.7 eV) in fenchone at three different photon energies, indicated. (a) Photoelectron spectrum of the carbonyl peak of the (1S,4R) enantiomer, recorded with right (solid line) and left (broken line) circularly polarized radiation at the magic angle, 54.7° to the beam direction. (b) The circular dichroism signal for fenchone for (1R,4S)-fenchone (×) and the (1S,4R)-fenchone (+) plotted as the raw difference $I_{lcp} - I_{rcp}$ of the 54.7° spectra, for example, as in the row above. (c) The asymmetry factor, Γ, obtained by normalizing the raw difference. In the lower rows, error bars are included, but are often comparable to size of plotting symbol: (1R,4S)-fenchone (×), (1S,4R)-fenchone (+). Data are taken from Ref. [38].

interaction terms cannot be automatically discounted at these photon energies [106–108].

1. Non-dipole interactions

Without needing quantitative calculations of terms of higher order than the electric dipole interaction it is possible to make some useful deductions about their likely signature in an angular distribution by considering just the symmetry properties of the angular coefficients that such interactions would provide. Ritchie has provided formulae, analogous to Eq. (A.15), for both the E1–E2 and E1–M1 electric dipole–magnetic dipole and electric dipole–electric quadrupole second-order interaction interference terms [109]. Somewhat similarly to the argument followed in Section III.A, symmetry properties of the angular coefficients under photon helicity and molecular enantiomer exchange may be deduced. These are summarized in Table II.

TABLE II

Symmetry Properties of Legendre Polynomial Coefficients in the Photoelectron Angular Distribution

Interaction	P_0	P_1	P_2	P_3
E1·E1	+p + E	−p − E	+p + E	
E1·M1	−p − E	+p + E	−p − E	
E1·E2	−p − E	±p + E	±p − E	−p + E

[a]The ±p indicates symmetry–antisymmetry of the Legendre coefficient with respect to a light polarization switch, ±E indicates the response to an enantiomer switch.

In the present context, where data are recorded at the magic angle, any result will be insensitive to the P_2 coefficient, so this need not be further discussed. Turning then to the other Legendre polynomial terms it is seen that the E1·M1 P_1 coefficient is symmetric with respect to a polarization switch and so would cancel from the experimental dichroism (difference) signal. Discussion of the E1·E2 interaction is rather more complicated due to the presence of first and second rank tensor components in the P_1 coefficient; these display alternate polarization symmetry, but this coefficient is clearly symmetric with respect to enantiomer exchange, as is the polarization antisymmetric P_3 coefficient. In either case, the contribution that persists in the PECD spectra would not vary with the enantiomer used, so the close mirroring seen between the PECD spectra of the two enantiomers would be broken. In the absence of any significant suggestion of that happening in the fenchone $C{=}O$ $1s$ data it seems reasonable to infer that E1·E2 terms are negligible.

It remains to consider the P_0 coefficients. For both the E1·M1 interaction and the E1·E2 interaction, this coefficient changes sign with a polarization or enantiomer switch; this is in fact the contribution that gives rise to CD in (non-angle resolved) absorption measurements. It is known that these terms are normally orders of magnitude smaller than the PECD dichroism discussed here, but it is also true that PECD measurements made at just one fixed angle would be incapable of distinguishing variations in the total cross-section from simple angular variations. It nevertheless seems safe to dismiss there being any significant E1·M1 terms since magnetic dipole transitions from the $1s$ shell are formally forbidden by selection rules. A significant E1·E2 contribution to the overall cross-section also seems unlikely when the two enantiomer curves closely mirror each other and provide no suggestion of a significant E1·E2 contribution to the P_1 coefficient.

The precision of the data is not such as to allow non-dipole interactions to be definitively ruled out, and more detailed study of this topic by careful measurement of the full angular *distribution*, as opposed to detection at a single angle, will be required to provide a complete probe. In the meantime a clear observation that enantiomer PECD curves have a mirror-image relationship seems sufficient to strongly suggest the dominance of the electric dipole approximations, and any success modeling experimental data with calculations built firmly from pure electric dipole interaction terms must surely confirm that.

B. Camphor

Camphor has served as a prototypical molecule for CD studies for a number of years. It has maintained that role in more recent investigations of photoelectron CD, where it has also quickly become the most studied system with papers describing PECD in the valence shell [36, 64, 65], the C $1s$ core region [56] and combined computational studies of both [57].

1. Core C 1s

The camphor XPS is extremely similar to that of fenchone, with the $C=O$ $1s$ being well resolved by being shifted a few electronvolts from the heavily overlapping main peak comprising all the other C $1s$ ionizations. Once again, this provides an opportunity to examine PECD in the ionization of a single, highly characterized $1s$ orbital. Figure 16 shows experimental $C=O$ $1s^{-1}$ data that have been obtained for the (R)- and (S)-enantiomers of camphor [56].

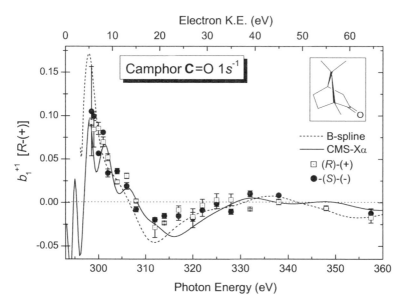

Figure 16. Carbonyl C $1s$ PECD from enantiomers of camphor. The experimentally derived $b_1^{\{1\}}$ data (Ref. [56]) for the (S)-enantiomer have been negated prior to plotting on expectation that they will then fall on the same trend line as the (R)-enantiomer data. The CMS-Xα and B-spline calculations (Ref. [57]) for the (R)-camphor enantiomer are included for comparison. The inset shows the (R)-camphor structure.

In this instance, the (S)-enantiomer data have been negated prior to plotting. From previous discussion of the antisymmetry of the $b_1^{\{\pm1\}}$ parameters under enantiomer exchange (e.g., Section III.A) it is recognized that it is then to be expected that the (R)- and (S)-enantiomer data should fall on the same experimental trend line. That they do indeed do so shows, as was argued in the Section IV.A for fenchone, that the behavior is at least qualitatively in accord with a pure electric dipole model. Furthermore, combining two distinct data sets [(R)- and (S)-enantiomers] in this manner provides a consistency check on the reproducibility of the PECD data. It seems good practice to include measurement of both enantiomers, where this is feasible, in an experimental study.

Rather than plotting the dichroism in the form of the asymmetry factor $\Gamma(54.7°)$, Eq. (10), this has been trivially reduced further to generate an estimate of $b_1^{\{+1\}}$ for plotting in Fig. 16. This facilitates a direct comparison in the figure with PECD calculations performed for the (R)-camphor enantiomer. Two such calculations are included: a CMS-Xα calculation and a B-spline calculation [57]. A strong degree of agreement can be seen between the calculations, and between either calculation and experiment. Hence, not only is the experimental dichroism in qualitative agreement with a pure electric dipole model, as exemplified by the exact antisymmetry exhibited by the two enantiomers, it is also in very convincing quantitative agreement with calculations conducted in this level of approximation.

The experimental dichroism is seen to have its greatest magnitude some 5 eV above threshold, where $b_1^{\{+1\}} \sim 0.10$. This corresponds to an asymmetry factor in the forward–backward scattering of $\gamma \sim 20\%$. Such a pronounced PECD asymmetry from a randomly oriented sample looks to comprehensively better the "amazingly high" 10% chiral asymmetry recorded with highly ordered nanocrystals of tyrosine enantiomer [25] or the "spectacular" 12.5% asymmetry reported from an oriented single crystal of a cobalt complex [28].

Unfortunately, experimental difficulties precluded measurements closer to threshold, and the B-spline calculation also does not properly span this near threshold region down to the onset [57]. However, the general trend rising above 5 eV is for the dichroism to become attenuated, easily rationalized as the ejected photoelectron displaying less sensitivity to the chiral molecular potential as it acquires more energy.

2. Valence Shell Ionization

The valence shell PES of camphor has a well-resolved band corresponding to the HOMO—nominally a carbonyl oxygen lone-pair orbital—and the next highest lying orbital can be distinguished as a shoulder on the second PES band [110]. It has already been seen (Fig. 1) that a PECD measurement at $h\nu = 10.3$ eV clearly distinguishes these two orbital ionizations. At higher photon energies, more orbital ionizations become energetically accessible, but while the PECD continues to show some variations that correlate with features in the PES (Ref. [36], Fig. 3]) an assignment indicates there is considerable overlap of individual orbital ionizations in the PES that precludes a detailed orbital by orbital analysis of the more strongly bound electrons.

Figure 17 shows the PECD, in the form of deduced $b_1^{\{+1\}}$ parameters, for the HOMO ionization of (R)- and (S)-enantiomers of camphor as a function of photon energy in the range from threshold to 26 eV. In addition to data recorded with the photoelectron imaging technique [36], this figure includes two data points obtained earlier with a somewhat different imaging arrangement [64] and a more substantial set of data for the 13–24-eV range obtained by an experiment using fixed direction detection at the magic angle [65]. These latter measurements were in fact made without the benefit of a polarization switching insertion device, but instead relied on using the out-of-plane synchrotron emission from a bending magnet beamline at BESSY I (Berlin). In this case, the authors found that observing the dichroism by switching enantiomers was more convenient than repeatedly switching polarization.

Implicitly, their data treatment assumes that the photoemission dissymmetry simply reverses direction with the enantiomer switch, equivalently to its

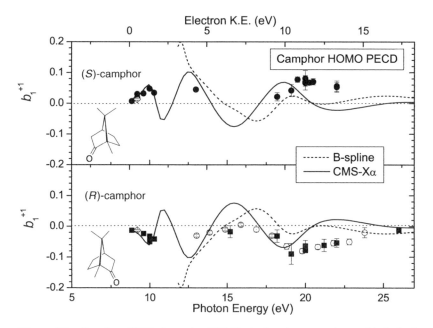

Figure 17. Camphor PECD from the HOMO (carbonyl oxygen lone-pair) ionization. Experimental data are: ● (S)-camphor, ■ (R)-camphor (both from Ref. [36]); ◇ (R)- and (S)-enantiomer data from Ref. [64]; ○ data from Ref [65]. Also shown as curves are CMS-Xα calculations (Ref. [36]) and B-spline calculations (Ref. [57]).

switching with polarization switch, as predicted in the electric dipole approximation. It can be seen from the enantiomer specific data plotted in Fig. 17 that the (R)- and (S)-enantiomer data do mirror each other, as previously seen and discussed. With this assumption secure, examination shows that the data reported from the BESSY I experiment is, after correction for the 54.7° detection direction, equivalent to the (R)-enantiomer $b_1^{\{+1\}}$ value, and has been plotted as such in the figure.

Overall, there is excellent agreement between the three different experiments, and between the enantiomer specific measurements made at the LURE and Elettra synchrotrons. The PECD reaches its maximum magnitude $|b_1^{\{+1\}}| \sim 0.9$ (equivalently $\gamma \sim 18\%$) at an electron kinetic energy of ~ 11 eV. This is directly comparable with the very pronounced asymmetry in the C $1s$ ionization (Section VI.B.1), but it is perhaps a little unintuitive to see that it occurs not at the start of a trend, gently declining with increasing energy, but rather as a secondary maximum located significantly above threshold. On the other hand, it may be remarked that the C=O $1s$ data (Fig. 16) also pass through a second maximum in magnitude (albeit reversed sign) some 20 eV above the ionization threshold. Hence, it seems that even photoelectrons departing with some tens of electronvolts kinetic energy retain a strong sensitivity to the chiral molecular potential.

Figure 18 presents in a similar fashion, data for the HOMO-1 ionization. Again there is excellent agreement between the (R)- and (S)-enantiomer data [36] and between these and the equivalent BESSY I data [65]. The magnitude of the asymmetry is not so great for this second orbital, which runs along the C—C—C bridge linking the two stereogenic centers. A visual comparison of these two outer-valence orbitals is provided in Fig. 19. It might, perhaps, be thought that the HOMO-1 orbital with its density surrounding both asymmetric carbons in this molecule would produce a more pronounced PECD than the HOMO, and certainly more than the spherically localized C=O $1s$ initial orbital. That it does not, reinforces the inference that final state scattering must be of great importance in transferring a sense of the molecular chirality to the outgoing electron angular distribution.

Also included in both Figs. 17 and 18 are the results of two calculations, performed with the CMS-Xα and B-spline methods. As discussed elsewhere these are in reasonable agreement, though less exactly so than for the C $1s$ ionization. [36, 57] Key characteristics, such as the magnitude and sign of the dichroism, are clearly captured. The CMS-Xα calculation provides close agreement with experiment for the HOMO in the threshold region. B-Spline results are not available close to the threshold region because the over-attractive LB94 potential used in these calculations causes calculated threshold behavior to fall into the discrete region of the spectrum [57]. Improved potentials should alleviate this problem in the future [79]. Both calculations reproduce the second

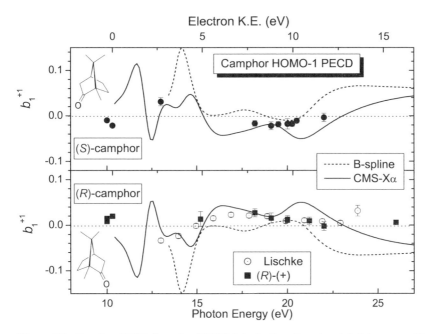

Figure 18   Camphor PECD from the HOMO-1 ionization. Experimental data are: ● (S)-camphor, ■ (R)-camphor (both from Ref. [36]); ○ data from Ref [65]. Also shown as curves are CMS-Xα calculations (Ref. [36]) and B-spline calculations (Ref. [57]).

maximum at ∼11 eV in the HOMO PECD, though both underestimate the width of this feature. It has been suggested that this structure may correlate with a shape resonance in camphor [57], but there is no real hint of such a continuum feature in either the core or the HOMO-1 PECD data. Let us also recall a

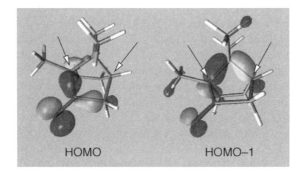

Figure 19.   Camphor outer-valence orbitals generated from a HF/cc-pVDZ calculation. The (R)- enantiomer is oriented with its carbonyl group toward bottom left in the figure. The two stereogenic centers (asymmetric carbons) in the molecule are indicated by arrows.

deduction that $b_1^{\{1\}}$ parameters may not be overtly sensitive to such continuum features [53, 60], as discussed in Section IV.D. Further work is required to clarify these points.

VII. CONCLUSIONS

The experimental work described in this chapter clearly demonstrates that chiral asymmetries in the forward–backward distribution of photoelectrons emitted from randomly oriented enantiomers when ionized with circularly polarized light can be spectacularly large (to borrow and apply a superlative from previous accounts of an unprecedented chiral asymmetry)—on the order of 20%. The theory discussed here, as implemented in two computational methods, is fully capable of predicting this and being applied to develop an understanding of a phenomenon that at times displays some counterintuitive properties. Doing so is very much an ongoing quest.

Summarizing the experimental data presented here, and the wider body of results now available, one can first identify the most general trend is for the photoelectron forward–backward asymmetry to show a reduction in the asymmetry factor as the energy increases, rationalized as a diminishing sensitivity of faster ejected electrons to the molecular scattering potential. There is also ample evidence for there being a strong dependence of detail on any measured PECD on the ionized orbital. In some cases, running across a photoelectron spectrum reveals structure in the PECD that is not readily evident in the total cross-section. A strong correlation of chiral effect with initial orbital can thus perhaps resolve otherwise hidden detail in the spectrum—rather like gas-phase magnetic CD measurements were able to resolve individual states contributing to the broad A-band continuum in methyl halides [111, 112].

The detail of the dependence on initial orbital is not, however, always what one might naiively expect. In studies of methyl oxirane [62], a correlation was noted between the extent of the initial orbital's chirality and the energy range over which PECD effects were measurable. It seems reasonable to infer that when the electron carries a greater sense of its environment's handedness into the photoionization process, then recognition of the molecular chirality will be more persistent in the final outcome. In seeming contradiction to this thought, in the camphor data considered here (Section IV.A.2) the orbital that has density surrounding both the chiral centers of the molecule produces the weakest PECD across the energy range. Stronger dichroism is observed from the more localized outermost valence orbital, and from the fully localized, spherical (therefore achiral) $C=O$ $1s$ orbital [36]. This latter observation, and others similar, leads to the inference of a dominant role for final-state scattering of the photoelectron off the chiral molecular potential in producing the chiral asymmetry observed in core level ionization.

The two computational methods outlined perform similarly and provide an excellent model for the camphor C 1*s* PECD. The description of valence ionization is more challenging for theory and greater discrepancies between theories and experiment emerge. Understanding and then improving on such differences provides a route forward, and so there is value in performing comparative studies, as has been done for camphor [57]. It has been speculated that differences between the performance in core and valence shell modeling may be due to a common neglect of electron response effects. Differences between the two computational methods in their description of the camphor valence shell could be attributed to the rather arbitrary spherical partitioning of the Xα potential about atomic centers, as this would seem better adapted to description of a spherical 1*s* initial orbital than to a more delocalized valence orbital. Against that, if the final-state scattering, delocalized over the entire molecule, continues to play a major role in valence ionization, then that aspect of the problem should be neither better nor worse described in the CMS-Xα core and valence region PECD models, and evidently works well in the former.

Clearly, there is considerable scope for further investigation of the orbital dependence, particularly in the valence shell, and interesting questions to pursue regarding the significance of localized versus delocalized, chiral versus achiral initial orbitals. Accompanying these themes there is scope for refinements in computations and in experimental technique—in particular in areas, such as detector technology, molecular beam sample sources, and the introduction of rapid polarization switching sources. As technical capabilities develop, resolution and sensitivity improve, so too will the questions that can be addressed advance. All this should lead toward greater precision in the data and its modeling.

At a fundamental level, it has been shown that PECD stems from interference between electric dipole operator matrix elements of adjacent continuum ℓ values, and that consequently the chiral $b_1^{\{1\}}$ parameters depend on the sine rather than the cosine of the relative scattering phases. Generally, this provides a unique probe of the photoionization dynamics in chiral species. More than that, this sine dependence invests the $b_1^{\{1\}}$ parameter with a greatly enhanced response to small changes in scattering phase, and it is believed that this accounts for an extraordinary sensitivity to small conformational changes, or indeed to molecular substitutions, that have only a minimal impact on the other photoionization parameters.

Such conformational dependence presents challenges and an opportunity. The challenges lie in properly accounting for its consequences. In many cases, exact conformational energetics and populations in a sample may be unknown, and the nature of the sample inlet may sometimes also mean that a Boltzmann distribution cannot be assumed. Introducing this uncertainty into the data modeling process produces some corresponding uncertainty in the theoretical interpretation of data at the present time. It is current practice for photoionization dynamics to be computed for fixed nuclear geometries, and this pronounced sensitivity to small structural changes suggests that large amplitude vibrations may need to be more explicitly incorporated in the modeling. The opportunity that arises is really just the converse; once there *is* confidence in the basic theoretical description of the process it is possible to deduce conformational details of an experimental sample from some kind of fitting process. Already, this has been convincingly demonstrated for 3-hydroxytetrahydrofuran [61] and other such demonstrations are expected shortly to be published. Whether or not the PECD is strongly affected by continuum dynamics, such as Cooper minima and resonances, remains an open question for further investigation.

What might be the wider implications of this newly investigated phenomenon in the broad context set out in the Introduction to this chapter? All the evidence now suggests that exceptional chiral asymmetries, of the order of 10%, may be quite routinely detected in PECD measurements. Already, it is clear that calculations of the phenomenon are sufficiently reliable to allow the absolute configuration of an unknown sample to be deduced by comparison of theory and experiment or perhaps to allow the estimation of enantiomeric excess. In a broader context, a PECD trace may help deconvolute an underlying structure from an otherwise congested ionization spectrum; the potential for this at the electronic state level has been demonstrated, and might even extend to reveal vibrational details. The possibilities for probing conformational, as well as electronic structure, have already been alluded to. Altogether, this suggests that PECD may be developed into a useful adjunct for gas-phase biomolecular spectroscopy—the more so since the instrumental source conditions required for photoelectron spectroscopy and PECD measurement also directly lend themselves to the accompanying performance of some form of mass spectrometry.

A persistent line of thought concerning possible explanations for the homochirality of terrestrial life has been the postulate that prebiotic molecular building blocks have probably been first formed in the interstellar medium and later transported to earth. Included among the suggestions for asymmetric processes that might give rise to the symmetry breaking exhibited as homochirality, opportunities for photochemical interactions with partially circularly polarized radiation in the interstellar medium have received some prominence [8, 113–115]. Given the quite remarkable asymmetries that have now been revealed, the PECD effect suggests itself as a candidate to be evaluated as a possibly significant asymmetric photophysical process. Accompanying the asymmetry in the recoil angular distribution of photoelectrons from target molecules in a CPL radiation field momentum conservation will require there be a corresponding asymmetry in the recoil angular distribution of the associated ions [80]. This might be dismissed first on the grounds that the large mass ratio between molecular ion and electron would render the ion recoil negligible, but recent laboratory studies have underlined that such photoion recoil asymmetries are clearly observable [116, 117]. The asymmetric drift of enantiomer ions generated from racemic precursor mixtures could hence contribute to some spatial differentiation of enantiomers. The results obtained so far indicate that at the astrophysically important Lyman-α photon energy, within a few electronvolts of the valence ionization threshold for many organic species, PECD asymmetries can be intense. But significant asymmetries have also been revealed several tens of electronvolts above thresholds, where the recoil momentum would also be greater.

APPENDIX A: THE PHOTOIONIZATION DIFFERENTIAL CROSS-SECTION

We consider the expression of the lab frame photoelectron angular distribution for a randomly oriented molecular sample. The frozen core, electric dipole approximation for the differential cross-section for electron emission into a solid angle about a direction $\hat{k}$ can be written as

$$\frac{d\sigma(\omega)}{d\hat{k}} = \frac{\alpha a_0^2 \omega k}{4\pi} |\langle \Psi_{\vec{k}}^{(-)}(\vec{r})|\hat{e} \cdot \vec{r}|\psi_i \rangle|^2 \tag{A.1}$$

$$= \frac{\alpha a_0^2 k}{\pi \omega} |\langle \Psi_{\vec{k}}^{(-)}(\vec{r})|\hat{e} \cdot \nabla|\psi_i \rangle|^2 \tag{A.2}$$

in, respectively, the dipole operator length and velocity forms, with α the fine-structure constant, a_0 the Bohr radius, $\Psi_{\vec{k}}^{(-)}(\vec{r})$ the electron continuum function (momentum $\vec{k}$), and ψ_i the initial, bound orbital. The photon energy is ω and its polarization $\hat{e}$.

Following normal practice, it is convenient to replace the continuum function $\Psi_{\vec{k}}^{(-)}$ in Eq. (A.1) or (A.2) with an incoming wave normalized partial wave expansion [39, 40, 118]:

$$\Psi_{\vec{k}}^{(-)}(\vec{r}) = \sum_{lm} i^l e^{-i\sigma_l} Y_{lm}^*(\hat{k}) \psi_{lm}^{(-)}(\vec{r}) \tag{A.3}$$

where $\sigma_l = \arg \Gamma(l + 1 - i/k)$ is the Coulomb phase, and the partial wave basis functions, $\psi_{lm}^{(-)}$, have a defined asymptotic angular momentum l, m. The bra–ket in Eq. (A.1) or (A.2) may then be expanded in matrix elements of the form:

$$M_{lm}^{\{v\}} = (-i)^l e^{-i\sigma_l} e_v^1 f_{lmv}^{(-)} Y_{lm}^*(\hat{k}') \tag{A.4}$$

where e_v^1 is a spherical tensor (vector) component of the photon electric vector and the $f_{lmv}^{(-)}$ are the radial transition amplitudes for $\psi_i \rightarrow \psi_{lm}^{(-)}$.

The electron and photon angular momentum projections, m, v, and the recoil direction, $\hat{k}'$, appearing in Eq. (A.3) are defined in the molecular frame, but our objective here is to obtain an expression for the lab frame distribution and its polarization dependence. Consequently, the relevant quantities are to be transformed to the lab frame by a rotation $\mathbf{R}$ expressed using rotation matrices $D_{mn}^j(\mathbf{R})$, namely $Y_{lm}^*(\hat{k}') = \Sigma_\mu [D_{\mu m}^l(\mathbf{R}) Y_{l\mu}(\hat{k})]^*$ and $e_p^1 = \Sigma_\nu D_{p\nu}^1(\mathbf{R}) e_\nu^1$. The transformed matrix element, $M_{lm}^{\{p\}}$, for a specific lab-frame polarization, p, and electron detection direction, $\hat{k}$ is thus written

$$M_{lm}^{\{p\}} = \sum_{\nu\mu} (-i)^l e^{-i\sigma_l} D_{p\nu}^1(\mathbf{R}) e_\nu^1 f_{lm\nu}^{(-)} [D_{\mu m}^l(\mathbf{R}) Y_{l\mu}(\hat{k})]^* \tag{A.5}$$

so that, with appropriate normalization, Eqs. (A.1) or (A.2) can be written

$$\frac{d\sigma(\omega; p, \mathbf{R})}{d\hat{k}} \sim \left| \sum_{lm} M_{lm}^{\{p\}} \right|^2 \tag{A.6}$$

$$= \sum_{\substack{lm\nu\mu \\ l'm'\nu'\mu'}} (-i)^{l-l'} e^{i(\sigma_l - \sigma_{l'})}$$

$$\times [D_{p\nu'}^1(\mathbf{R}) e_{\nu'}^1 f_{l'm'\nu'}^{(-)}]^* D_{\mu'm'}^{l'}(\mathbf{R}) Y_{l'\mu'}(\hat{k})$$

$$\times D_{p\nu}^1(\mathbf{R}) e_\nu^1 f_{lm\nu}^{(-)} [D_{\mu m}^l(\mathbf{R}) Y_{l\mu}(\hat{k})]^*. \tag{A.7}$$

Now the pairs of rotation matrix products in Eq. (A.7) can be replaced with Clebsch–Gordan series

$$[D_{p\nu'}^1(\mathbf{R})]^* D_{p\nu}^1(\mathbf{R}) = (-1)^{p-\nu'} D_{-p-\nu'}^1(\mathbf{R}) D_{p\nu}^1(\mathbf{R})$$

$$= \sum_j \langle 1 - p, 1p|j0\rangle\langle 1 - \nu', 1\nu|j\nu - \nu'\rangle$$

$$\times D_{0\nu-\nu'}^j(\mathbf{R})(-1)^{p-\nu'} \tag{A.8}$$

and similarly

$$D_{\mu'm'}^{l'}(\mathbf{R})[D_{\mu m}^l(\mathbf{R})]^* = \sum_k \langle l'\mu', l - \mu|k\mu' - \mu\rangle\langle l'm', l - m|km' - m\rangle$$

$$\times D_{\mu'-\mu m'-m}^k(\mathbf{R})(-1)^{\mu-m} \tag{A.9}$$

The four rotation matrices of Eq. (A.7) thus reduce to a product of two having the same argument, $\mathbf{R}$. The orthonormality of these causes a further simplification when an integration over all lab frame orientations, **R**, is performed to accommodate random molecular orientation:

$$\int D^j_{0\nu-\nu'}(\mathbf{R})D^k_{\mu'-\mu m'-m}(\mathbf{R})d\mathbf{R} = (-1)^{\nu'-\nu}\frac{8\pi^2}{(2j+1)}\delta_{jk}\delta_{0,\mu'-\mu}\delta_{\nu'-\nu,m'-m} \quad (\text{A.10})$$

Hence, for a randomly oriented sample Eq. (A.7) becomes

$$\frac{d\sigma(\omega;p)}{d\hat{k}} = 8\pi^2 \sum_{\substack{lm\nu\mu \\ l'm'\nu'j}} (-i)^{l-l'}e^{i(\sigma_l-\sigma_{l'})}(-1)^{p-m-\nu}(-1)^{\mu}\frac{1}{(2j+1)}$$

$$\times \langle 1-p,1p|j0\rangle\langle 1-\nu',1\nu|j\nu-\nu'\rangle\langle l'\mu,l-\mu|j0\rangle$$

$$\times \langle l'm',l-m|jm'-m\rangle$$

$$\times [e^1_{\nu'}f^{(-)}_{l'm'\nu'}]^*e^1_{\nu}f^{(-)}_{lm\nu}Y_{l'\mu}(\hat{k})Y^*_{l\mu}(\hat{k})\delta_{\nu'-\nu,m'-m} \quad (\text{A.11})$$

The sum on μ can be eliminated by examining the μ-containing subexpression

$$\zeta \equiv \sum_{\mu}(-1)^{\mu}\langle l'\mu,l-\mu|j0\rangle Y_{l'\mu}(\hat{k})Y^*_{l\mu}(\hat{k}) \quad (\text{A.12})$$

By reexpressing the spherical harmonics in the form of D rotation matrices they may be effectively substituted by a Clebsch–Gordan series yielding

$$\zeta = \frac{\sqrt{(2l'+1)(2l+1)}}{4\pi}\sum_{\mu k}(-1)^{l'+l+k}\langle l'-\mu,l\mu|j0\rangle$$

$$\times \langle l'-\mu l\mu|k0\rangle\langle l'0,l0|k0\rangle D^k_{00}(\hat{k}) \quad (\text{A.13})$$

M-sum unitarity of the first two Clebsch–Gordan coefficients now means that the sum over μ reduces to a simple delta function:

$$\zeta = \frac{\sqrt{(2l'+1)(2l+1)}}{4\pi}(-1)^{l'+l+j}\sum_{k}\langle l'0,l0|k0\rangle D^k_{00}(\hat{k})\delta_{jk}$$

$$= \frac{\sqrt{(2l'+1)(2l+1)}}{4\pi}(-1)^{l'+l+j}\langle l'0,l0|j0\rangle P_j(\cos\theta) \quad (\text{A.14})$$

Substituting this ζ subexpression back into Eq. (A.11) finally yields

$$\frac{d\sigma(\omega;p)}{d\hat{k}} = 2\pi \sum_{j} \sum_{lmv} (-i)^{l-l'} e^{i(\sigma_l - \sigma_{l'})} (-1)^{p-m-v} \frac{\sqrt{(2l'+1)(2l+1)}}{(2j+1)}$$

$$l'm'v'$$

$$\times \langle 1-p, 1p|j0 \rangle \langle 1-v', 1v|jv-v' \rangle$$

$$\times \langle l'0, l0|j0 \rangle \langle l'-m', lm|jm-m' \rangle$$

$$\times \left[e_{v'}^{1} f_{l'm'v'}^{(-)} \right]^{*} e_{v}^{1} f_{lmv}^{(-)} \delta_{v'-v,m'-m} P_{j}(\cos\theta) \qquad (A.15)$$

APPENDIX B: PECD CONVENTIONS

Because the sense, or sign, of chiral asymmetry in the forward–backward electron scattering asymmetry depends on the helicity of the photon and of the molecule, it is essential that these variables are properly specified in any study to permit meaningful comparisons to be made. Discussing and comparing quantitative asymmetry factors, γ [Eq. (8)] and dichroism [Eq. (9)] likewise requires agreement on the convention adopted in the definition of these terms.

- Molecular enantiomers. The absolute configuration at the stereogenic center of an enantiomer is easily specified (where known) using the labels (*R*)- and (*S*)- derived according to the Cahn–Ingold–Prelog sequence rules [105] taught in all undergraduate organic chemistry classes. The labels (+),(−) are also widely used to indicate the direction of optical rotation by a chiral sample, but this is phenomenological designation that does not provide an absolute configuration. Both (*R*)- and (*S*)- are therefore more meaningful when comparing theory with theory or with experiment.

- Light polarization. The polarization state of the ionizing radiation appears in the theoretical expressions [e.g., Eq. (12)] as the photon helicity, *p*, describing the spin projection of the photon along its propagation direction; $p = +1$ is thus the right-handed helicity index. Unfortunately, the optical convention describes the handedness of the radiation by the observed direction of rotation of the electric vector with time viewed looking toward the source of the light. A clockwise direction, labeled right circularly polarized (rcp) light then corresponds to a negative (or left-handed) helicity. The potential confusion has been compounded by inconsistencies in the application of the optical convention in the literature. The Stokes parameters provide an alternative characterization of the polarization, with s_3 describing the circular polarization component. For clarity we specify

$$\text{rcp (optical convention)} \equiv (s_3 = +1) \equiv (p = -1)$$
$$\text{lcp (optical convention)} \equiv (s_3 = -1) \equiv (p = +1)$$

- Circular dichroism convention. In this chapter, dichroism and asymmetry factors are defined as $I_{\text{lcp}} - I_{\text{rcp}}$ or $I_{p=+1} - I_{p=-1}$, which the author believes to be the more common, though not universal, choice.

There have sometimes been practical difficulties deducing absolute handedness of the circular polarization produced using multi-order quarter waveplates where the exact order or retardation is not known. In the VUV and SXR regions, such devices are anyway not available, but the use of insertion devices (IDs) in synchrotron storage rings to produce CPL rather simplifies things. It is relatively straightforward to deduce the sense of the corkscrew motion induced in the electron beam by the magnetic array of the ID, and hence to assign the absolute sense of the resulting circularly polarized synchrotron radiation.

Acknowledgments

The very considerable contributions to the experimental investigations that have been made by my collaborators are gratefully acknowledged: Laurent Nahon, Uwe Hergenhahn and his group—especially Emma Rennie and Oliver Kugeler who contributed greatly to first initiating the SXR PECD experiments. As a member of our group, Gustavo Garcia contributed to all experimental activities, and continues to do so in his current position. Another ex-student, Chris Harding, has played an invaluable role in both experimental and computational aspects of this research. On the theoretical side there have been stimulating interactions with Piero Declava and especially Mauro Stener, and their efforts in theoretical modeling have helped provide a solid base for interpreting PECD measurements.

References

1. R. S. Chivukula, *J. Phys. Conf. Ser.* 37–28 (2006).
2. O. Naviliat-Cuncic, T. A. Girard, J. Deutsch, and N. Severijns, *J. Phys. G: Nucl. Part. Phys.* **17** (6), 919, (1991).
3. J. van Klinken, *J. Phys. G: Nucl. Part. Phys.* **22** (9), 1239, (1996).
4. Lia Addadi and Steve Weiner, *Nature* (London) **411** (6839), 753 (2001).
5. P. Cintas, *Angew. Chem.-Int. Ed.* (Engl) **41** (7), 1139, (2002).
6. T. Asami, R. H. Cowie, and K. Ohbayashi, *Am. Nat.* **152** (2), 225 (1998).
7. M. Schilthuizen and A. Davison, *Naturwissenschaften* **92** (11), 504 (2005).
8. W. A. Bonner, *Orig. Life Evol. Biosph.* **25**(1–3), 175 (1995).
9. W. A. Bonner, *Orig. Life Evol. Biosph.* **21** (2), 59 (1991).

10. A. Brack, **24** (4), 417 (1999).

11. J. L. Bada, *Proc. Natl. Acad. Sci. U. S. A.* **98** (3), 797 (2001).

12. J. L. Bada, M. A. Sephton, P. Ehrenfreund, R. A. Mathies, A. M. Skelley, F. J. Grunthaner, A. P. Zent, R. C. Quinn, J. L. Josset, F. Robert, O. Botta, and D. P. Glavin, *Astron. Geophys.* **46** (6), 26 (2005).

13. A. J. MacDermott, L. D. Barron, A. Brack, T. Buhse, A. F. Drake, R. Emery, G. Gottarelli, J. M. Greenberg, R. Haberle, R. A. Hegstrom, K. Hobbs, D. K. Kondepudi, C. McKay, S. Moorbath, F. Raulin, M. Sandford, D. W. Schwartzman, W. H. P. Thiemann, G. E. Tranter, and J. C. Zarnecki, *Planet Space Sci.* **44** (11), 1441 (1996).

14. C. J. Welch and J. I. Lunine, *Enantiomer* **6**(2–3), 69 (2001).

15. E. Brenna, C. Fuganti, and S. Serra, *Tetrahedron: Asymm.* **14** (1), 1 (2003).

16. A. M. Rouhi, *Chem. Eng. News* **82** (24), 47, (2004).

17. A. M. Rouhi, *Chem. Eng. News* **81** (18), 45, (2003).

18. L. D. Barron, *Molecular light scattering and optical activity.* Cambridge University Press, Cambridge, 2nd ed., 2004.

19. Nina Berova, Kji Nakanishi, and Robert, Woody, *Circular dichroism principles and applications*, Wiley–VCH, New York, 2nd ed., 2000.

20. Alison Rodger and Bengt Nordén, *Circular dichroism and linear dichroism.* Oxford University Press, Oxford, UK, 1997.

21. Werner Kuhn and E. Braun, *Z. Physik. Chem.* **8**(Abt. B), 445 (1930).

22. Gerald D. Fasman, (ed.), *Circular Dichroism and the Conformational Analysis of Biomolecules*, Plenum, New York, 1996.

23. Balavoin G, Moradpou A, and H B. Kagan, *J. Am. Chem. Soc.* **96** (16), 5152 (1974).

24. J. Paul, A. Dorzbach, and K. Siegmann, *Phys. Rev. Lett.* **79** (16), 2947 (1997).

25. J. Paul and K. Siegmann, *Chem. Phys. Lett.* **304**, 23 (1999).

26. L. Alagna, T. Prosperi, S. Turchini, J. Goulon, A. Rogalev, C. Goulon-Ginet, C. R. Natoli, R. D. Peacock, and B. Stewart. *Phys. Rev. Lett.* **80** (21), 4799 (1998).

27. Robert D. Peacock and Brian Stewart, *J. Phys. Chem. B* **105** (2), 351 (2001).

28. B. Stewart, R. D. Peacock, L. Alagna, T. Prosperi, S. Turchini, J. Goulon, A. Rogalev, and C. Goulon-Ginet, *J. Am. Chem. Soc.* **121** (43), 10233 (1999).

29. Julie Fidler, P. Mark Rodger, and Alison Rodger, *J. Chem. Soc. Perkin Trans. 2* **2**, 235 (1993).

30. Magdalena Pecul, Domenico Marchesan, Kenneth Ruud, and Sonia Coriani. *J. Chem. Phys.* **122** (2), 024106/1 (2005).

31. Friedhelm Pulm, Jorg Schramm, Horst Lagier, and Josef Hormes, *Enantiomer* **3**(4–5), 315 (1998).

32. T. Muller, K. B. Wiberg, and P. H. Vaccaro, *J. Phys. Chem. A* **104** (25), 5959 (2000).

33. S. Turchini, N. Zema, S. Zennaro, L. Alagna, B. Stewart, R. D. Peacock, and T. Prosperi, *J. Am. Chem. Soc.* **126** (14), 4532 (2004).

34. B. Ritchie, *Phys. Rev. A* **13**, 1411 (1976).

35. I. Powis, *J. Chem. Phys.* **112** (1), 301 (2000).

36. L. Nahon, G. A. Garcia, C. J. Harding, E. A. Mikajlo, and I. Powis, *J. Chem. Phys.* **125**, 114309, (2006).

37. G. A. Garcia, L. Nahon, C.H. Harding, and I. Powis, ChemPhysPhysChem (submitted).

38. C. J. Harding, *Photoelectron Circular Dichroism in Gas Phase Chiral Molecules*, Ph.D. thesis, University of Nottingham, UK, 2005.

39. D. Dill, *J. Chem. Phys.* **65**, 1130 (1976).

40. D. Dill and J. L. Dehmer, *J. Chem. Phys.* **61**, 692 (1974).

41. D. Loomba, S. Wallace, D. Dill, and J. L. Dehmer, *J. Chem. Phys.* **75**, 4546 (1981).

42. H. Park and R. N. Zare, *J. Chem. Phys.* **104** (12), 4554 (1996).

43. J. Cooper and R. N. Zare, Photoelectron angular distributions. In Sydney Geltman, Kalayana T. Mahanthappa, and Wesley Emil Brittin (ed.), *Lectures in theoretical physics: Vol. 11c. Atomic collision processes*, Gordon and Breach, New York, 1969, pp. 317–337.

44. O. Gessner, Y. Hikosaka, B. Zimmermann, A. Hempelmann, R. R. Lucchese, J. H. D. Eland, P. M. Guyon, and U. Becker, *Phys. Rev. Lett.* **88**, 193002 (2002).

45. T. Jahnke, Th. Weber, A. L. Landers, A. Knapp, S. Schössler, J. Nickles, S. Kammer, O. Jagutzki, L. Schmidt, A. Czasch, T. Osipov, E. Arenholz, A. T. Young, R. Díez Muiño, D. Rolles, F. J. García de Abajo, C. S. Fadley, M. A. Van Hove, S. K. Semenov, N. A. Cherepkov, Rösch J., M. H. Prior, H. Schmidt-Böcking, C. L. Cocke, and R. Dörner, *Phys. Rev. Lett.* **88**:073002, (2002).

46. G. Schönhense and J. Hormes, Photoionization of oriented systems and circular dichroism. In U. Becker and D. A. Shirley (eds.), *VUV and Soft X-Ray Photoionization*, Chapt. 17, Plenum, New York, 1996, pp. 607–652.

47. C. Westphal, M. Bansmann, M. Getzlaff, G. Schönhense, N. A. Cherepkov, M. Braunstein, V. McKoy, and P. L. Dubs, *Surf. Sci.* **253**, 205 (1991).

48. N. A. Cherepkov, *Chem. Phys. Lett.* **87** (4), 344 (1982).

49. J. W. Kim, M. Carbone, J. H. Dil, M. Tallarida, R. Flammini, M. P. Casaletto, K. Horn, and M. N. Piancastelli, *Phys. Rev. Lett.* **95** (10), 107601 (2005).

50. M. Polcik, F. Allegretti, D. I. Sayago, G. Nisbet, C. L. A. Lamont, and D. P. Woodruff, *Phys. Rev. Lett.* **92** (23), 236103 (2004).

51. I. Powis, *J. Phys. Chem. A* **104** (5), 878 (2000).

52. C. J. Harding and I. Powis, *J. Chem. Phys.* **125**, 234306 (2006).

53. M. Stener, G. Fronzoni, D. Di Tommaso, and P. Decleva, *J. Chem. Phys.* **120** (7), 3284 (2004).

54. J. W. Davenport. *Theory of Photoemission from Molecules in the Gas Phase and on Solid Surfaces*. Ph.D. thesis, University of Pennsylvania, 1976.

55. Chris J. Harding, Elisabeth A. Mikajlo, Ivan Powis, Silko Barth, Sanjeev Joshi, Volker Ulrich, and Uwe Hergenhahn, *J. Chem. Phys.* **123** (23), 234310 (2005).

56. U. Hergenhahn, E. E. Rennie, O. Kugeler, S. Marburger, T. Lischke, I. Powis, and G. Garcia, *J. Chem. Phys.* **120** (10), 4553 (2004).

57. M. Stener, D. Di Tommaso, G. Fronzoni, P. Decleva, and Ivan Powis, *J. Chem. Phys.* **124**, 024326 (2006).

58. K. H. Johnson. *Adv. Quant. Chem.* **7**, 143, (1973).

59. D. Toffoli, M. Stener, G. Fronzoni, and P. Decleva, *Chem. Phys.* **276** (1), 25 (2002).

60. D. Di Tommaso, M. Stener, G. Fronzoni, and P. Decleva, *ChemPhysChem* **7** (4), 924 (2006).

61. A. Giardini, D. Catone, S. Stranges, M. Satta, M. Tacconi, S. Piccirillo, S. Turchini, N. Zema, G. Contini, T. Prosperi, P. Decleva, D. Di Tommaso, G. Fronzoni, M. Stener, A. Filippi, and M. Speranza, *ChemPhysChem*, **6** (6), 1164 (2005).

62. S. Stranges, S. Turchini, M. Alagia, G. Alberti, G. Contini, P. Decleva, G. Fronzoni, M. Stener, N. Zema, and T. Prosperi, *J. Chem. Phys.* **122**, (2005).

63. S. Turchini, N. Zema, G. Contini, G. Alberti, M. Alagia, S. Stranges, G. Fronzoni, M. Stener, P. Decleva, and T. Prosperi, *Phys. Rev. A* **70** (1), art. no.014502 (2004).

64. G. A. Garcia, L. Nahon, M. Lebech, J. C. Houver, D. Dowek, and I. Powis, *J. Chem. Phys.* **119** (17), 8781 (2003)

65. T. Lischke, N. Böwering, B. Schmidtke, N. Müller, T. Khalil, and U. Heinzmann, *Phys. Rev. A* **70** (2): art. no. 022507 (2004).

66. G. Fronzoni, M. Stener, and P. Decleva, *J. Chem. Phys.* **118** (22), 10051 (2003).

67. M. Stener and P. Decleva, *J. Chem. Phys.* **112** (24), 10871 (2000).

68. M. Stener, G. Fronzoni, D. Toffoli, and P. Decleva, *Chem. Phys.* **282** (3), 337 (2002).

69. A. G. Csaszar, *J. Phys. Chem.* **100** (9), 3541 (1996).

70. P. D. Godfrey, S. Firth, L. D. Hatherley, R. D. Brown, and A. P. Pierlot, *J. Am. Chem. Soc.* **115** (21), 9687 (1993).

71. R. Kaschner and D. Hohl, *J. Phys. Chem. A* **102** (26), 5111 (1998).

72. T. Egawa, Y. Kachi, T. Takeshima, H. Takeuchi, and S. Konaka, *J. Mol. Struct.* **658** (3), 241 (2003).

73. G. G. Hoffmann, *J. Mol. Struct.* **661**, 525 (2003).

74. M. Mineyama and T. Egawa, *J. Mol. Struct.* **734**(1–3), 61 (2005).

75. D. P. Chong, P. Aplincourt, and C. Bureau, *J. Phys. Chem. A* **106** (2), 356 (2002).

76. G. Cavigliasso and D. P. Chong, *J. Chem. Phys.* **111** (21), 9485 (1999).

77. Yuji Takahata and Delano P. Chong, *J. Elec. Spec. Rel. Phenom.* 133: 69–76, 2003.

78. D. P. Chong, *J. Chem. Phys.* **103** (5), 1842, (1995).

79. M. Stener, D. Toffoli, G. Fronzoni, and P. Decleva, *J. Chem. Phys.* **124** (11), 114306 (2006).

80. N. Böwering, T. Lischke, B. Schmidtke, N. Müller, T. Khalil, and U. Heinzmann, *Phys. Rev. Lett.* **86** (7), 1187 (2001).

81. J. A. Clarke, *The science and technology of undulators and wigglers*, Oxford series on synchrotron radiation 4. Oxford University Press, Oxford, UK, 2004.

82. Hideo Onuki and Pascal Elleaume (eds.). *Undulators, Wigglers and Their Applications.* Taylor and Francis CRC Press, London, 2003.

83. H. Onuki, N. Saito, and T. Saito, *Appl. Phys. Lett.* **52** (3), 173 (1988).

84. S. Sasaki, *Nucl. Instrum. Methods Phys. Res. Sect. A-Accel. Spectrom. Dect. Assoc. Equip.* **347** (1–3), 83 (1994).

85. M. R. Weiss, R. Follath, K. J. S. Sawhney, F. Senf, J. Bahrdt, W. Frentrup, A. Gaupp, S. Sasaki, M. Scheer, H. C. Mertins, D. Abramsohn, F. Schäfers, W. Kuch, and W. Mahler, *Nucl. Instrum. Methods Phys. Res. A* **467** (1), 449 (2001).

86. J. Bahrdt, W. Frentrup, A. Gaupp, M. Scheer, W. Gudat, G. Ingold, and S. Sasaki, *Nucl. Instrum. Methods Phys. Res. Sect. A-Accel. Spectrom. Dect. Assoc. Equip.* **467**, 21 (2001).

87. F. Schäfers, H. C. Mertins, A. Gaupp, W. Gudat, M. Mertin, I. Packe, F. Schmolla, S. Di Fonzo, G. Soullie, W. Jark, R. Walker, X. Le Cann, R. Nyholm, and M. Eriksson, *Appl. Opt.* **38** (19), 4074 (1999).

88. K. Rabinovitch, L. R. Canfield, and R. P. Madden, *Appl. Opt.* **4** (8), 1005, (1965).

89. L. Nahon, F. Polack, B. Lagarde, R. Thissen, C. Alcaraz, O. Dutuit, and K. Ito, *Nucl. Instrum. Methods Phys. Res. Sect. A-Accel. Spectrom. Dect. Assoc. Equip.* **467**, 453 (2001).

90. L. Nahon, M. Corlier, P. Peaupardin, F. Marteau, O. Marcouille, P. Brunelle, C. Alcaraz, and P. Thiry, *Nucl. Instrum. Methods Phys. Res. A* **396** (1–2), 237 (1997).

91. L. Nahon, R. Thissen, C. Alcaraz, M. Corlier, P. Peaupardin, F. Marteau, O. Marcouille, and P. Brunelle, *Nucl. Instrum. Methods Phys. Res. A* **447** (3), 569 (2000).

92. B. Mercier, M. Compin, C. Prevost, G. Bellec, R. Thissen, O. Dutuit, and L. Nahon, *J. Vac. Sci. Tech. A* **18** (5), 2533 (2000).

93. L. Nahon and C. Alcaraz, *Appl. Opt.* **43** (5), 1024 (2004).

94. R. P. Walker and B. Diviacco, *Rev. Sci. Instrum.* **63** (1), 332 (1992).

95. A. Derossi, F. Lama, M. Piacentini, T. Prosperi, and N. Zema, *Rev. Sci. Instrum.* **66** (2), 1718 (1995).

96. D. Desiderio, *Synchrotron Radiation News* 12:38, 1999. ISSN 0894-0886.

97. C. Bordas, F. Paulig, H. Helm, and D. L. Huestis, *Rev. Sci. Inst.* **67** (6), 2257 (1996).

98. B. J. Whitaker (ed.) *Imaging in Molecular Dynamics.* Cambridge University Press, Cambridge, 2003.

99. G. A. Garcia, L. Nahon, and I. Powis, *Rev. Sci. Instrum.* **75** (11), 4989 (2004).

100. V. Dribinski, A. Ossadtchi, V. A. Mandelshtam, and H. Reisler, *Rev. Sci. Inst.* **73** (7), 2634 (2002).

101. A. T. J. B. Eppink and D. H. Parker, *Rev. Sci. Inst.* **68** (9), 3477 (1997).

102. G. A. Garcia, L. Nahon, C. J. Harding, E. A. Mikajlo, and I. Powis, *Rev. Sci. Instrum.* **76** (5), 053302 (2005).

103. G. A. Garcia, L. Nahon, and I. Powis, *Int. J. Mass. Spectro.* **225** (3), 261, (2003).

104. D. Ceolin, G. Chaplier, M. Lemonnier, G. A. Garcia, C. Miron, L. Nahon, M. Simon, N. Leclercq, and P. Morin, *Rev. Sci. Instrum.* **76** (4), 043302, (2005).

105. R. S. Cahn, C. Ingold, and V. Prelog, *Angew. Chem.-Int. Ed.* **5** (4), 385, (1966).

106. O. Hemmers, H. Wang, P. Focke, I. A. Sellin, D. W. Lindle, J. C. Arce, J. A. Sheehy, and P. W. Langhoff, *Phys. Rev. Lett.* **87** (27), art.no. 273003, (2001).

107. O. Hemmers, R. Guillemin, and D. W. Lindle, *Radiat. Phys. Chem.* **70**(1–3), 123 (2004).

108. O. Hemmers, M. Blackburn, T. Goddard, P. Glans, H. Wang, S. B. Whitfield, R. Wehlitz, I. A. Sellin, and D. W. Lindle, *J. Elec. Spec. Rel. Phen.* **123**(2–3), 257 (2002).

109. B. Ritchie, *Phys. Rev. A*, **14** 359 (1976).

110. E. E. Rennie, I. Powis, U. Hergenhahn, O. Kugeler, G. Garcia, T. Lischke, and S. Marburger, *J. Elec. Spec. Rel. Phenom.* **125** (3), 197 (2002).

111. A. Gedanken and M. D. Rowe, *Chem. Phys. Lett.* **34** (1), 39 (1975).

112. A. Gedanken, *Chem. Phys. Lett.* **137** (5), 462 (1987).

113. A. Jorissen and C. Cerf, *Orig. Life Evol. Biosph.* **32** (2), 129 (2002).

114. J. Bailey, A. Chrysostomou, J. H. Hough, T. M. Gledhill, A. McCall, S. Clark, F. Menard, and M. Tamura, *Science* **281**(5377), 672 (1998).

115. J. Bailey, *Orig. Life Evol. Biosph.* **31**(1–2), 167 (2001).

116. M. L. Lipciuc, J. B. Buijs, and M. H. M. Janssen, *Phys. Chem. Chem. Phys.* **8** (2), 219 (2006).

117. M. L. Lipciuc and M. H. M. Janssen, *Phys. Chem. Chem. Phys.* **8** (25), 3007 (2006).

118. N. Chandra, *J. Phys. B: At. Mol. Phys.* **20**, 3405 (1987).

SPECTROSCOPY OF THE POTENTIAL ENERGY SURFACES FOR C–H AND C–O BOND ACTIVATION BY TRANSITION METAL AND METAL OXIDE CATIONS

R. B. METZ

Department of Chemistry, University of Massachusetts Amherst, Amherst, MA 01003, USA

CONTENTS

Advances in Chemical Physics, Volume 138, edited by Stuart A. Rice
Copyright © 2008 John Wiley & Sons, Inc.

I. INTRODUCTION

The potential energy surface (PES) determines the course of a chemical reaction, and characterizing PESs for reactions has been a central focus of chemical dynamics [1]. Reaction studies, especially detailed measurements of reaction cross-sections as a function of collision energy or the reactivity of state selected reactants to produce products whose quantum states are measured, provide a great deal of indirect information on the PES for a reaction. However, even relatively simple reactions may involve several intermediates, connected by multiple transition states, and a particular experiment will only be sensitive to some features of the PES. Spectroscopic studies complement bimolecular experiments as they provide a direct probe of a region of the PES. Spectroscopy of reactive PESs presents particular challenges, as the stationary points on the surface correspond to intermediates (local minima) that are often short lived and to transition states (local maxima) that are inherently transient. Our studies of the PES of transition metal ion-molecule reactions are part of a tradition of transition state spectroscopy, which has primarily been used to study reactions of neutrals. Early studies by Brooks, Stwalley, and co-workers looked at far wing absorption by reacting complexes during the course of a harpoon reaction [2–4]. These experiments are extremely challenging, and interpreting the results is complicated by the averaging over impact parameter and orientation inherent in bimolecular reactions. Photoinitiating a reaction from a precursor with a similar geometry to the region of the PES of interest provides a powerful alternative. One advantage is that photoexcitation provides a start time to the reaction, which makes possible time-resolved studies, such as those of Zewail and co-workers [5,6]. A second advantage is that the energy, orientation, and angular momentum (impact parameter) are constrained, and energy-resolved studies can give well-resolved spectra rich in information, although one is limited to studying regions of the PES with good Franck–Condon overlap with the initial complex. In work on neutral metal atom reactions, Polanyi and co-workers [7] and Soep, Visticot, and co-workers [8,9] photoexcite van der Waals clusters, such as Ca (XH), where X is a halide to study the Ca* + XH → CaX + H harpoon reaction. Kleiber and co-workers have studied photodissociation of complexes of alkaline earth cations, Al^+ and Zn^+ with hydrogen, hydrocarbons, and simple organic molecules, observing several examples where photoexcitation leads to H–H, C–H, or C–C bond activation [10–13]. Photodetachment of a stable negative ion can give information on the corresponding neutral, which may be an intermediate or transition state of a neutral reaction. Brauman and co-workers measured photodetachment cross-sections of hydrogen-bound negative ions to study the transition state of hydrogen-transfer reactions [14]. Photoelectron spectroscopy gives more detailed information. Lineberger and co-workers studied isomerization of vinylidene to acetylene from photoelectron

and PtO_2. Followup photoelectron spectroscopy studies are planned, to measure vibrational frequencies and low lying electronic states of the cations.

The PES for even a simple transition metal ion reaction is quite complex; experiments alone cannot characterize it and electronic structure theory is required. Calculations of reactive PESs are inherently challenging, as they involve partially made and –broken bonds. Calculations involving transition metals are even more difficult, due to the presence of several unpaired electrons and large number of low lying electronic states. Transition metal reactions frequently involve multiple spin states. Figure 1 shows a fairly common case, in which ground state, high spin M^+ reacts to form ground-state, high spin MA^+, yet the reaction proceeds through low spin intermediates. Modeling such a reaction requires accurately calculating the high and low spin surfaces and their relative energies, as well as the coupling between the surfaces. Spectroscopy of the reactants and reaction intermediates provides a rigorous test of the accuracy of computational methods.

II. EXPERIMENTAL METHODS

We use laser photofragment spectroscopy to study the vibrational and electronic spectroscopy of ions. Our photofragment spectrometer is shown schematically in Fig. 2. Ions are formed by laser ablation of a metal rod, followed by ion molecule reactions, cool in a supersonic expansion and are accelerated into a dual TOF mass spectrometer. When they reach the reflectron, the mass-selected ions of interest are irradiated using one or more lasers operating in the infrared (IR), visible, or UV. Ions that absorb light can photodissociate, producing fragment ions that are mass analyzed and detected. Each of these steps will be discussed in more detail below, with particular emphasis on the ions of interest.

A. Ion Production

Ions are produced in a Smalley-type laser ablation source [40,41]. The frequency-doubled output (532 nm) of a pulsed Nd:YAG laser is loosely focused

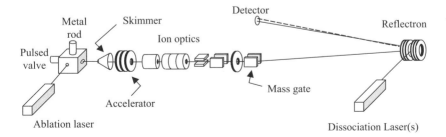

Figure 2. Time-of-flight (TOF) photofragment spectrometer.

onto a rotating and translating metal rod. Ablation produces transition metal ions M^+, which then react with an appropriate precursor entrained in a pulse of gas introduced through a piezoelectric pulsed valve [42]. Simple metal–ligand complexes M^+L where the ligand is a stable molecule [e.g., V^+ (CO_2)] are produced using 0.05–5% ligand seeded in an inert carrier, such as helium or argon at a backing pressure of 1–5 atm.

Most of the ions we study have covalent bonds to the metal (e.g., MO^+, $[HO-M-CH_3]^+$). These ions are synthesized via ion molecule reactions, and our choice of synthetic precursors is guided by the extensive literature on ion molecule reactions, typically carried out under single-collision conditions in ion cyclotron resonance (ICR) spectrometers [18, 43]. Although conditions in our ablation source are very different from those in an ICR, we generally observe the same products (other than, of course, the ablation source also produces cluster ions). Thus, for example, FeO^+ is produced by reaction of Fe^+ with nitrous oxide [44]. Studying the intermediates of a reaction presents a special challenge, as they are all isomers and thus cannot be separated in a mass spectrometer. In our study of the intermediates of the $FeO^+ + CH_4$ reaction [45] (Section III.C) it was critical to find precursors that selectively form each intermediate. Our choices were based on the extensive study of Schröder et al., who produced several of the intermediates by reacting Fe^+ with a variety of neutral molecules in an ICR [46]. They identified the resulting ions based on fragments produced after collision-induced dissociation (CID). Again, we are generally able to produce the desired intermediate using the same reaction. For example, the $[H_2C=Fe-OH_2]^+$ intermediate is synthesized by the reaction of Fe^+ with acetic acid or, with less efficiency and specificity, n-propanol. Reaction of Fe^+ with methanol efficiently produces $[HO-Fe-CH_3]^+$; acetic acid and n-propanol also give modest yields of this isomer. In each case, we characterize the ion formed through its dissociation pathways and electronic and vibrational photodissociation spectrum [45].

Once formed, ions travel through a short tube and supersonically expand into the source vacuum chamber. This cools the ions, reducing spectral congestion due to transitions from excited rotational and vibrational states. The molecular beam is then skimmed and ions pass into the differential pumping chamber. We have measured rotational temperatures of 8 K for FeO^+ [47] and 12 K for $V^+(OCO)$ [48] although the supersonic expansion is not at equilibrium, so the rotational-state distribution has a small component at a higher temperature, which is not unusual for an ablation source [49]. At these low temperatures, weakly bound cluster ions are readily produced. Vibrational spectroscopy of $V^+(OCO)_n$ and $Fe^+(CH_4)_n$ is discussed in Sections IV and V. When using argon (Ar) as the carrier gas, cluster ions containing argon can be formed. As a result, we study vibrations of the of $[HO-Fe-CH_3]^+$ insertion intermediate by measuring the vibrational spectrum of $[HO-Fe-CH_3]^+(Ar)_n$ ($n=1,2$)

(Section III.C). Using a rotational temperature to characterize an ion source can be misleading, as the reactions used to form the ions of interest can be quite exothermic, producing vibrationally and even electronically excited ions. These degrees of freedom are more difficult to cool than rotations. Transitions from vibrationally excited molecules provide very useful information, if they can be identified and analyzed. Hot FeO^+ (produced using 3% N_2O in helium) has a $v' = 1 \leftarrow v'' = 1$ sequence band that disappears when 15% N_2 is added to the carrier gas [47]. This sequence band allows us to measure the ground-state vibrational frequency $v''_0 = 838 \pm 4 \, cm^{-1}$. Similarly, by adjusting the delay between the ablation laser and the pulsed valve, we can observe [48] the $v'_3 = 0 \leftarrow v''_3 = 1$ hot band in $V^+(OCO)$, which gives the frequency of the metal–ligand stretch in the ground electronic state $v''_3 = 210 \, cm^{-1}$.

After the skimmer, ions are extracted along the beam axis using a pulsed electric field, then accelerated to 1800-V kinetic energy. This is a coaxial version of the classic orthogonal Wiley–McLaren TOF mass spectrometer [50]. It is convenient to have the source and flight tube grounded, so, after acceleration, the ions are rereferenced to ground potential [51]. This is accomplished by having the rereferencing tube at -1800 V potential as the ion cloud enters, then pulsing it to ground prior to the ions' exit using a potential switch [52]. An Einzel lens and deflectors guide the ions through an aperture into the detector chamber. A final deflector allows the ion beam to traverse the 5° angle through the reflectron and to the detector. When the deflector is off, <0.1% of the incident ions reach the detector. Applying a pulsed voltage to the deflector allows only ions within a few mass units of the ion of interest to reach the detector, forming an effective mass gate. This is essential as, without the mass gate, peaks from lighter ions, particularly the large M^+ peak, distort the baseline for the rest of the mass spectrum.

Ions are photodissociated at the turning point [53] of the reflectron [54, 55] with pulsed lasers. Photofragment ions and undissociated parent ions (dashed line in Fig. 2) reaccelerate out of the reflectron and strike a 40-mm diameter dual microchannel plate detector. Masses of parent and fragment ions are determined from their flight times. Two laser systems were employed for the studies described here. Electronic spectra were obtained using the unfocused output of a pulsed, tunable Continuum dye laser pumped by a Continuum Nd: YAG laser. With mixing and doubling crystals, this laser system is tunable from 220 to >900 nm with < 0.08-cm^{-1} line width. A LaserVision IR OPO/OPA pumped by an injection seeded Nd:YAG laser produces light in the near- and mid-IR for vibrational spectra. The IR laser system uses a 532-nm pumped OPO, followed by a 1064-nm pumped OPA. In the mid-IR, it is tunable from ~2100 to >4000 cm^{-1}, with 0.3-cm^{-1} line width. It produces ~3 mJ/pulse near 2400 cm^{-1} and ~10 mJ/pulse near 3800 cm^{-1}. The lasers operate at 20-Hz repetition rate.

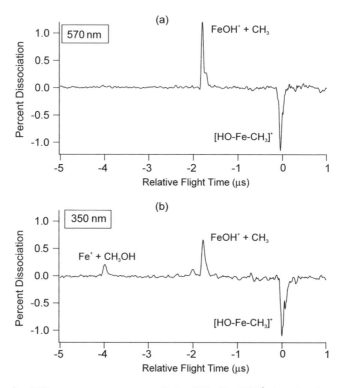

Figure 3. Difference mass spectrum of the $[HO-Fe-CH_3]^+$ insertion intermediate at photolysis wavelengths of 570 nm (a) and 350 nm (b). Simple Fe—C bond fission is observed at both wavelengths, but photolysis at 350 nm also triggers the half reaction to produce $Fe^+ +$ methanol (CH_3OH).

The ion signal is amplified, collected on a digital oscilloscope or a gated integrator, and averaged using a LabView-based program. Subtracting mass spectra collected with the dissociation laser blocked from those when it is unblocked produces a *difference mass spectrum*. As shown in Fig. 3, this allows immediate identification of the dissociation channels active at a particular wavelength, along with their relative importance. Mass resolution is $m/\Delta m \sim 200$ for fragment ions, although the large parent ion signal can make it difficult to detect H atom loss from heavy ions, such as $AuCH_2^+$.

The shape of the fragment peak in the difference mass spectrum can be affected by kinetic energy release and the dissociation rate. Significant kinetic energy release leads to broadening, while slow dissociation leads to tailing. We have not yet observed broadening in photofragments of singly charged ions, probably because the long-range attraction between the fragments and the large number of available product quantum states both favor low kinetic energy release. We do, however, observe significant broadening in singly charged fragments formed by photodissociation of dications, where the Coulomb repulsion between the fragments leads to kinetic energy releases of $80–170\,kJ\,mol^{-1}$ [56–58]. If an ion is photoexcited, then travels some distance through the reflectron before it dissociates, its flight time will lie between those of the parent ion and prompt-dissociating fragments. The fragment peak in the difference mass spectrum will then exhibit an exponential tail whose time constant depends on the dissociation rate. In our apparatus, tailing should be observed for dissociation lifetimes in the range $\sim 50\,ns$ to $\sim 3\,\mu s$. We have not observed tailing for any singly charged, metal-containing ions, indicating photodissociation lifetimes $< 50\,ns$. Larger ions do show tailing, for example, the ethylbenzene radical cation near 450 nm [59] and $Co^{2+}(CH_3OH)_4$ at 570 nm [58] from work in our group, and $Fe_3coronene^+$ at 532 nm [60] from work by Duncan's group.

Monitoring the yield of a particular fragment ion as a function of laser wavelength and normalizing to parent ion signal and laser fluence yields the *photodissociation spectrum*. This is the absorption spectrum of those ions that photodissociate to produce the fragment being monitored. The photo-dissociation spectrum is obtained by monitoring the fragment ion signal with a gated integrator or, if the fragment ion signal is very small, measuring the area under the fragment peak in the difference mass spectrum using numerical integration and then normalizing to the parent signal and laser fluence.

B. Electronic Spectroscopy

The ions we study typically have several unpaired electrons and consequently have many excited electronic states. This high density of states, along with peak broadening due to photoinduced reactions, fast internal conversion and rapid dissociation can lead to broad, featureless electronic photodissociation spectra. These cases provide a strong impetus for vibrational spectroscopy (described below). However, broad spectra do contain useful thermodynamic information, as one-photon dissociation of internally cold ions requires the photon energy to exceed the strength of the bond being broken. The photodissociation onset thus provides an upper limit to the true, thermodynamic bond strength [30–32, 61]. The high density of electronic states in ions with a coordinatively unsaturated transition metal center means that they are likely to absorb widely in the visible and near UV. As a result of this broad absorption and strong coupling between states near the dissociation limit, these ions often photodissociate at the thermodynamic threshold, and the photodissociation onsets give bond strengths precise to 5 kJ/mol or better [25, 62–64]. Photodissociation studies can thus complement bond strengths measured with collisional methods [32] such as endothermic reactions or collision induced dissociation in guided ion beams. In general, collisional methods are preferred for measuring bond strengths, as they have higher throughput and are more generally applicable because they do not rely on specific absorption properties of the molecule. The exception is those ions for which the photodissociation spectrum exhibits vibrational structure that completely converges to a diabatic dissociation limit. This has been seen for VAr^+ and $CoAr^+$; the dissociation limit can be determined with spectroscopic accuracy, giving bond strengths with <0.1-kJ mol^{-1} precision [65–67]. The methods used to measure ion thermochemistry are compared in an excellent review by Ervin [30].

Fortunately, many of the ions we study predissociate, giving photodissociation spectra with resolved vibrational progressions and, in some cases, residual rotational structure. Electronic spectroscopy of the FeO^+ reactant and the $[HO{-}Fe{-}CH_3]^+$ intermediate of the $FeO^+ + CH_4 \rightarrow Fe^+ + CH_3OH$ reaction is discussed in Section III, while the electronic spectroscopy and coupling between high and low spin states of $V^+(OCO)$ is covered in Section IV. Analysis of the electronic photodissociation spectrum gives detailed information on the excited-state potential energy surface(s), coupling between electronic states, and the geometry of the ion.

If an ion predissociates, then the rotational congestion and vibrational sequence bands can determine the resolution of the photodissociation spectrum. Most of the ions we study are quite rotationally cold, irrespective of the source conditions. However, varying the source conditions can greatly affect the vibrational temperature of the ions. As noted above, vibrational hot bands can be very useful, characterizing low frequency vibrations in the ground electronic state. More commonly, the peaks in the spectrum are simply broadened due to overlapping vibrational sequence bands. Helium, the typical buffer gas, is rather poor at cooling vibrations, so we often add $<10\%$ of a polyatomic molecule to enhance vibrational and electronic cooling. Adding O_2 eliminates an electronic hot band in PtO^+ and gives a much sharper dissociation onset in $CoCH_2^+$, again due to cooling metastable excited electronic states [63,68]. We have used CF_4, which is rather inert, for vibrational cooling in $Au^+(C_2H_4)$ [69]. The unimolecular dissociation rate of ethylbenzene cation is strongly dependent on the available energy. To produce vibrationally cold ions, we form them first by charge transfer from Pt^+, which is nearly thermoneutral, and also include CO_2 in the carrier gas [59]. In producing cluster ions, more concentrated mixes are not always better. We produce cold $V^+(OCO)_5$ using 5% CO_2 in helium. This same mix produces $V^+(OCO)$, which is fairly hot, as, in order to get good yields of the smaller clusters, we have to adjust the delay between the ablation laser and pulsed valve to have lower gas pressure in the ablation region. Using only 0.1% CO_2 in helium produces plenty of $V^+(OCO)$, and it is cold [48].

Photofragment spectroscopy is extremely sensitive, but it has the disadvantage that one is only sensitive to absorption that leads to photodissociation. For single-photon experiments, this means that one is restricted to studying electronic states that lie above the dissociation limit, and the lifetime of these states usually determines the resolution of the spectrum. We, along with several other groups, have developed techniques to measure the spectroscopy of states that lie below the dissociation limit, with laser-limited resolution. We have used resonance-enhanced photodissociation (REPD) spectroscopy to study the electronic spectrum of FeO^+ near $14,000 \, cm^{-1}$. One photon promotes the molecule to an excited electronic state; absorption of a second photon leads to dissociation. We obtain the rotationally resolved spectrum, with 0.05-cm^{-1} resolution, which allows us to measure the rotational constants (and bond lengths) in the ground and excited electronic states. The results are discussed in more detail in Section III.B. Brucat and co-workers used REPD to study charge-transfer transitions in CoO^+ [49]. One disadvantage of REPD is the low dissociation yield. For FeO^+, we obtain 0.1% dissociation at a peak, which restricts us to studying ions that can be produced in abundance. A second consideration is power broadening. For the FeO^+ study, the UV laser was not focused and was attenuated to avoid saturating the more intense transitions. Using two different wavelengths helps to avoid this problem, as the fluence of the laser exciting the bound–bound transition can be low, while that of the second laser is much higher. In addition, one can access a much broader range of electronic states, as one is not restricted to those that lie above 50% of the dissociation energy. An excellent example of two-color resonance enhanced photodissociation is the study of vibronic transitions in Ni_2^+ by Brucat and co-workers [70].

Another method to measure electronic spectra of ions below the dissociation limit is to photoexcite molecules relatively weakly bound to a ligand. For example, photofragmentation of $FeCH_2^+(H_2O)$ in the visible probes electronic states of the $FeCH_2^+$ chromophore that lie below $D_0(Fe^+–CH_2) = 28500 \, cm^{-1}$ [31, 63]. Absorption of $14,000$–$20,000$-cm^{-1} photons leads to loss of H_2O. The resulting spectrum is vibrationally resolved, but the peaks are $\sim 300 \, cm^{-1}$ wide. The REPD of $FeCH_2^+$ shows a similar vibrational progression, shifted $1740 \, cm^{-1}$ to the red. [45]. This shift is due to the different binding energies of H_2O to the ground and excited states of $FeCH_2^+$. The peaks in the REPD spectrum are also $300 \, cm^{-1}$ wide, probably due to rapid internal conversion of the excited state. In their photofragment study of the $ZrO^+(CO_2)$ and $ZrO^+(N_2)$ complexes, obtained by monitoring loss of CO_2 or N_2, Brucat and co-workers observe beautifully resolved photodissociation spectra from $14,900$ to $17,700 \, cm^{-1}$, with progressions in the Zr–O stretch, as well as the ZrO–ligand stretch and rock [71]. This technique provides a way to study the electronic spectroscopy of ZrO^+ far below its dissociation limit. Similarly, we observe a well-resolved spectrum in $TiO^+(CO_2)$ from $14,000$ to $17,300 \, cm^{-1}$, which probes a $^2\Pi \leftarrow \, ^2\Delta$ transition in TiO^+ [72].

C. Vibrational Spectroscopy

Electronic spectroscopy of jet-cooled molecules is an excellent tool to characterize excited electronic states of molecules, but it rarely gives information on vibrations in the *ground* electronic state. Vibrational (IR) spectroscopy is the tool of choice to study bonding in the ground electronic state. In electrostatically bound entrance channel complexes, such as $M^+(CH_4)$, vibrational spectroscopy illuminates the mechanism of C—H bond activation by measuring how interaction with the metal center affects bonds in the reactant. Similarly, vibrational spectroscopy of reaction intermediates elucidates the mechanism of ligand activation by the metal center. An additional advantage is that vibrational spectroscopy gives well-resolved spectra even in quite complex molecules, such as M^+ bound to several ligands, where the electronic spectrum is usually featureless [73]. As a result, vibrational spectroscopy of ions has chiefly been applied to studying *noncovalent interactions* in solvated cluster ions [73–92]. Vibrational spectroscopy has been used to reveal the structure of products and intermediates of ion molecule reactions [93, 94] as well as of intracluster reactions [76, 95]. We have used vibrational spectroscopy to study covalently bound reaction intermediates, such as $[HO—Fe—CH_3]^+$ (Section III.C), and noncovalent complexes, such as $V^+(OCO)$ and $Fe^+(CH_4)_n$ (Sections IV and V). These studies were carried out using an IR OPO/OPA laser system that is pumped by an injection-seeded Nd:YAG laser. The IR laser system uses a 532-nm pumped OPO and a 1064-nm pumped OPA to produce tunable light from 2100 to $>4000\ cm^{-1}$, with 0.2-cm^{-1} line width and $\sim$3 mJ/pulse near 2400 cm^{-1} and 8 mJ/pulse near 3500 cm^{-1}. Free electron lasers, such as FELIX (Netherlands) and CLIO (France), produce very high fluences down to $\sim$600 cm^{-1}. This wide spectral range has allowed the characterization of many ions that have several potential isomers [74, 81, 93, 94, 96]. Free electron lasers have disadvantages: beam time is limited and the spectra tend to be broad due to the 2% bandwidth of the light and high fluences required for efficient IR multiphoton dissociation.

Photofragment spectroscopy requires that the ion dissociate, which is a challenge as most of the ions we wish to study have significant binding energies: for example, $V^+(OCO)$ is bound by 6000 cm^{-1} [33]. Groups in this field have developed a toolbox of complementary techniques to measure IR spectra of strongly bound ions [74]. No one method is perfect or completely general, but at least one method should work on any given ion. Because absorption of several IR photons can dissociate even strongly bound ions, *resonance enhanced IR multiphoton dissociation (IRMPD)* has proven a useful tool for studying the spectroscopy of ions since early work on organic ions using CO_2 lasers [97]. Transition metal containing systems have also been studied using CO_2 lasers [98–101]. The high fluences available from IR OPOs and free-electron lasers make IRMPD an attractive technique. Efficient IRMPD requires fairly rapid intramolecular vibrational redistribution (IVR), to ensure that vibrationally excited molecules continue to absorb at the resonant wavelength and also to efficiently transfer energy from the vibration excited to the dissociation coordinate. Small molecules and clusters tend to have high binding energies and low vibrational density of states (and hence small IVR rates); larger molecules are better candidates for IRMPD due to their higher densities of states and more rapid IVR. For example, Duncan and co-workers observe no signal in IRMPD of $Fe^+(CO_2)$, weak signal from $Fe^+(CO_2)_2$, and strong signals in larger clusters, attributing this to slow IVR in the small clusters [102]. Similarly, $[HO-Fe-CH_3]^+$ is not a good candidate for IRMPD studies. We observe very small amounts of multiphoton dissociation: only 0.1% dissociation in the O—H stretching region. Also, power broadening and preferential photodissociation of hotter ions in the beam lead to a broad, poorly resolved spectrum.

"*Argon tagging*" is a version of the "spectator spectroscopy" developed by Lee and co-workers [103] in which the ion of interest is clustered with one or more argon atoms. The IR absorption by the ion core leads to vibrational predissociation and loss of the tag. The choice of argon as the tag is a compromise: Argon has sufficiently weak binding to ensure one-photon dissociation and to minimize perturbations of the vibrational spectrum, yet binds sufficiently strongly to allow production of usable quantities of tagged ions. Argon tagging has been extensively developed by Johnson and co-workers for the spectroscopy of negative ions [78, 104–107] and has been used by Lisy and co-workers to study $Cs^+(H_2O)$ and $Li^+(H_2O)$ [86,88] and Duncan and co-workers to study many ions, including $M^+(CO_2)_n$ (M = Al, Fe, Mg, Ni, V) [76, 77, 102, 108–112]. We have used this technique to measure the O—H and C—H stretching vibrations of $[HO-Fe-CH_3]^+$ (Section III.C). Neon-tagged ions are more challenging to produce, but the weaker binding is an advantage in studying low frequency vibrations and also gives smaller perturbations [79]. Helium tagging leads to the smallest perturbations and has been used to measure vibrational spectra of vanadium oxide cations using a free-electron laser [113].

Vibrationally mediated photodissociation (VMP) can be used to measure the vibrational spectra of small ions, such as $V^+(OCO)$. Vibrationally mediated photodissociation is a double resonance technique in which a molecule first absorbs an IR photon. Vibrationally excited molecules are then selectively photodissociated following absorption of a second photon in the UV or visible [114–120]. With neutral molecules, VMP experiments are usually used to measure the spectroscopy of regions of the excited-state potential energy surface that are not Franck–Condon accessible from the ground state and to see how different vibrations affect the photodissociation dynamics. In order for VMP to work, there must be some wavelength at which vibrationally excited molecules have an electronic transition and photodissociate, while vibrationally unexcited molecules do not. In practice, this means that the ion has to have a vibrationally resolved photodissociation spectrum due to predissociation [as is the case for $V^+(OCO)$] or that the photodissociation spectrum have a sharp onset (e.g., at the bond strength, as for $FeCH_2^+$, $AuCH_2^+$ and several other ions we have studied) [25, 63, 121]. This is not always the case, or the electronic transition may be weak, or lie at an inconvenient wavelength, so VMP is less generally applicable to ion spectroscopy than IRMPD or tagging. However, VMP has the great advantage that it can be used to measure vibrational spectra of unperturbed ions with resolution limited only by the laser line width. Section IV discusses our use of VMP to measure the OCO antisymmetric stretch in $V^+(OCO)$ and to see how exciting vibronic transitions involving this vibration affect the photodissociation dynamics.

III.   METHANE–METHANOL CONVERSION BY FeO^+

Methane cannot be liquefied by pressure alone, it must also be cooled, which makes it awkward to transport. Therefore there has been a great deal of effort directed toward direct conversion of methane to an easily transportable and more synthetically useful liquid, such as a larger hydrocarbon or methanol [122–144]. Direct conversion of methane to methanol is also of great fundamental interest as the simplest alkane oxidation. Although no direct, efficient methane–methanol conversion scheme has yet been developed [122], significant advances have been made using iron-based catalysts. Wang and Otsuka have achieved high catalytic selectivity for direct oxidation of methane to methanol using an $FePO_4$ catalyst with N_2O and H_2/O_2 as the oxidizing agents, but the reaction yield is low [125, 126]. Other approaches that have achieved modest success include direct oxidation by N_2O in a plasma [127], oxidation of methane to a methyl ester using a platinum catalyst [128, 129], and direct methane–methanol conversion using an iron-doped zeolite [130]. Methanogenic bacteria efficiently convert methane to methanol. The reaction is catalyzed by the enzyme methane monooxygenase (MMO), which contains non-heme iron centers at the active site [131–133].

In 1990, Schröder and Schwarz reported that gas-phase FeO^+ directly converts methane to methanol under thermal conditions [21]. The reaction is efficient, occuring at $\sim$20% of the collision rate, and is quite selective, producing methanol 40% of the time ($FeOH^+ + CH_3$ is the other major product). More recent experiments have shown that NiO^+ and PtO^+ also convert methane to methanol with good efficiency and selectivity [134]. Reactions of gas-phase transition metal oxides with methane thus provide a simple model system for the direct conversion of methane to methanol. These systems capture the essential chemistry, but do not have complicating contributions from solvent molecules, ligands, or multiple metal sites that are present in condensed-phase systems.

Bond activation by transition metals is a complex process, as it often involves making and breaking several bonds and can occur on multiple, coupled potential energy surfaces. Detailed understanding of the mechanism requires many experiments, as each is sensitive to only part of the potential energy surface, but experiments alone are not sufficient: Accurate calculations are also required. Studies of small model systems that retain the essential chemistry are necessary, as they are the systems for which we can carry out the most detailed experiments and highest level calculations, allowing us to assess the reliability of competing theoretical methods. These methods can then be used to predict mechanisms for more complex, condensed-phase reactions for which we have limited experimental data and to help develop improved catalysts.

Methane-to-methanol conversion by gas-phase transition metal oxide cations has been extensively studied by experiment and theory: see reviews by Schröder, Schwarz, and co-workers [18, 23, 134, 135] and by Metz [25, 136]. We have used photofragment spectroscopy to study the electronic spectroscopy of FeO^+ [47, 137], NiO^+ [25], and PtO^+ [68], as well as the electronic and vibrational spectroscopy of intermediates of the $FeO^+ + CH_4$ reaction.[45, 136] We have also used photoionization of FeO to characterize low lying, low spin electronic states of FeO^+ [39]. Our results on the iron-containing molecules are presented in this section.

A. Reaction and Computational Studies: Mechanism

Figure 4 shows a schematic potential energy surface for the conversion of methane to methanol by FeO^+. The sextet (high spin) reaction path is indicated by a solid line and the quartet (low spin) path is dotted. The minor pathway leading to $FeCH_2^+ + H_2O$ is not shown. The relative energies of reactants and products are based on experiment [138]. The energies of intermediates are based on our calculations [45, 139] at the B3LYP/6-311+G(d,p) level, and the energies of transition states are relative to the previous intermediate, as calculated by Yoshizawa et al. [140]. Recent calculations in our group at the CCSD(T)/6-311+G(3df,p) and B3LYP/6-311+G(3df,p) level give similar results, but predict that the quartet and sextet states of the insertion intermediate have very similar energies [141]. Our calculations are an extension of computational studies at the B3LYP/6-311G(d,p) level by Yoshizawa et al. on methane activation by FeO^+ [142–144] and the other first-row MO^+ [140, 145]. Schröder et al. [46] and Fiedler et al. [146] have also carried out calculations on methane–methanol conversion by FeO^+ and the late first-row transition metals, respectively.

The mechanism that has been developed for the conversion of methane to methanol by FeO^+ is an excellent example of the synergy between experiment and theory. This mechanism includes two key concepts: concerted reaction involving the critical $[HO—Fe—CH_3]^+$ insertion intermediate and two-state reactivity. The reaction proceeds as follows: electrostatic interaction between FeO^+ and methane produces the $[OFe\cdots CH_4]^+$ entrance channel complex.

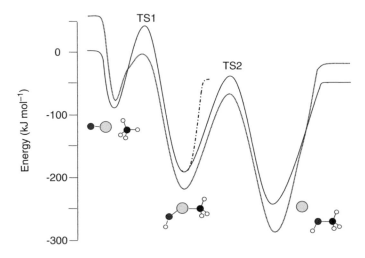

Figure 4. Schematic potential energy surface for the reaction of FeO$^+$ with methane. The solid line indicates the sextet surface; the quartet surface is shown with a dotted line, in each case leading to the production of Fe$^+$ + CH$_3$OH. The dashed line leads to formation of FeOH$^+$ + CH$_3$. The pathway leading to the minor FeCH$_2^+$ + H$_2$O channel is not shown. Schematic structures are shown for the three minima: the [OFe···CH$_4$]$^+$ entrance channel complex, [HO—Fe—CH$_3$]$^+$ insertion intermediate, and Fe$^+$(CH$_3$OH) exit channel complex. See text for details on the calculations on which the potential energy surface is based.

Depending on the level of theory, the iron coordinates to methane in an η_2 or η_3 configuration, and weakens the proximate C—H bonds by accepting electron density from C—H σ bonding orbitals and backdonating electron density to C—H antibonding orbitals. At transition state TS1 the strong C—H bond in methane is being replaced by two bonds: a strong O—H bond and a fairly weak Fe—C bond. Although both the reactants and products are high spin, at thermal energies the reaction occurs through low spin intermediates [143,144], as the high spin TS1 lies significantly above the reactants. This "two-state reactivity" has been extensively studied by Shaik and co-workers, especially in the exothermic, but very inefficient, FeO$^+$ + H$_2$ → Fe$^+$ + H$_2$O reaction [147–150]. The *efficiency* of the reaction is determined by the likelihood that reactants will cross TS1. This is determined by the energy of the quartet TS1, as well as by the probability that the initially formed sextet entrance channel complex will undergo a spin change to the quartet state. Shiota and Yoshizawa have calculated [144] the spin–orbit coupling in the entrance channel complex to be a modest ~130 cm^{-1}. This and the lifetime of the [OFe···CH$_4$]$^+$ entrance channel complex determine the likelihood of crossing to the quartet surface. The low efficiency of the FeO$^+$ + H$_2$ reaction is due to the short lifetime of the entrance channel complex. The TS1 leads to the key insertion intermediate

$[HO-Fe-CH_3]^+$, which can dissociate to produce $FeOH^+ + CH_3$ or can undergo migration of a methyl group via TS2 to produce the iron–methanol exit channel complex $[Fe(CH_3OH)]^+$, which subsequently dissociates. The *selectivity* of the reaction between methanol and methyl radical products is primarily determined by the energy of TS2 relative to methyl radical products. Because methyl radical is produced by simple bond fission of the insertion intermediate, it is entropically favored over the methanol channel, which occurs through the tighter transition state TS2. Thus, if TS2 is at an energy close to or above methyl products, the reaction will overwhelmingly produce $MOH^+ + CH_3$, as is observed for MnO^+ [134, 145]. For FeO^+, TS2 lies somewhat below methyl radical products, so the two pathways are competitive at thermal energies, but increased translational energy strongly favors the methyl radical pathway [151]. Producing quartet $Fe^+ + CH_3OH$ from sextet reactants is endothermic, so a second spin change is required to produce exothermic sextet products. Previously, most discussions assumed that this occurs in the $Fe^+(CH_3OH)$ exit channel complex. However, Shiota and Yoshizawa calculate [144] that the spin–orbit coupling in this complex is only $0.3\ cm^{-1}$. They suggest that the second spin change occurs in the insertion intermediate, which has a spin–orbit coupling of $\sim 20\ cm^{-1}$. Our vibrational spectroscopy experiments on the insertion intermediate support this idea, as our results suggest that the quartet and sextet states of $[HO-Fe-CH_3]^+$ are both formed in the molecular beam and can likely interconvert. In order to characterize FeO^+ we have measured the electronic spectroscopy of predissociative and bound states, as well as photoionization of neutral FeO. We have also studied the electronic and vibrational spectroscopy of the $[HO-Fe-CH_3]^+$ intermediate.

B.   Spectroscopy of FeO^+

1.   Electronic Spectroscopy of Predissociative States

In 1986 Freiser and co-workers measured the photodissociation spectrum of FeO^+ in an ion cyclotron resonance spectrometer using a lamp–monochromator as their light source. They observed a gradual onset at $\sim 420\ nm$ leading to a sharp peak near $350\ nm$ ($28,600\ cm^{-1}$) whose width was determined by the 10-nm resolution of their instrument [61]. More recent guided ion beam measurements place $D_0(Fe^+-O)$ at $28,000 \pm 400\ cm^{-1}$ [31]. We produce FeO^+ by reacting Fe^+ with N_2O and measure its photodissociation spectrum from $\sim 28,000$ to $> 30,000\ cm^{-1}$. Figure 5 shows the origin of the $^6\Sigma \leftarrow X\ ^6\Sigma$ band, which lies just above the dissociation limit. The partially resolved rotational structure in the peak can be simulated to obtain rotational constants for the upper state, using the known (see below) constants for the ground state [47,137]. The simulation (dashed lines) gives a bond length $r_0' = 1.664\ \text{Å}$ and the spin–spin splitting constant $\lambda' = 0.60\ cm^{-1}$ for the upper state. The resolution is limited to

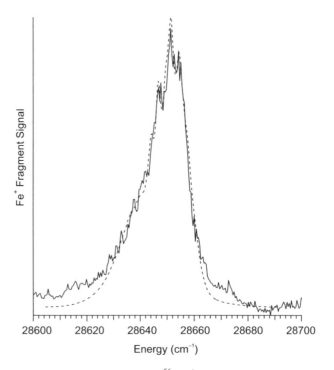

Figure 5. Photodissociation spectrum of $^{56}FeO^+$. The 0–0 vibrational transition of the $^6\Sigma \leftarrow X^6\Sigma$ band (solid line) and best-fit simulation (dashed line) are shown.

$1.5\ cm^{-1}$ by the 3.5-ps predissociation lifetime of the excited state. The $v' = 1$ peak is observed $662\ cm^{-1}$ higher in energy and is $40\ cm^{-1}$ wide, dissociating in 140 fs. By producing the ions using helium as the carrier gas at low backing pressure, we minimize vibrational cooling and observe the $v' = 1 \leftarrow v'' = 1$ transition at $28,473\ cm^{-1}$, which allows us to measure the vibrational frequency in the ground electronic state $v_0'' = 838\ cm^{-1}$ [47].

2. Resonance-Enhanced Photodissociation: FeO$^+$ States Below the Dissociation Limit

Unfortunately, predissociation of the excited-state limits the resolution of our photodissociation spectrum of FeO$^+$. One way to overcome this limitation is by resonance enhanced photodissociation. Molecules are electronically excited to a state that lies below the dissociation limit, and photodissociate after absorption of a second photon. Brucat and co-workers have used this technique to obtain a rotationally resolved spectrum of CoO$^+$ from which they derived rotational constants for the $^5\Delta_4$ ground state and the $^5\Pi_3$ and $^5\Phi_5$ excited states [49]. We carried out time-dependent density functional theory calculations (TD–DFT) using the B3LYP hybrid density functional to see if FeO^+ has excited electronic states in the relevant energy range. The TD–DFT results, shown in Fig. 6, predict an excited $^6\Sigma$ state near 27500 cm^{-1}, with a bond length similar to the ground state. These predictions are in excellent agreement with our observed $^6\Sigma$–$^6\Sigma$ transition. At lower energy, the calculations predict three $^6\Pi$ states with equilibrium bond lengths significantly longer than the ground state. A vertical transition to the lowest $^6\Pi$ state should occur at about one-half of the dissociation energy, making this state a good candidate for resonance enhanced $(1+1)$ photodissociation (REPD) studies. Figure 7 shows a portion of the REPD

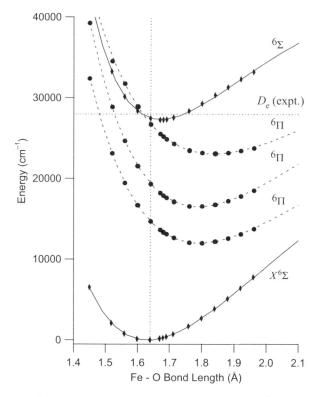

Figure 6. Calculated potential energy curves for sextet states of FeO^+. The ground electronic state and excited states accessible by allowed electronic transitions from the ground state are shown. Points are calculated using TD–DFT at the B3LYP/6-311G(d,p) level. Solid lines are Σ states and dashed lines are Π states, the vertical dashed line indicates r_e for the ground state. The experimental value of the dissociation energy D_e is also shown for reference.

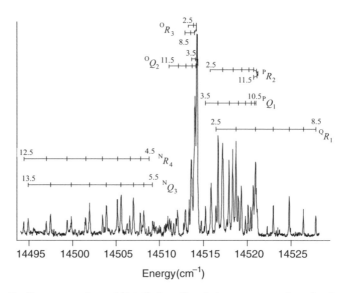

Figure 7. Resonance enhanced $(1 + 1)$ photodissociation spectrum and rotational assignments of the $^6\Pi_{7/2}$ ($v' = 8$) $\leftarrow$ $^6\Sigma$ ($v'' = 0$) band of $^{56}FeO^+$. Numbers indicate J'' for each line; the subscripts indicate the spin component of the $^6\Sigma$ state (F_1 is $\Sigma = \frac{5}{2}$); the superscripts indicate ΔN for the transition.

spectrum. The resolution is limited only by the $0.05\ cm^{-1}$ line width of the laser, leading to clearly resolved rotational structure. Based on the energy shift observed for the minor $^{54}FeO^+$ isotopomer, the transition is from $v'' = 0$ of the $^6\Sigma$ ground state to $v' = 8$ of the $^6\Pi_{7/2}$ excited state. The $^6\Sigma$ ground state has six spin–spin sublevels and we observe transitions from four of them to $v' = 8, 9$ of the $^6\Pi_{7/2}$ excited state. The observed transitions have been fit to a detailed Hamiltonian to obtain rotational constants for the ground and $^6\Pi_{7/2}$ excited states [137]. The rotational constant for the ground state gives $r_0'' = 1.643 \pm 0.001$ Å. Other molecular parameters determined for the $^6\Sigma^+$ ground state are the spin–spin coupling constant $\lambda = -0.126\ cm^{-1}$ and the spin–rotational coupling constant $\gamma = -0.033\ cm^{-1}$. Detailed spectroscopy of FeO^+ and similar systems provide a demanding test of electronic structure methods. Measurements of the ground-state bond length and the ground- and excited-state vibrational frequencies test calculations of the potentials, while measurements of the spin–spin and spin–orbit coupling test calculations of the interactions between states. Observation of transitions to high vibrational levels of the $^6\Pi$ state is consistent with the TD–DFT calculations, which predict that the low lying $^6\Pi$ states have charge-transfer character, with much longer bond length than the $^6\Sigma$

ground state. In general, we find that TD–DFT calculations of excited electronic states of open-shell metal-containing diatomics are surprisingly accurate, with mean errors of 0.03 Å in bond lengths, 2000 cm^{-1} in electronic excitation energies and 50 cm^{-1} in vibrational frequencies [137].

3. Photoionization of FeO: Low lying Quartet States of FeO$^+$

Transition metal containing ions often have low lying electronic states with different spin multiplicity from the ground state. Electronic transitions to these states from the ground state are spin forbidden and often occur at awkward wavelengths, making their study difficult. Photoionization of the neutral provides an alternate route to access these states. For example, electronic structure calculations predict that FeO$^+$ has a $^6\Sigma$ ground state (which our experiments confirm) and several low lying quartet states, $^4\Pi$, $^4\Delta$, and $^4\Phi$ within ~1 eV of the ground state, depending on the level of theory [145, 146, 152, 153]. The quartet states have not been optically observed. All of these states can be accessed by photoionization of neutral FeO (X, $^5\Delta$). We have measured photoionization efficiencies for FeO (Fig. 8) and also of CuO at the Advanced Light Source, obtaining ionization energies for the metal oxides with

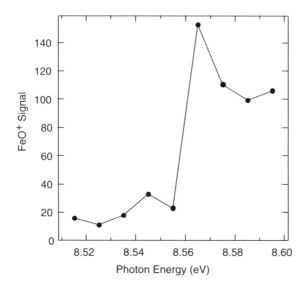

Figure 8. Photoionization efficiency curve for ^{56}FeO near the ionization onset, corresponding to production of FeO$^+$ X, $^6\Sigma$.

0.01-eV precision [39]. This allows us to refine the $Fe^+ - O$ bond strength using

$$D_0(M^+ - O) - D_0(M - O) = IE(M) - IE(MO)$$

Recent photofragment imaging experiments give a very precise value $D_0(Fe - O) = 4.18 \pm 0.01$ eV [154], which implies $D_0(Fe^+ - O) = 3.52 \pm 0.02$ eV. This value is consistent with, and slightly more precise than, the guided ion beam value of 3.47 ± 0.06 eV [31]. For most transition metal oxides, bond strengths for the ions are significantly more precise than for the neutrals, so measuring the molecule's ionization energy leads to improved values of the neutral bond strength. In this vein, we have also recently measured ionization energies of PtC, PtO, and PtO_2.

Photoionization can also access excited electronic states of the ion that are difficult to study by optical methods. The photoionization yield of FeO increases dramatically 0.36 eV above the ionzation energy. This result corresponds to the threshold for producing low spin quartet states of FeO^+. These states had not been previously observed, as transitions to them are spin forbidden and occur at inconveniently low energy. Because the $FeO^+ + CH_4$ reaction occurs via low spin intermediates, accurately predicting the energies of high and low spin states is critical.

C.   Spectroscopy of the $[HO-Fe-CH_3]^+$ Insertion Intermediate

1.   Electronic Spectroscopy

The potential energy surface for the $FeO^+ + CH_4$ reaction has four intermediates: an $[OFe \cdots CH_4]^+$ entrance channel complex, $[HO-Fe-CH_3]^+$ insertion intermediate and $Fe^+(CH_3OH)$ exit channel complex, as shown in Fig. 4. In addition, there is an $[H_2O-Fe=CH_2]^+$ exit channel complex that leads to the minor $FeCH_2^+ + H_2O$ channel. As these intermediates are, at least, local minima, once they are produced and cooled, they should be stable in the absence of collisions and they can be studied. Studying these ions is complicated by the fact that they are all isomers and thus cannot be separated in a mass spectrometer. So, we use ion molecule reactions to synthesize a particular intermediate and then characterize it from its photodissociation pathways and photodissociation spectrum. In an elegant series of ICR experiments, Schröder et al. synthesized the intermediates of the $FeO^+ + CH_4$ reaction by reacting Fe^+ with a variety of organic precursors. The intermediates were characterized based on fragment ions produced by collision-induced dissociation [46]. Although the reaction conditions in our ion source are very different from those in an ICR, we find that the reactions used by Schröder et al. generally form the same intermediates in our source. We have produced the $[HO-Fe-CH_3]^+$ and $[H_2O-Fe=CH_2]^+$

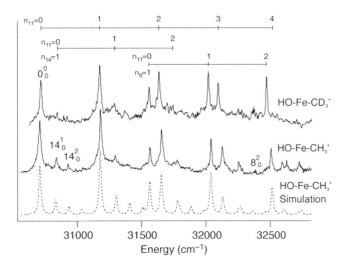

Figure 9. Photodissociation spectra of the insertion intermediate of the $FeO^+ + CH_4$ reaction. Top: $[HO-Fe-CD_3]^+$, middle: $[HO-Fe-CH_3]^+$, bottom (dashed): Franck–Condon simulation of the $[HO-Fe-CH_3]^+$ spectrum. The spectrum shows a long progression in the Fe-C stretch ($v_{11}' = 478$ cm^{-1}) and short progressions in the Fe–O stretch ($v_8' = 861$ cm^{-1}) and O–Fe–C bend ($v_{14}' = 132$ cm^{-1}).

intermediates, cooled them in a supersonic expansion, and measured their electronic and vibrational spectra. Here we will focus on the $[HO-Fe-CH_3]^+$ insertion intermediate, results on $[H_2O-Fe=CH_2]^+$ have been presented elsewhere [45, 139].

Photodissociation of the insertion intermediate produces $Fe^+ + CH_3OH$ and $FeOH^+ + CH_3$ in a 44:56 ratio at each photodissociation resonance peak. Nonresonant photodissociation leads to less Fe^+ product, as shown in Fig. 3. So, in a half-collision experiment, photoexcitation of the $[HO-Fe-CH_3]^+$ intermediate triggers the $FeO^+ + CH_4$ reaction, leading to the same products as are observed in the bimolecular reaction. The photodissociation spectra of $[HO-Fe-CH_3]^+$ and $[HO-Fe-CD_3]^+$ obtained by monitoring $FeOH^+$ are shown in Fig. 9. The data shown was obtained from the insertion intermediate produced by reacting Fe^+ with CH_3OH and CD_3OH. As this reaction can also produce the $Fe^+(CH_3OH)$ exit channel complex, we also synthesize the ions by reacting Fe^+ with acetic acid, which produces $[HO-Fe-CH_3]^+$ and $[H_2O-Fe=CH_2]^+$. Monitoring the Fe^+ or $FeOH^+$ fragment, we get the same spectrum as shown in Fig. 9, confirming that it is due to the insertion intermediate. If we monitor $FeCH_2^+$, we get a very different photodissociation spectrum due to $[H_2O-Fe=CH_2]^+$. The spectrum of the insertion intermediate is vibrationally resolved and the peaks show some tailing to lower energy due to unresolved rotational structure. As these intermediates are probably in the sextet state [45], the peaks are due to vibrations in an electronically excited sextet state of the intermediate. The longest vibrational progression observed is in the Fe—C stretch ($v_{11}' = 478\,\mathrm{cm}^{-1}$) and there are short progressions in the Fe—O stretch ($v_8' = 861\,\mathrm{cm}^{-1}$) and O—Fe—C bend ($v_{14}' = 132\,\mathrm{cm}^{-1}$). This assignment is supported by isotope shifts in the spectrum of $[\mathrm{HO-Fe-CD_3}]^+$ and extensive hybrid density functional theory (B3LYP) calculations [45]. The Franck–Condon simulation shown in dashed lines in the figure predicts that electronic excitation leads to a 0.13-Å change in the Fe—C bond length, a 0.05 Å change in the Fe—O bond length, and O—Fe—C angle change of 4°. The low frequency bend is the primary motion required for the molecule to get from the insertion intermediate to transition state TS2. While the electronic spectrum allows us to study an excited electronic state of $[\mathrm{HO-Fe-CH_3}]^+$, it gives little information on the structure and bonding in the ground state of the complex. This information is best obtained from vibrational spectroscopy.

2. Vibrational Spectroscopy

To characterize bonding in the insertion intermediate, we have measured its vibrational spectrum. As noted in Section II, using photofragment spectroscopy to measure vibrational spectra of strongly bound ions is challenging, as one IR photon does not have sufficient energy to break a bond. This is certainly the case for $[\mathrm{HO-Fe-CH_3}]^+$, which is bound by $\sim$130 kJ mol^{-1} (three IR photons at 3600 cm^{-1}) relative to Fe$^+$ + CH$_3$OH [45]. Infrared multiphoton dissociation is inefficient ($<$0.1% dissociation at 10 mJ/pulse) and gives a very broad spectrum. Instead, we measure the vibrational spectrum of argon-tagged molecules $[\mathrm{HO-Fe-CH_3}]^+(\mathrm{Ar})_n$ ($n = 1,2$) in the O—H stretching region (Fig. 10) [141]. The spectrum of $[\mathrm{HO-Fe-CH_3}]^+(\mathrm{Ar})$ peaks at 3646 cm^{-1} and has a shoulder at $\sim$3633 cm^{-1}, while the spectrum of $[\mathrm{HO-Fe-CH_3}]^+(\mathrm{Ar})_2$ is significantly narrower, peaking at 3660 cm^{-1}. For comparison, the O—H stretching frequency in methanol is 3681 cm^{-1}. The shoulder in the $[\mathrm{HO-Fe-CH_3}]^+(\mathrm{Ar})$ spectrum suggests the presence of two isomers or spin states of the molecule, and this feature persists when we change source conditions. The same spectrum is obtained with methanol and acetic acid as the precursor, confirming that we are studying the insertion intermediate, rather than the exit channel complex. We have extended our previous calculations on the FeO$^+$ + CH$_4$ reaction to B3LYP and CCSD(T) with larger basis sets, and the calculations predict that the quartet and sextet states of $[\mathrm{HO-Fe-CH_3}]^+$ are at very similar energies. Calculations on the argon-tagged complexes predict that the O—H stretching frequency of sextet $[\mathrm{HO-Fe-CH_3}]^+$ lies 11 cm^{-1} above that of the quartet state, but that the frequencies are very similar

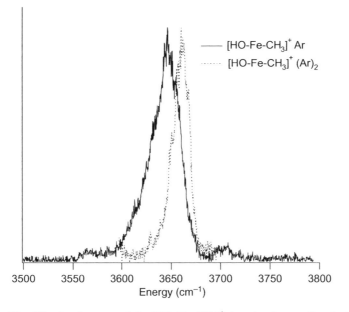

Figure 10. Vibrational spectra of the $[HO{-}Fe{-}CH_3]^+$ insertion intermediate in the O—H stretching region. Spectra are obtained by monitoring loss of argon from IR resonance enhanced photodissociation of the argon-tagged complexes $[HO{-}Fe{-}CH_3]^+(Ar)_n$ ($n = 1,2$).

for the argon-tagged complexes. So, it is possible that both the sextet and quartet states of the insertion intermediate are formed in our experiment and that the significant spin–orbit coupling calculated by Shiota and Yoshizawa [144] allows them to interconvert in the source. We have also measured the vibrational spectrum in the C—H stretching region, which is significantly more difficult, as the C—H stretches are at least an order of magnitude weaker than the O—H stretch. We observe peaks at 2880 and 2860 cm^{-1} for $[HO{-}Fe{-}CH_3]^+(Ar)$ and narrower peaks at 2885 and 2860 cm^{-1} for $[HO{-}Fe{-}CH_3]^+(Ar)_2$. The calculations predict these to be the symmetric C—H stretch, which is the most intense and occurs at the lowest energy. They also suggest that the higher frequency vibrations are due to the sextet state. The vibrational spectrum is sensitive to both spin states, as both have intense O—H stretching vibrations. In the electronic photodissociation spectrum, we observe a structured spectrum near 31,000 cm^{-1}. The TD-DFT calculations suggest that this is due to the quartet state, as the sextet is not predicted to absorb in this energy region, while the quartet is predicted to have several electronic states with modest absorption from 29,300 to 35,900 cm^{-1}. Some of these states also likely lead to the nonresonant photodissociation observed in this region.

IV. C–O BOND ACTIVATION BY V$^+$: SPECTROSCOPY AND DISSOCIATION DYNAMICS OF V$^+$(OCO)$_n$

The reaction between V$^+$ cation and CO$_2$ is quite interesting, as it demonstrates the effect of spin on an exothermic reaction, and how spin effects differ between a bimolecular reaction and a photoinduced half-reaction. It also shows how photoexcitation can be used to influence the products of the chemical reaction. The V$^+$ + CO$_2$ reaction is exothermic:

$$V^+(^5D) + CO_2 \rightarrow VO^+(^3\Sigma) + CO(^1\Sigma) \qquad \Delta H = -52\,kJ\,mol^{-1} \qquad (1)$$

However, the reaction does not occur at thermal energies (the reaction rate is <0.1% of the collision rate) [26], but is observed at higher collision energies [33]. This is unusual: Most exothermic ion molecule reactions have an appreciable rate at thermal energies, as the attractive electrostatic forces between the reactants depress reaction barriers so that they lie below the reactants. Vanadium cation forms an electrostatic entrance channel complex with CO$_2$; calculations show that the complex is linear: V$^+$(OCO), with a $^5\Delta$ or $^5\Sigma$ ground state [48,155], and guided ion beam experiments [33] show that it is bound by 72 kJ mol^{-1} relative to V$^+$ + CO$_2$. Collision-induced dissociation [33] of V$^+$(OCO) leads exclusively to V$^+$ + CO$_2$ at low energies, with VO$^+$ only observed at collision energies > 8 eV. This result is surprising, as VO$^+$ production is energetically favored, but spin forbidden

$$V^+(OCO) \begin{cases} V^+ (^5D) + CO_2 (^1\Sigma) & \Delta H = 72 \pm 4\ kJ\ mol^{-1} \qquad (2a) \\ \\ VO^+ (^3\Sigma) + CO (^1\Sigma) & \Delta H = 20 \pm 10\ kJ\ mol^{-1} \qquad (2b) \end{cases}$$

A. Electronic Spectroscopy of V$^+$(OCO)

Brucat and co-workers studied photodissociation of V$^+$(OCO) in the visible [156]. The excited-state predissociates, giving a vibrationally resolved spectrum with partially resolved rotational structure with progressions in the V$^+$–OCO stretch and rock, and in the OCO bend. Dissociation occurs via both the reactive VO$^+$ and nonreactive V$^+$ channels, with a VO$^+$/V$^+$ ratio of 0.4 at the vibronic origin. They also observed interesting mode selectivity in the photodissociation: Exciting the rocking vibration enhances the reactive channel by 50%. Because the reactive channel requires breaking a C–O bond, we felt that the OCO antisymmetric stretch would be similar to the reaction coordinate. So, we proposed that exciting vibronic transitions involving the antisymmetric stretch, perhaps in combination with V^+–OCO stretch and rock, would enhance the reactive channel even further. Brucat and co-workers did not observe bands involving the antisymmetric stretch vibration, presumably due to small Franck–Condon factors. To overcome this problem, we used vibrationally mediated photodissociation to access transitions involving the antisymmetric stretch vibration, v_1. An IR photon vibrationally excites molecules to $v_1'' = 1$, then a visible photon promotes the molecules to vibrational levels in the excited electronic state with $v_1' = 1$, whereupon they predissociate, and we measure the products. This experiment would also measure the antisymmetric stretch frequencies in the ground and excited electronic states, which had not been previously determined.

B. Vibrational Spectroscopy and Vibrationally Mediated Photodissociation

As noted in the experimental section, vibrational predissociation is the simplest way to measure the vibrational spectra of ions. Vibrational predissociation requires that the photon energy exceed the bond strength. The OCO antisymmetric stretch frequency in V^+(OCO) should be similar to the value in bare CO_2, 2349 cm^{-1}. The V^+–OCO bond strength [33] is 6050 ± 320 cm^{-1}, so three IR photons are required to dissociate the molecule, ruling out vibrational predissociation. V^+(OCO) is also a poor candidate for IRMPD studies, as it is small and thus has a low density of vibrational states. We first looked at vibrational spectroscopy of V^+(OCO)$_5$, which had previously been studied by Duncan and co-workers [112]. They find that, for V^+, the first four CO_2 ligands bind to the metal; additional ligands are in the second solvent shell. This results in weak binding, and our IR resonance enhanced photodissociation spectrum of V^+(OCO)$_5$ is shown in Fig. 11. The antisymmetric stretch vibrational frequency for outer-shell CO_2 is nearly unchanged from its value in free CO_2. For inner-shell CO_2, binding to V^+ shifts the antisymmetric stretch 26 cm^{-1} to the blue. This shift should be even larger for the V^+(OCO) cluster, as the metal–ligand interaction is stronger, and Duncan and co-workers observe a peak at 2378 cm^{-1} for the argon-tagged complex V^+(OCO)Ar, but the signal for this ion is quite weak [112].

The V^+(OCO) ion has a structured electronic photodissociation spectrum, which allows us to measure its *vibrational* spectrum using vibrationally mediated photodissociation (VMP). This technique requires that the absorption spectrum (or, in our case, the photodissociation spectrum) of vibrationally excited molecules differ from that of vibrationally unexcited molecules. The photodissociation spectrum of V^+(OCO) has an extended progression in the V^+–OCO stretch, indicating that the ground and excited electronic states have different equilibrium V^+–OCO bond lengths. Thus, the OCO antisymmetric stretch frequency v_1 should be different in the two states, and the

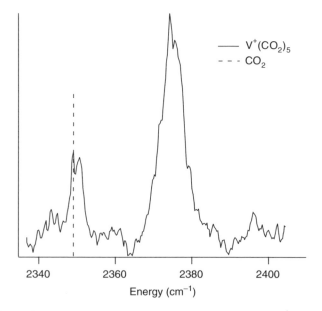

Figure 11. Infrared resonance enhanced photodissociation spectrum of $V^+(OCO)_5$ obtained by monitoring loss of CO_2. The antisymmetric stretch of outer-shell CO_2 is near 2349 cm^{-1} (the value in free CO_2, indicated by the dashed vertical line). The vibration shifts to 2375 cm^{-1} for inner-shell CO_2.

$v_1' = 1 \leftarrow v_1'' = 1$ transition will be at a different energy than the $v_1' = 0 \leftarrow v_1'' = 0$. We use this fact to measure the vibrational spectrum of $V^+(OCO)$ in a *depletion* experiment (Fig. 12a). A visible laser is set to the $v_1' = 0 \leftarrow v_1'' = 0$ transition at 15,801 cm^{-1}, producing fragment ions. A tunable IR laser fires before the visible laser. Absorption of IR photons removes population from the ground state, which is observed as a decrease in the fragment ion signal. This technique is a variation of ion-dip spectroscopy, in which ions produced by $1 + 1$ REMPI are monitored as an IR laser is tuned. Ion-dip spectroscopy has been used by several groups to study vibrations of neutral clusters and biomolecules [157–162].

We observe 8% depletion of the fragment signal at 2391.5 cm^{-1}, which establishes the antisymmetric stretch frequency in the $V^+(OCO)$ ground electronic state and that *at least* 8% of the ions are vibrationally excited. This value is only a lower limit because vibrationally excited molecules may also absorb at 15,801 cm^{-1} and photodissociate. The major drawback to the depletion experiment is that we are trying to observe a small decrease in the fragment ion signal, which is not particularly large or stable. Obtaining good signal-to-noise ratio thus requires extensive signal averaging, limiting the number of data points we can measure. An *enhancement* experiment, in

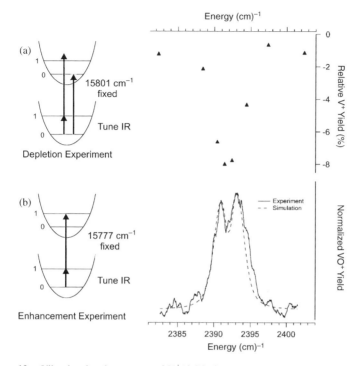

Figure 12. Vibrational action spectra of $V^+(OCO)$ in the OCO antisymmetric stretch region. (a) Spectrum obtained by monitoring *depletion* in the V^+ photofragment produced by irradiation at the vibronic origin at $15{,}801 \text{ cm}^{-1}$. The IR absorption near 2391.5 cm^{-1} removes molecules from $v_1'' = 0$, leading to an 8% reduction in the fragment yield. (b) Spectrum obtained by monitoring *enhancement* in the VO^+ photofragment signal as the IR laser is tuned, with the visible laser fixed at $15{,}777 \text{ cm}^{-1}$ (the $v_1' = 1 \leftarrow v_1'' = 1$ transition). The simulated spectrum gives a more precise value of the OCO antisymmetric stretch vibration in $V^+(OCO)$ of 2392.0 cm^{-1}.

which tuning the IR laser onto a vibrational resonance increases the fragment ion yield from a near-zero background level would greatly improve the signal to noise. In order to do this, we first find a electronic transition originating from $v_1'' = 1$. The IR laser is set to excite the OCO antisymmetric stretch and the visible laser is tuned in the vicinity of the $v_1' = 0 \leftarrow v_1'' = 0$ band. Figure 13 shows that we observe the $v_1' = 1 \leftarrow v_1'' = 1$ band at $15{,}777 \text{ cm}^{-1}$, 24 cm^{-1} to the red of the $0 \leftarrow 0$ band. Now, we can measure the vibrational spectrum with much improved signal-to-noise (and using a much smaller energy spacing) by setting the visible laser to the $v_1' = 1 \leftarrow v_1'' = 1$ electronic transition and tuning the IR laser while monitoring the fragment ion yield. The resulting vibrational action spectrum is shown in Fig. 12(b). The resolution of the spectrum is limited only by the

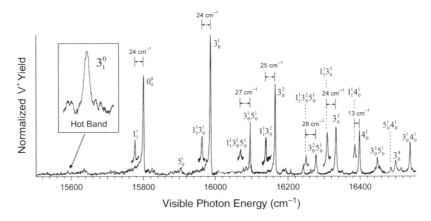

Figure 13. Photodissociation spectrum of $V^+(OCO)$, with assignments. Insets and their assignments show the photodissociation spectrum of molecules excited with one quantum of OCO antisymmetric stretch, v_1'' at 2390.9 cm^{-1}. These intensities have been multiplied by a factor of 2. The shifts show that v_1' (excited state) lies $\sim 24 \text{ cm}^{-1}$ below v_1'' (ground state), and that there is a small amount of vibrational cross-anharmonicity. The box shows a hot band at $15{,}591 \text{ cm}^{-1}$ that is shifted by 210 cm^{-1} from the origin peak and is assigned to the V^+–OCO stretch in the ground state.

IR laser line width. As a result, the spectrum clearly shows P and R branches, establishing that $V^+(OCO)$ is linear. The simulated spectrum (dashed line) gives an improved value for the OCO antisymmetric stretch frequency in the ground electronic state $v_1'' = 2392.0 \text{ cm}^{-1}$. The simulation assumes a $^5\Sigma$ ground state and uses $B'' = B' = 0.057 \text{ cm}^{-1}$, the value given by our density functional theory calculations [48]. Because the rotational constants are so small, the simulated spectrum is not sensitive to B' or B'', but $B'-B''$ affects the relative intensities of the P and R branches, and the separation between the branches is due to the 12 K rotational temperature of the ions. This is the first use of vibrationally mediated photodissociation for spectroscopy of an ion. The $v_1' = 1 \leftarrow v_1'' = 1$ band lies 24 cm^{-1} below $v_1' = 0 \leftarrow v_1'' = 0$, so the OCO antisymmetric stretch frequency in the excited electronic state is $v_1' = 2392-24 = 2368 \text{ cm}^{-1}$. Table I summarizes the vibrational frequencies of the ground and excited states of $V^+(OCO)$ and compares the values to those in CO_2. Electronic excitation of $V^+(OCO)$ lengthens the metal–ligand bond, resulting in a smaller blue shift. More generally, based on our calculations [163], the blue shift in the OCO antisymmetric stretch on binding to V^+ correlates with the amount of charge transferred to the CO_2, which means that this is an electronic effect, rather than simply a mechanical effect due to the proximity of the V^+ to the ligand. Larger $V^+(OCO)_n$ clusters have smaller blue shifts [112], due to longer metal–ligand bonds and less charge transfer to each ligand.

TABLE I
Vibrational Frequencies (cm^{-1}) of $V^+(OCO)$ and CO_2

	ν_1 OCO Antisymmetric Stretch	ν_2 OCO Symmetric Stretch	ν_3 Stretch $V^+-(OCO)$	ν_4 OCO Bend	ν_5 $V^+-(OCO)$ Rock
$V^+(OCO)$ Ground State	2392.0		210		
$V^+(OCO)$ [15.8] State (Visible)	2368	1340	186	597	105
CO_2	2349.16	1333		667.38	

C. Mode Selective Photodissociation of $V^+(OCO)$

Figure 13 shows the photodissociation spectrum of $V^+(OCO)$ in the visible and the insets show the photodissociation spectrum of vibrationally excited molecules containing one quantum of OCO antisymmetric stretch (ν_1). The shift between a vibronic band and the corresponding $\nu_1' = 1 \leftarrow \nu_1'' = 1$ sequence band is $\sim24\ cm^{-1}$ in each case, indicating that there is little cross-anharmonicity between ν_1 and the other vibrations observed. To characterize how the OCO antisymmetric stretch, OCO bend and metal–ligand stretch and rock affect the products of photodissociation of $V^+(OCO)$, we used difference spectra to measure the relative yield of the nonreactive V^+ (and CO_2) versus the reactive VO^+ (and CO) channels at each band assigned in Fig. 13. The results, shown in Fig. 14, indicate that there is mode selectivity in the reaction. Data $< 16,600\ cm^{-1}$ was obtained from one-photon dissociation. Compared to the origin band, one quantum of $V^+-(OCO)$ rock (ν_5') enhances the reactive channel, while one quantum of $V^+-(OCO)$ stretch (ν_3') or CO_2 bend (ν_4') slightly decreases it. These results agree with those previously obtained by Brucat and co-workers [156]. We also find that exciting $3\nu_3' + \nu_5'$ enhances reactivity by $\sim70\%$, a result that was not reported earlier, presumably due to the low total dissociation yield for this peak. Brucat and co-workers also observed vibronic features at $17,000$–$17,500\ cm^{-1}$ that lead to VO^+/V^+ branching ratios of ~1.2, showing a trend toward higher VO^+ production at higher energy. We used vibrationally mediated photodissociation to measure how the OCO antisymmetric stretch vibration ν_1, alone or in combination with other vibrations, affects the mode selectivity. We excite the peak of the P branch of ν_1 at $2390.9\ cm^{-1}$, and then photoexcite vibrationally excited molecules in the visible. The VO^+/V^+ branching ratio for these vibronic transitions is shown in the high energy portion of Fig. 14, where the energy is the total energy of the IR and

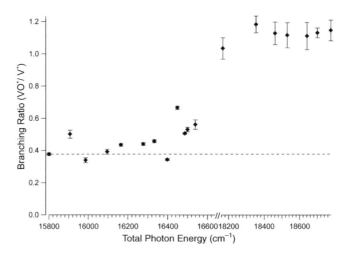

Figure 14. Mode selectivity in photodissociation of $V^+(OCO)$. The ratio of the reactive ($VO^+ + CO$) to nonreactive ($V^+ + CO_2$) product is measured at the peaks of the vibronic bands labeled in Fig. 13. The data below $16,600 \text{ cm}^{-1}$ is from bands accessed by one-photon excitation; data at higher energy was obtained by vibrationally mediated photodissociation exciting the OCO antisymmetric stretch.

visible photons. Exciting the OCO antisymmetric stretch enhances the reactive channel by almost a factor of 3. Exciting combination bands of the antisymmetric stretch with several other vibrations further increases the VO^+ yield slightly, but combination bands involving the antisymmetric stretch all give similar branching ratios [163].

The photodissociation and guided ion beam experiments on $V^+(OCO)$ give interesting and, at first glance, conflicting results. Photodissociation of $V^+(OCO)$ leads to significant amounts of the reactive, spin forbidden VO^+ product, and its yield increases significantly for specific vibrations in the excited electronic state [156, 163]. On the other hand, collision-induced dissociation of $V^+(OCO)$ at energies $< 8 \text{ eV}$ leads exclusively to spin-allowed V^+ [33]. We also find that photodissociation in the near IR, near 7000 cm^{-1} only produces V^+ [163]. To investigate these observations, we carried out extensive electronic structure calculations. We first characterized the stationary points on the $V^+ + CO_2$ potential energy surface (Fig. 15). The ground state of V^+ is high spin (^5D), and the ground state of the $V^+(OCO)$ complex is also a quintet. Triplet $OV^+(CO)$ is the most stable species on the potential energy surface, so insertion of V^+ into the C—O bond is exothermic. However, there is a substantial barrier to insertion, that explains why the $V^+ + CO_2 \rightarrow VO^+ + CO$ reaction does not occur at thermal energies [26]. The transition state to insertion is tight and quite bent. So, in collision-induced dissociation of $V^+(OCO)$, the

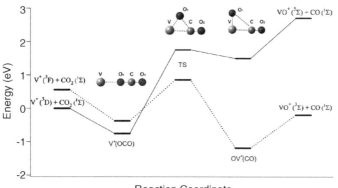

Figure 15. Calculated potential energy surface and geometries of intermediates of the $V^+ + CO_2$ reaction. The energy of the lowest energy state for the quintet (solid lines) and triplet (dotted lines) stationary points are shown. Energies are calculated at the CCSD(T)/6-311+G(3df) level, at the B3LYP/6-311+G(d) geometry and include zero-point energy at the B3LYP/6-311+G(d) level.

$V^+ + CO_2$ channel dominates as it is spin allowed and occurs via a loose transition state. In contrast, production of VO^+ is spin forbidden and goes *via* a tight transition state that lies higher in energy than $V^+ + CO_2$.

Interpreting the photodissociation experiments requires further calculations to characterize the ground and excited triplet and quintet states of $V^+(OCO)$, and the spin–orbit coupling between the states. We find that there are several triplet states that cross the excited quintet states accessed in the photodissociation experiments. These crossings occur at bent V^+–OCO geometries and extended V^+–OCO bond lengths. Also, the excited quintet states have modest spin–orbit couplings of $\leq 90\ cm^{-1}$ to nearby triplet states [163]. These observations help to explain the photodissociation results. The photodissociation spectrum of $V^+(OCO)$ in the near-IR is broad and poorly resolved [48], suggesting that dissociation is rapid. This gives little time for the excited quintet state to convert to triplet states, so quintet V^+ is the only channel observed.

In contrast, the photodissociation spectrum of $V^+(OCO)$ in the visible shows vibrational and partial rotational structure. The observed $< 3.5\ cm^{-1}$ line width, indicates excited state lifetimes of at least 1.4 ps. During this time, the molecules have an opportunity to cross to triplet surfaces, from which they can dissociate to form VO^+ ($^3\Sigma$). The calculations also help to explain the observed mode selectivity. Exciting vibrations that lead to regions of the excited state potential with higher spin–orbit coupling should enhance production of VO^+. Triplet states of $V^+(OCO)$ cross the excited quintet state at long V^+–OCO bond lengths and at large V^+–OCO bend angles. Thus, exciting the V^+–OCO stretch (v_3') and the V^+–OCO rock (v_5') should enhance VO^+ production, as is observed. We also find that exciting the OCO antisymmetric stretch v_1' leads to significant increase in the VO^+ product. If $V^+(OCO)$ dissociated to produce $VO^+ + CO$ via a spin-allowed, collinear transition state, then the OCO antisymmetric stretch would be similar to the reaction coordinate, and exciting v_1' should enhance VO^+ production. The calculations suggest that the reaction mechanism is much more complex: The transition states leading to VO^+ are not linear, and second, producing VO^+ requires intersystem crossing to the triplet surface, which is enhanced at nonlinear geometries. Thus, the enhanced reactivity we observe following excitation of the OCO antisymmetric stretch is likely due to overall enhancement of VO^+ production with increasing energy, combined with a mode selective effect due to stretching the C—O bond, which is required for reaction, even at nonlinear geometries.

V. PREREACTIVE COMPLEXES: VIBRATIONAL SPECTROSCOPY OF $Fe^+(CH_4)_n$

Most of the third-row transition metal cations react with methane under thermal conditions to produce $MCH_2^+ + H_2$ [164]. The corresponding reactions of first- and second-row M^+ are endothermic [18]. Even in these cases, bonding in $M^+(CH_4)$ is not merely electrostatic, but includes significant covalency due to donation from C—H bonding orbitals into empty or partially empty $4s$ and $3d$ orbitals on the metal, along with back donation into C—H antibonding orbitals. These interactions weaken the C—H bonds and significantly lower the C—H stretching frequencies, so vibrational spectroscopy of $M^+(CH_4)_n$ entrance channel complexes reveals the amount of donation and backdonation. For our first studies [165] of vibrational spectroscopy of these complexes we selected $Fe^+(CH_4)_n$. The complexes are easy to produce, and the $Fe^+ + CH_4$ reaction is 123-kJ mol^{-1} endothermic [31, 63, 166], and there is a substantial barrier to C—H insertion [167] so, at the conditions in our ion source, we should only produce the entrance channel complex. Later studies will look at vibrational spectra of the entrance channel complex of the exothermic $FeO^+ + CH_4$ reaction, which is more challenging to produce.

Sequential binding energies of methane to Fe^+ have been measured by Schultz and Armentrout using collision-induced dissociation [168] and more recently by Zhang, Kemper, and Bowers by cluster equilibrium [169]. According to the equilibrium measurements, the first methane binds by 77 kJ mol^{-1} and the second binds by 108 kJ mol^{-1}. Stronger binding of the second ligand is often observed in electrostatic Fe^+ complexes. The first excited state of Fe^+ ($3d^7$, ^4F) interacts more strongly with ligands than the ^6D ($3d^6 4s$) ground state, as the $4s$ orbital is empty, reducing repulsive interactions. Calculations predict that $Fe^+(CH_4)$ has a quartet ground state [169], so its measured binding energy includes the 24 kJ mol^{-1} Fe^+ ^6D–^4F promotion energy. Larger clusters (at least up to $n = 6$) are calculated to remain quartets, so this promotion energy is no longer an issue. Binding energies for the third and fourth methane are much smaller, 23 and 20 kJ mol^{-1}. This large change in binding energy with cluster size suggests that the binding in these complexes is not simply electrostatic. We measured the resonance enhanced dissociation spectra of $Fe^+(CH_4)_n$ ($n = 3,4$) in the C—H stretching region (Fig. 16). The CH_4 binding energies are sufficiently small that absorption of one

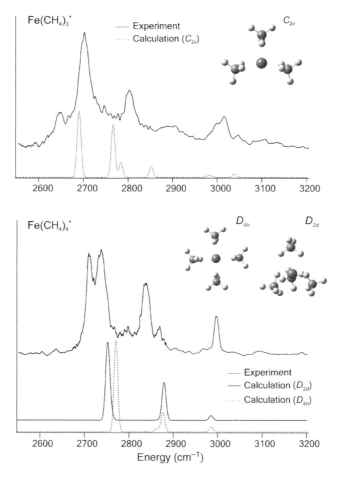

Figure 16. Experimental and calculated IR resonance enhanced photodissociation spectra of $Fe^+(CH_4)_3$ and $Fe^+(CH_4)_4$. Experimental spectra were obtained by monitoring loss of CH_4. Calculated spectra are based on vibrational frequencies and intensities calculated at the B3LYP/ 6-311+G(d,p) level. Calculated frequencies are scaled by 0.96. The calculated spectra have been convoluted with a 10-cm^{-1} full width at half-maximum (FWHM) Gaussian. The D_{2d} and D_{4h} geometries of $Fe^+(CH_4)_4$ are calculated to have very similar energies, and it appears that both isomers are observed in the experiment.

IR photon leads to dissociation. The $Fe^+(CH_4)_3$ cluster has an intense absorption at 2703 cm^{-1}, with smaller peaks at 2648, 2803, 2895, and 3015 cm^{-1}. Compared to the C—H stretching frequencies in bare CH_4 (2917-cm^{-1} symmetric stretch and 3019-cm^{-1} antisymmetric stretch), the large peak is red-shifted $> 200\ cm^{-1}$. The red shift is smaller in the $Fe^+(CH_4)_4$ cluster, which has a doublet at 2711/ 2737 cm^{-1} and large peaks at 2838 and 2998 cm^{-1}. Zhang et al. calculated structures of low lying isomers of $Fe^+(CH_4)_n$ and their binding energies using the B3LYP method with the 6-31G(d,p) basis on carbon and hydrogen and the TZVP and augmented Wachter's basis on iron [169]. For several of the clusters they identify multiple isomers with similar calculated energies, with the energy difference depending on the basis set. In principle, the supersonic expansion in our experiment should cool the molecules to the minimum-energy structure, so to identify this structure we compare measured and calculated vibrational spectra. Thus, we reoptimized the geometries and calculated energies and vibrational frequencies and intensities at the B3LYP/6-311+G(d,p) level. Calculated frequencies have been scaled by 0.96, which predicts C—H stretching frequencies for CH_4 within 15 cm^{-1} of experiment. Due to the high symmetry of the complexes, the calculated vibrational spectra consist of only a few peaks. The calculated spectrum for $Fe^+(CH_4)_3$ is in good agreement with experiment, although it does not reproduce the smaller peak at 2648 cm^{-1}. This may be due to an isomer in which one of the methanes is even closer to the metal. For the larger $Fe^+(CH_4)_4$ cluster, neither of the low energy isomers has a calculated vibrational spectrum with a doublet near 2700 cm^{-1}. However, the calculations reproduce the doublet observed at 2711/2737 cm^{-1} if both the D_{2d} and D_{4h} isomers are included. Our calculations predict that the D_{4h} isomer is more stable by a mere 21 cm^{-1}, so the presence of two isomers in the beam is not unreasonable.

The calculated C—H stretching frequencies are very sensitive to the C—H bond lengths, which in turn depend on the metal–methane distance. In $Fe^+(CH_4)_3$ each methane is bound to the metal in a η_2 configuration. The calculated C—H bond length r_{CH} is 1.114 Å for hydrogens pointing toward the metal, 0.023 Å longer than in bare CH_4, while $r_{CH} = 1.089$ Å for the hydrogens pointing away from the metal. In $Fe^+(CH_4)_4$ the D_{4h} structure has four equal Fe—C bond lengths, with the proximate C—H bonds extended by 0.017 Å. The D_{2d} isomer has two short and two longer Fe—C distances; the C—H bonds are extended by 0.019 Å for the nearer methanes. Zhang et al. note that systematic measurements for many metals, along with theoretical studies, were required to establish the relative importance of four factors to binding of M^+ to H_2 [169]. These are donation from bonding orbitals on the ligand to empty $4s$ and partially empty $3d$ orbitals on the metal, backdonation from partially filled $3d$ orbitals to σ^* orbitals on the ligand, metal–ligand repulsion due to partially filled $3d$ and $4s$ orbitals, and simple electrostatic attraction. Systematic studies of the vibrational spectroscopy of $M^+(CH_4)_n$, along with thermodynamic measurements, should allow us to characterize metal–methane bonding in even greater detail.

VI. FUTURE PROSPECTS

Characterizing the potential energy surfaces of an ion molecule reaction requires using a variety of experimental techniques. The studies described above use vibrational and electronic spectroscopy as well as photoelectron spectroscopy, each of which is sensitive to particular features of the potential. Advances in techniques and equipment are extending the applicability of each of these techniques. Our current IR laser system is limited to the range of 2100 to >4000 cm^{-1}. Free electron lasers have a much broader tunability down to $\sim$600 cm^{-1} and outstanding fluences, but they have modest line width and beam time on these multiuser facilities is limited. Difference frequency mixing in the new nonlinear optical materials LiInS$_2$ and AgGaSe$_2$ is extending the tunability of laboratory infrared lasers down to 900 and 600 cm^{-1}, respectively [170]. This region of the IR spectrum has recently been used in elegant laboratory studies of vibrational spectra and structure of H$_3$O$^+$(H$_2$O)$_n$ [79, 80]. Using AgGaSe$_2$ to obtain light at these longer wavelengths will allow us to directly probe M—OH and M=CH$_2$ stretches, as well as C=C and C=O stretches and H—C—H bends, which are beyond the tuning range of our current IR laser. Improved techniques for vibrational photofragment spectroscopy will also extend the systems that can be studied. Techniques to measure the vibrational spectra of unperturbed ions with <1-cm^{-1} resolution are particularly attractive. Vibrationally mediated photodissociation is one such method, and we are exploring complementary methods.

Photoelectron spectroscopy of negative ions has been used to probe the transition state of neutral reactions [16, 17], so photoelectron spectroscopy of appropriate neutral complexes could be a powerful tool for studying the potential energy surfaces of reactions of transition metal cations. As discussed above, we have measured photoionization efficiencies of FeO, CuO, PtC, PtO, and PtO$_2$ at the Chemical Dynamics beamline at the Advanced Light Source. We have also obtained preliminary photoelectron spectra of FeO. These photoelectron imaging experiments are significantly more challenging as the source has to produce good yields of the neutral of interest, with excellent selectivity. Photoelectrons from other neutrals can obscure features from the molecule of interest. Photoelectron–photoion coincidence experiments eliminate most interferences, but have far lower signal. Yang and co-workers have elegantly applied ZEKE photoelectron spectroscopy to studying metal–ligand complexes with ionization energies $<\sim$6 eV [171]. Covalently bound complexes of late transition metal ions tend to have ionization energies >8 eV, in the vacuum UV, where current laboratory laser systems have much lower fluences. A possible solution to this problem is resonant multiphoton ionization, where the resonance selects the neutral of interest, and all photon energies lie $<6\,\text{eV}$ [172, 173].

Studying photoelectron spectroscopy of intermediates of metal ion reactions requires synthesizing neutrals with a geometry similar to the ion of interest. Several third-row transition metal cations react with methane at thermal energies, but no neutral metal atoms do. However, calculations show that neutral Pt readily inserts into the C–H bond in methane, producing a H–Pt–CH$_3$ intermediate that lies $1.4\,\text{eV}$ below the reactants [174]. It is fairly long lived, which leads to the observed high termolecular rate for the Pt + CH$_4$ clustering reaction [174–176]. We plan to measure the photoionization of H–Pt–CH$_3$, which would give the energy of this insertion intermediate relative to Pt + CH$_4$, as the corresponding value has been measured for the cation [35]. Subsequent photoelectron spectroscopy of H–Pt–CH$_3$ would measure vibrational frequencies of [H–Pt–CH$_3$]$^+$, particularly the Pt–C stretch, complementing vibrational spectroscopy, which is better suited to higher frequency vibrations.

Acknowledgments

Financial support from the National Science Foundation under award numbers CHE-0308439 and CHE-0608446 is gratefully acknowledged. The photofragment studies described in this chapter are the work of many talented group members: graduate students Fernando Aguirre, Gokhan Altinay, Murat Citir, John Husband, Kay Stringer and Chris Thompson and undergraduate Peter Ferguson. Photoionization experiments at the Advanced Light Source (ALS) were carried out with the expert assistance of Drs. Musahid Ahmed, Dr. Johannes Abate, Dr. Leonid Belau, and Dr. Christophe Nicolas and Prof. Stephen Leone. The Chemical Dynamics beamline at the ALS is supported by the Director of the Office of Energy Research, Office of Basic Energy Sciences, Chemical Sciences Division of the U. S. Department of Energy under Contract No. DE-AC02-05CH11231.

References

1. R. D. Levine and R. B. Bernstein, *Molecular Reaction Dynamics and Chemical Reactivity.* Oxford University Press, New York, 1987.
2. P. R. Brooks, *Chem. Rev.* **88**, 407 (1988).
3. M. D. Barnes, P. R. Brooks, R. F. Curl, P. W. Harland, and B. R. Johnson, *J. Chem. Phys.* **96**, 3559 (1992).
4. P. D. Kleiber, W. C. Stwalley, and K. M. Sando, *Annu. Rev. Phys. Chem.* **44**, 13 (1993).
5. J. C. Polanyi and A. H. Zewail, *Acc. Chem. Res.* **28**, 119 (1995).
6. A. H. Zewail, *J. Phys. Chem.* **100**, 12701 (1996).
7. A. J. Hudson, H. B. Oh, J. C. Polanyi, and P. Piecuch, *J. Chem. Phys.* **113**, 9897 (2000).
8. A. Keller, R. Lawruszczuk, B. Soep, and J. P. Visticot, *J. Chem. Phys.* **105**, 4556 (1996).
9. J. M. Mestdagh, B. Soep, M. A. Gaveau, and J. P. Visticot, *Int. Rev. Phys. Chem.* **22**, 285 (2003).
10. P. D. Kleiber and J. Chen, *Int. Rev. Phys. Chem.* **17**, 1 (1998).
11. W. Y. Lu, T. H. Wong, and P. D. Kleiber, *Chem. Phys. Lett.* **347**, 183 (2001).
12. W. Y. Lu, Y. Abate, T. H. Wong, and P. D. Kleiber, *J. Phys. Chem. A* **108**, 10661 (2004).

13. W.-Y. Lu, P. D. Kleiber, M. A. Young, and K.-H. Yang, *J. Chem. Phys.* **115**, 5823 (2001).

14. C. R. Moylan, J. A. Dodd, C. C. Han, and J. I. Brauman, *J. Chem. Phys.* **86**, 5350 (1987).

15. K. M. Ervin, J. Ho, and W. C. Lineberger, *J. Chem. Phys.* **91**, 5974 (1989).

16. R. B. Metz, S. E. Bradforth, and D. M. Neumark, *Adv. Chem. Phys.* **81**, 1 (1992).

17. D. M. Neumark, *Phys. Chem. Chem. Phys.* **7**, 433 (2005).

18. K. Eller and H. Schwarz, *Chem. Rev.* **91**, 1121 (1991).

19. J. C. Weisshaar, *Acc. Chem. Res.* **26**, 213 (1993).

20. P. B. Armentrout, in *Organometallic Bonding and Reactivity: Fundamental Studies*, J. M. Brown and P. Hofmann (eds.), Springer-Verlag, Berlin, 1999, vol. IV, pp. 1–45.

21. D. Schröder and H. Schwarz, *Angew. Chem. Intl. Ed. Engl.* **29**, 1433 (1990).

22. D. Schröder and H. Schwarz, *Helv. Chim. Acta* **75**, 1281 (1992).

23. D. Schröder, H. Schwarz, and S. Shaik, *Struct. Bonding* **97**, 91 (2000).

24. D. K. Böhme and H. Schwarz, *Angew. Chem., Int. Ed. Engl.* **44**, 2336 (2005).

25. R. B. Metz, *Int. Rev. Phys. Chem.* **23**, 79 (2004).

26. G. K. Koyanagi and D. K. Bohme, *J. Phys. Chem. A* **110**, 1232 (2006).

27. S. S. Yi, M. R. A. Blomberg, P. E. M. Siegbahn, and J. C. Weisshaar, *J. Phys. Chem. A* **102**, 395 (1998).

28. R. J. Noll, S. S. Yi, and J. C. Weisshaar, *J. Phys. Chem. A* **102**, 386 (1998).

29. E. L. Reichert, S. S. Yi, and J. C. Weisshaar, *Int. J. Mass. Spectrom.* **196**, 55 (2000).

30. K. M. Ervin, *Chem. Rev.* **101**, 391 (2001).

31. P. B. Armentrout and B. L. Kickel, in *Organometallic Ion Chemistry*, B. S. Freiser (ed.), Kluwer Academic Publishers, Dordrecht, The Netherlands, 1996, p. 1.

32. P. B. Armentrout, *Int. J. Mass Spectrom.* **227**, 289 (2003).

33. M. R. Sievers and P. B. Armentrout, *J. Chem. Phys.* **102**, 754 (1995).

34. C. L. Haynes, Y. M. Chen, and P. B. Armentrout, *J. Phys. Chem.* **100**, 111 (1996).

35. X. G. Zhang, R. Liyanage, and P. B. Armentrout, *J. Am. Chem. Soc.* **123**, 5563 (2001).

36. P. B. Armentrout, *Ann. Rev. Phys. Chem.* **41**, 313 (1990).

37. L. Sanders, S. D. Hanton, and J. C. Weisshaar, *J. Chem. Phys.* **92**, 3498 (1990).

38. H. Schwarz, *Int. J. Mass Spectrom.* **237**, 75 (2004).

39. R. B. Metz, C. Nicolas, M. Ahmed, and S. R. Leone, *J. Chem. Phys.* **123**, 114313 (2005).

40. T. G. Dietz, M. A. Duncan, D. E. Powers, and R. E. Smalley, *J. Chem. Phys.* **74**, 6511 (1981).

41. P. J. Brucat, L.-S. Zheng, C. L. Pettiette, S. Yang, and R. E. Smalley, *J. Chem. Phys.* **84**, 3078 (1986).

42. D. Proch and T. Trickl, *Rev. Sci. Instrum.* **60**, 713 (1989).

43. *Organometallic Ion Chemistry*, B. S. Freiser (ed.), Kluwer Academic Publishers, Dordrecht, The Netherlands, 1996.

44. R. L. Hettich and B. S. Freiser, *J. Am. Chem. Soc.* **108**, 2537 (1986).

45. F. Aguirre, J. Husband, C. J. Thompson, K. L. Stringer, and R. B. Metz, *J. Chem. Phys.* **116**, 4071 (2002).

46. D. Schröder, A. Fiedler, J. Hrušák, and H. Schwarz, *J. Am. Chem. Soc.* **114**, 1215 (1992).

47. J. Husband, F. Aguirre, P. Ferguson, and R. B. Metz, *J. Chem. Phys.* **111**, 1433 (1999).

48. M. Citir, G. Altinay, and R. B. Metz, *J. Phys. Chem. A* **110**, 5051 (2006).

49. A. Kamariotis, T. Hayes, D. Bellert, and P. J. Brucat, *Chem. Phys. Lett.* **316**, 60 (2000).

50. W. C. Wiley and I. H. McLaren, *Rev. Sci. Instrum.* **26**, 1150 (1955).

51. L. A. Posey, M. J. DeLuca, and M. A. Johnson, *Chem. Phys. Lett.* **131**, 170 (1986).

52. R. E. Continetti, D. R. Cyr, and D. M. Neumark, *Rev. Sci. Instrum.* **63**, 1840 (1992).

53. D. S. Cornett, M. Peschke, K. Laitting, P. Y. Cheng, K. F. Willey, and M. A. Duncan, *Rev. Sci. Instrum.* **63**, 2177 (1992).

54. V. I. Karataev, B. A. Mamyrin, and D. V. Shmikk, *Sov. Phys.-Tech. Phys.* **16**, 1177 (1972).

55. B. A. Mamyrin, V. I. Karataev, D. V. Shmikk, and V. A. Zagulin, *Sov. Phys.-JETP* **37**, 45 (1973).

56. C. J. Thompson, J. Husband, F. Aguirre, and R. B. Metz, *J. Phys. Chem. A* **104**, 8155 (2000).

57. K. P. Faherty, C. J. Thompson, F. Aguirre, J. Michne, and R. B. Metz, *J. Phys. Chem. A* **105**, 10054 (2001).

58. C. J. Thompson, K. P. Faherty, K. L. Stringer, and R. B. Metz, *Phys. Chem. Chem. Phys.* **7**, 814 (2005).

59. K. L. Stringer, M. Citir, and R. B. Metz, to be submitted (2008).

60. J. W. Buchanan, J. E. Reddic, G. A. Grieves, and M. A. Duncan, *J. Phys. Chem. A* **102**, 6390 (1998).

61. R. L. Hettich, T. C. Jackson, E. M. Stanko, and B. S. Freiser, *J. Am. Chem. Soc.* **108**, 5086 (1986).

62. L. M. Russon, S. A. Heidecke, M. K. Birke, J. Conceicao, M. D. Morse, and P. B. Armentrout, *J. Chem. Phys.* **100**, 4747 (1994).

63. J. Husband, F. Aguirre, C. J. Thompson, C. M. Laperle, and R. B. Metz, *J. Phys. Chem. A* **104**, 2020 (2000).

64. J. Husband, F. Aguirre, C. J. Thompson, and R. B. Metz, *Chem. Phys. Lett.* **342**, 75 (2001).

65. D. Lessen and P. J. Brucat, *J. Chem. Phys.* **91**, 4522 (1989).

66. D. Lessen, R. L. Asher, and P. Brucat, *Int. J. Mass Spec. Ion. Proc.* **102**, 331 (1990).

67. R. L. Asher, D. Bellert, T. Buthelezi, and P. J. Brucat, *Chem. Phys. Lett.* **227**, 277 (1994).

68. C. J. Thompson, K. L. Stringer, M. McWilliams, and R. B. Metz, *Chem. Phys. Lett.* **376**, 588 (2003).

69. K. L. Stringer, M. Citir, and R. B. Metz, *J. Phys. Chem. A* **108**, 6996 (2004).

70. D. Bellert, T. Buthelezi, V. Lewis, K. Dezfulian, D. Reed, T. Hayes, and P. J. Brucat, *Chem. Phys. Lett.* **256**, 555 (1996).

71. D. Bellert, T. Buthelezi, T. Hayes, and P. J. Brucat, *Chem. Phys. Lett.* **276**, 242 (1997).

72. K. M. Gunawardhana and R. B. Metz, to be submitted (2008).

73. M. A. Duncan, *Int. Rev. Phys. Chem.* **22**, 407 (2003).

74. M. A. Duncan, *Int. J. Mass Spectrom.* **200**, 545 (2000).

75. R. S. Walters, P. R. von Schleyer, C. Corminboeuf, and M. A. Duncan, *J. Am. Chem. Soc.* **127**, 1100 (2005).

76. N. R. Walker, R. S. Walters, G. A. Grieves, and M. A. Duncan, *J. Chem. Phys.* **121**, 10498 (2004).

77. G. Gregoire, J. Velasquez, and M. A. Duncan, *Chem. Phys. Lett.* **349**, 451 (2001).

78. E. A. Woronowicz, W. H. Robertson, G. H. Weddle, M. A. Johnson, E. M. Myshakin, and K. D. Jordan, *J. Phys. Chem. A* **106**, 7086 (2002).

79. N. I. Hammer, E. G. Diken, J. R. Roscioli, M. A. Johnson, E. M. Myshakin, K. D. Jordan, A. B. McCoy, X. Huang, J. M. Bowman, and S. Carter, *J. Chem. Phys.* **122**, 244301 (2005).

80. J. M. Headrick, E. G. Diken, R. S. Walters, N. I. Hammer, R. A. Christie, J. Cui, E. M. Myshakin, M. A. Duncan, M. A. Johnson, and K. D. Jordan, *Science* **308**, 1765 (2005).

81. R. C. Dunbar, *Int. J. Mass Spectrom.* **200**, 571 (2000).

82. R. L. Grimm, J. B. Mangrum, and R. C. Dunbar, *J. Phys. Chem. A* **108**, 10897 (2004).

83. T. J. Selegue and J. M. Lisy, *J. Phys. Chem.* **96**, 4143 (1992).

84. J. M. Lisy, *Int. Rev. Phys. Chem.* **16**, 267 (1997).

85. O. M. Cabarcos, C. J. Weinheimer, and J. M. Lisy, *J. Chem. Phys.* **108**, 5151 (1998).

86. T. D. Vaden, B. Forinash, and J. M. Lisy, *J. Chem. Phys.* **117**, 4628 (2002).

87. T. D. Vaden, C. J. Weinheimer, and J. M. Lisy, *J. Chem. Phys.* **121**, 3102 (2004).

88. T. D. Vaden, J. M. Lisy, P. D. Carnegie, E. D. Pillai, and M. A. Duncan, *Phys. Chem. Chem. Phys.* **8**, 3078 (2006).

89. E. J. Bieske and O. Dopfer, *Chem. Rev.* **100**, 3963 (2000).

90. D. A. Wild, R. L. Wislon, P. S. Weiser, and E. J. Bieske, *J. Chem. Phys.* **113**, 10154 (2000).

91. D. A. Wild, Z. M. Loh, P. P. Wolynec, P. S. Weiser, and E. J. Bieske, *Chem. Phys. Lett.* **332**, 531 (2000).

92. G. H. Wu, J. G. Guan, G. D. C. Aitken, H. Cox, and A. J. Stace, *J. Chem. Phys.* **124**, 201103 (2006).

93. P. Maitre, S. Le Caer, A. Simon, W. Jones, J. Lemaire, H. N. Mestdagh, M. Heninger, G. Mauclaire, P. Boissel, R. Prazeres, F. Glotin, and J. M. Ortega, *Nucl. Instrum. Meth. A* **507**, 541 (2003).

94. A. Simon, W. Jones, J. M. Ortega, P. Boissel, J. Lemaire, and P. Maitre, *J. Am. Chem. Soc.* **126**, 11666 (2004).

95. R. S. Walters, T. D. Jaeger, and M. A. Duncan, *J. Phys. Chem. A* **106**, 10482 (2002).

96. J. Oomens, B. G. Sartakov, G. Meijer, and G. Von Helden, *Int. J. Mass Spectrom.* **254**, 1 (2006).

97. L. R. Thorne and J. L. Beauchamp, in *Gas Phase Ion Chemistry*, M. T. Bowers (ed.), Academic Press, Orlando, FL, 1984, vol. **3**, pp. 41.

98. S. K. Shin and J. L. Beauchamp, *J. Am. Chem. Soc.* **112**, 2057 (1990).

99. P. I. Surya, L. M. Roth, D. R. A. Ranatunga, and B. S. Freiser, *J. Am. Chem. Soc.* **118**, 1118 (1996).

100. P. I. Surya, D. R. A. Ranatunga, and B. S. Freiser, *J. Am. Chem. Soc.* **119**, 3351 (1997).

101. G. Dietrich, S. Kruckeberg, K. Lutzenkirchen, L. Schweikhard, and C. Walther, *J. Chem. Phys.* **112**, 752 (2000).

102. G. Gregoire and M. A. Duncan, *J. Chem. Phys.* **117**, 2120 (2002).

103. M. Okumura, L. I. Yeh, J. D. Meyers, and Y. T. Lee, *J. Chem. Phys.* **85**, 2328 (1986).

104. P. Ayotte, G. H. Weddle, J. Kim, and M. A. Johnson, *J. Am. Chem. Soc.* **120**, 12361 (1998).

105. P. Ayotte, C. G. Bailey, J. Kim, and M. A. Johnson, *J. Chem. Phys.* **108**, 444 (1998).

106. P. Ayotte, J. Kim, J. A. Kelley, S. B. Nielsen, and M. A. Johnson, *J. Am. Chem. Soc.* **121**, 6950 (1999).

107. S. A. Corcelli, J. A. Kelley, J. C. Tully, and M. A. Johnson, *J. Phys. Chem. A* **106**, 4872 (2002).

108. R. S. Walters, N. R. Brinkmann, H. F. Schaefer, and M. A. Duncan, *J. Phys. Chem. A* **107**, 7396 (2003).

109. R. S. Walters and M. A. Duncan, *Aust. J. Chem.* **57**, 1145 (2004).

110. G. Gregoire, N. R. Brinkmann, D. van Heijnsbergen, H. F. Schaefer, and M. A. Duncan, *J. Phys. Chem. A* **107**, 218 (2003).

111. N. R. Walker, R. S. Walters, M. K. Tsai, K. D. Jordan, and M. A. Duncan, *J. Phys. Chem. A* **109**, 7057 (2005).

112. N. R. Walker, R. S. Walters, and M. A. Duncan, *J. Chem. Phys.* **120**, 10037 (2004).

113. K. R. Asmis, G. Meijer, M. Mrümmer, C. Kaposta, G. Santambrogio, L. Wöste, and J. Sauer, *J. Chem. Phys.* **120**, 6461 (2004).

114. R. C. Dunbar, in *Gas Phase Ion Chemistry*, M. T. Bowers (ed.), Academic Press, Orlando, FL, 1984, vol. **3**, pp. 129.

115. F. F. Crim, *Annu. Rev. Phys. Chem.* **44**, 397 (1993).

116. A. Bach, J. M. Hutchison, R. J. Holiday, and F. F. Crim, *J. Chem. Phys.* **116**, 4955 (2002).

117. A. Callegari and T. R. Rizzo, *Chem. Soc. Rev.* **30**, 214 (2001).

118. I. Bar and S. Rosenwaks, *Int. Rev. Phys. Chem.* **20**, 711 (2001).

119. C. Tao and P. J. Dagdigian, *Chem. Phys. Lett.* **350**, 63 (2001).

120. O. Votava, D. F. Plusquellic, and D. J. Nesbitt, *J. Chem. Phys.* **110**, 8564 (1999).

121. F. Aguirre, J. Husband, C. J. Thompson, and R. B. Metz, *Chem. Phys. Lett.* **318**, 466 (2000).

122. J. Haggin, *Chem. & Eng. News* **71**, 27 (1993).

123. R. H. Crabtree, *Chem. Rev.* **95**, 987 (1995).

124. B. K. Warren and S. T. Oyama, *Heterogeneous Hydrocarbon Oxidation*. American Chemical Society, Washington, D.C., 1996.

125. Y. Wang and K. Otsuka, *J. Chem. Soc. Chem. Commun.* **1994**, 2209 (1994).

126. Y. Wang, K. Otsuka, and K. Ebitani, *Catal. Lett.* **35**, 259 (1995).

127. H. Matsumoto, S. Tanabe, K. Okitsu, Y. Hayashi, and S. L. Suib, *J. Phys. Chem. A* **105**, 5304 (2001).

128. R. A. Periana, D. J. Taube, S. Gamble, H. Taube, T. Satoh, and H. Fujii, *Science* **280**, 560 (1998).

129. R. A. Periana, G. Bhalla, W. J. Tenn, K. J. H. Young, X. Y. Liu, O. Mironov, C. J. Jones, and V. R. Ziatdinov, *J. Mol. Catal. A* **220**, 7 (2004).

130. R. Raja and P. Ratnasamy, *Appl. Catal. A General* **158**, L7 (1997).

131. J. Haggin, *Chem. & Eng. News* **72**, 24 (1994).

132. A. C. Rosenzweig, C. A. Frederick, S. J. Lippard, and P. Nordlund, *Nature (London)* **366**, 537 (1993).

133. L. J. Shu, J. C. Nesheim, K. Kauffmann, E. Munck, J. D. Lipscomb, and L. Que, *Science* **275**, 515 (1997).

134. D. Schröder and H. Schwarz, *Angew. Chem. Int. Ed. Engl.* **34**, 1973 (1995).

135. H. Schwarz and D. Schröder, *Pure Appl. Chem.* **72**, 2319 (2000).

136. R. B. Metz, in *Research Advances in Physical Chemistry*, R. M. Mohan (ed.), Global, Trivandrum, 2001, vol. 2, pp. 35.

137. F. Aguirre, J. Husband, C. J. Thompson, K. L. Stringer, and R. B. Metz, *J. Chem. Phys.* **119**, 10194 (2003).

138. H. Y. Afeefy, J. F. Liebman, and S. E. Stein, in *NIST Chemistry WebBook, NIST Standard Reference Database Number 69*, W. G. Mallard and P. J. Linstrom (eds.), National Institute of Standards and Technology, Gaithersburg MD 20899, July 2001.

139. F. Aguirre, Electronic spectroscopy of Au CH_2^+ and intermediates involved in the conversion of methane to methanol by FeO^+ Ph.D. Dissertation, Dept. of Chemistry. University of Massachusetts, 2002.

140. K. Yoshizawa, Y. Shiota, and T. Yamabe, *J. Am. Chem. Soc.* **120**, 564 (1998).

141. G. Altinay, M. Citir, and R. B. Metz, to be submitted (2008).

142. K. Yoshizawa, Y. Shiota, and T. Yamabe, *Chem. Eur. J.* **3**, 1160 (1997).

143. K. Yoshizawa, Y. Shiota, and T. Yamabe, *J. Chem. Phys.* **111**, 538 (1999).

144. Y. Shiota and K. Yoshizawa, *J. Chem. Phys.* **118**, 5872 (2003).

145. Y. Shiota and K. Yoshizawa, *J. Am. Chem. Soc.* **122**, 12317 (2000).

146. A. Fiedler, D. Schröder, S. Shaik, and H. Schwarz, *J. Am. Chem. Soc.* **116**, 10734 (1994).

147. D. Schröder, A. Fiedler, M. F. Ryan, and H. Schwarz, *J. Phys. Chem.* **98**, 68 (1994).

148. M. Filatov and S. Shaik, *J. Phys. Chem. A* **102**, 3835 (1998).

149. D. Danovich and S. Shaik, *J. Am. Chem. Soc.* **119**, 1773 (1997).

150. D. Schröder, S. Shaik, and H. Schwarz, *Acc. Chem. Res.* **33**, 139 (2000).

151. D. Schröder, H. Schwarz, D. E. Clemmer, Y. Chen, P. B. Armentrout, V. Baranov, and D. K. Bohme, *Int. J. Mass Spectrom. Ion Proc.* **161**, 175 (1997).

152. A. Fiedler, J. Hrusák, W. Koch, and H. Schwarz, *Chem. Phys. Lett.* **211**, 242 (1993).

153. Y. Nakao, K. Hirao, and T. Taketsugu, *J. Chem. Phys.* **114**, 7935 (2001).

154. D. A. Chestakov, D. H. Parker, and A. V. Baklanov, *J. Chem. Phys.* **122**, 084302 (2005).

155. M. Sodupe, V. Branchadell, M. Rosi, and C. W. Bauschlicher, Jr., *J. Phys. Chem. A* **101**, 7854 (1997).

156. D. E. Lessen, R. L. Asher, and P. J. Brucat, *J. Chem. Phys.* **95**, 1414 (1991).

157. R. H. Page, Y. R. Shen, and Y. T. Lee, *J. Chem. Phys.* **88**, 4621 (1988).

158. R. N. Pribble and T. S. Zwier, *Science* **265**, 75 (1994).

159. B. Brutschy, *Chem. Rev.* **100**, 3891 (2000).

160. T. Ebata, A. Iwasaki, and N. Mikami, *J. Phys. Chem. A* **104**, 7974 (2000).

161. N. A. Macleod and J. P. Simons, *Phys. Chem. Chem. Phys.* **5**, 1123 (2003).

162. E. Nir, I. Hunig, K. Kleinermanns, and M. S. de Vries, *ChemPhysChem* **5**, 131 (2004).

163. M. Citir and R. B. Metz, *J. Chem. Phys.*, accepted to appear in vol 127 (2007).

164. K. K. Irikura and J. L. Beauchamp, *J. Phys. Chem.* **95**, 8344 (1991).

165. M. Citir and R. B. Metz, to be submitted (2008).

166. R. H. Schultz and P. B. Armentrout, *Organometallics* **11**, 828 (1992).

167. M. Hendrickx, K. Gong, and L. Vanquickenborne, *J. Chem. Phys.* **107**, 6299 (1997).

168. R. H. Schultz and P. B. Armentrout, *J. Phys. Chem.* **97**, 596 (1993).

169. Q. Zhang, P. R. Kemper, and M. T. Bowers, *Int. J. Mass Spectrom.* **210**, 265 (2001).

170. W. D. Chen, E. Poullet, J. Burie, D. Boucher, M. W. Sigrist, J. J. Zondy, L. Isaenko, A. Yelisseyev, and S. Lobanov, *Appl. Opt.* **44**, 4123 (2005).

171. D. S. Yang, *Coord. Chem. Rev.* **214**, 187 (2001).

172. V. Goncharov, L. A. Kaledin, and M. C. Heaven, *J. Chem. Phys.* **125**, 133202 (2006).

173. M. C. Heaven, *Phys. Chem. Chem. Phys.* **8**, 4497 (2006).

174. J. J. Carroll, J. C. Weisshaar, P. E. M. Siegbahn, C. A. M. Wittborn, and M. R. A. Blomberg, *J. Phys. Chem.* **99**, 14388 (1995).

175. J. J. Carroll and J. C. Weisshaar, *J. Phys. Chem.* **100**, 12355 (1996).

176. M. L. Campbell, *J. Chem. Soc. Faraday Trans.* **94**, 353 (1998).

STABILIZATION OF DIFFERENT CONFORMERS OF WEAKLY BOUND COMPLEXES TO ACCESS VARYING EXCITED-STATE INTERMOLECULAR DYNAMICS

DAVID S. BOUCHER and RICHARD A. LOOMIS

Department of Chemistry, Washington University in St. Louis One Brookings Drive, CB 1134, St. Louis, Missouri, 63130, USA

CONTENTS

Advances in Chemical Physics, Volume 138, edited by Stuart A. Rice
Copyright © 2008 John Wiley & Sons, Inc.

I. INTRODUCTION

For 30 years, intermolecular interactions and half-collision, dissociation dynamics have been interrogated by stabilizing multiple entities in weakly bound complexes. These interactions are most often investigated on inter-molecular potential energy surfaces associated with the species in their ground-electronic state or with one moiety having electronic and vibrational excitation. The intermolecular orientations and the energy levels that can be accessed on the excited sate are dictated by both the Franck–Condon factors and selection rules for these transitions. Consequently, it is desirable to stabilize ground-state complexes with different average geometries to interrogate the dependence of the excited-state interactions and dynamics on the intermolecular orientation and energetics. This chapter reviews work performed predominantly in our laboratory that has focused on stabilizing and spectroscopically characterizing different conformers of the rare gas–dihalogen complexes. Also summarized are the varying photoinitiated excited-state dynamics that have been and can be investigated by taking advantage of the different Franck–Condon windows from the conformers.

The primary objective of studying chemical reaction dynamics is the full disclosure of the energy-transfer processes that occur between reactants and products. A critical aspect of this endeavor is revealing the features and contours of the multidimensional potential energy surfaces (PESs) that dictate the outcome of the reaction(s) of interest. Understanding the nature of a PES not only serves to further our understanding of an encounter between a specific set of reactants, but reveals the roles that the intermolecular interactions have on the overall reaction mechanism that in turn helps us to expand the knowledge of the more fundamental dynamical processes underlying the encounter.

A great deal of experimental and theoretical effort has been exerted to characterize intermolecular PESs for a wide range of atomic and molecular systems. Much of this effort has involved the inversion of experimental data obtained from molecular beam scattering experiments [1, 2]. In conjunction with high-level theoretical methodologies, the scattering and reaction cross-sections and their dependence on entrance-channel parameters, such as the internal energies, relative velocities and momenta of the reactants, and the distribution of final product states have proven to be particularly useful for developing detailed PESs [3, 4]. In addition to the dependence of chemical reaction dynamics on the energetics of the reactants in the entrance channel, the specific interreactant orientations will also dictate probabilities for reaction and propensities for forming specific product channels. While there has been a great deal of success manipulating the energetics of reactive encounters, influence over the orientations of reactants in the entrance channel has proven to be a challenge. As a result, the available experimental information regarding the orientation dependence of reactions is still somewhat limited.

In principle, a convenient way of controlling reactant orientation is to trap the reactants in a weakly bound, nonreactive complex within a molecular beam or free-jet expansion [5–9]. Then preferred geometries of the prereactive complexes with well-defined energies and angular momentum can be accessed by vibrationally or electronically photoexciting one of the moieties. By doing so, the use of weakly bound complexes reduces the full-collisional problem to a much more amenable, half-collision, reaction event, or photodissociation problem. The stereospecificity and state-selectivity also make them an ideal venue for studying reaction dynamics, characterizing multidimensional PESs, and testing various theoretical methodologies. One corresponding limitation in this methodology was believed to be that the weakly bound complexes tend to be stabilized only in the lowest energy intermolecular vibrational level, and thus limited spatial and energy regions of the excited-state PES can be accessed.

Over the past three decades, considerable attention has been paid to the triatomic complexes, $Rg \cdots XY$, consisting of a rare gas atom, Rg, weakly bound to a dihalogen chromophore, XY, and a large body of data has been compiled regarding the structure and dynamics of these systems [10–16]. These complexes are particularly attractive because they can be subjected to both detailed theoretical and experimental study. The dihalogens, and thus complexes containing them, are easily probed using a variety of standard spectroscopic techniques, including laser-induced fluorescence (LIF), absorption spectroscopy, and two-laser, pump–probe action spectroscopy. Also, the structural simplicity of these triatomic complexes makes them amenable to high level theoretical calculations, involving both classical dynamics and full quantum-mechanical treatments. Furthermore, members of this class of weakly bound complexes exhibit intriguing and varying dynamical behavior during photo-initiated dissociation that has made them prototypical systems for studying long-range intermolecular forces, intermolecular PESs, energy redistribution and unimolecular reaction dynamics.

The most often investigated dynamical process in rare gas–dihalogen complexes is vibrational predissociation (VP) [14]. During VP a complex is prepared in an intermolecular vibrational level associated with a specific vibrational level of the dihalogen most often in an electronically excited state. The vibrational energy of the excited dihalogen bond is sufficient to rupture the weak van der Waals bond, and separate rare gas atom and dihalogen molecule products are formed with varying amounts of translational and orbital energy. The molecular products also have distributions of vibrational and rotational excitation that can be directly measured using established two-laser, pump–probe spectroscopy [14, 17]. These photodissociation events may then be treated as half-collision events that begin from preferred orientations on the excited intermolecular PES.

The mismatch between the high frequency dihalogen bond and the low frequency intermolecular bond, and the weak coupling of these two modes, gives rise to interesting VP behavior that cannot be readily described using standard statistical models [14]. The observed VP rates are much slower than those predicted by Rice–Ramspergen–Kassel–Marcus (RRKM) theory. Instead, the restrictions on the observed product states are more readily described using energy gap arguments, which is itself a signature of nonstatistical behavior. Deviations from statistical behavior imply that the dynamical attributes of the internal energy flow play a significant role in determining the VP rates. Within the scope of unimolecular reaction dynamics, the inherent nonstatistical energy disposal within these complexes and the competition that may take place between energy redistribution and fragmentation provide a venue for investigating the propensities of the complex to break apart via specific photodissociation product channels. The final distribution of product states associated with each channel reflects the partitioning of energy over the product degrees of freedom and may reveal detailed information about the anisotropy of the intermolecular PES and the strength of the interactions between the initial and final states.

Although the simpler triatomic $Rg \cdots XY$ complexes have been the subject of more extensive experimental and theoretical scrutiny, the higher order $Rg_n \cdots XY$, $n \geq 2$, complexes, or "clusters", have also received a great deal of attention [18–23]. As additional rare gas atoms are complexed with a dihalogen constituent, multibody interactions and the increasing number of internal degrees of freedom adds to the complexity of the energy-transfer processes within the higher order clusters. For example, intramolecular vibrational energy redistribution (IVR) within the cluster has been shown to become more prevalent as the sizes of the clusters increase due to the competition that develops between the increasing number of van der Waals vibrational modes for energy flow out of the dihalogen moiety [24–26]. It is well established that by regulating the gas mixture composition and conditions of the supersonic expansion used to stabilize these complexes the size distribution of the higher order complexes can be controlled to some extent [27]. This affords the experimentalist with a convenient means of systematically varying the sizes of the complexes. In general, studying complexes of varying size and investigating the effects of size evolution on the complexity of the energy redistribution in these systems could help bridge the gap between state-specific interactions and macroscopic solvent effects.

It is important to emphasize that while the electronically excited states of the triatomic and higher order rare gas–dihalogen complexes exhibit so much interesting behavior, it is the geometry of the ground-state conformers that ultimately dictates the excited-state dynamics that can be interrogated. For vibronic excitation, the Franck–Condon windows associated with different ground-state configurations provide the access to regions of the excited-state PESs that may result in different intermolecular interactions and energy-transfer processes. In this way, the ground-state complexes can themselves be regarded as precursors or launch states for the reactive excited-state. Therefore, the full characterization of the excited-state dynamics in these complexes hinges on the ability to stabilize and excite multiple ground-state conformers. Finally, it is highly desirable to have control over the populations of the different ground-state conformers that are stabilized in an expansion so that each can be preferentially investigated.

This chapter describes work done by our research group aimed at stabilizing and investigating the different conformers of ground-state $Rg \cdots XY$ complexes and how these populations can be manipulated. Specifically, Section II contains an overview of the geometries and spectroscopy of the different ground-state conformers of rare gas–dihalogen complexes, $Rg_n \cdots XY$, $n \geq 1$, focusing primarily on the $He_n \cdots ICl$ and $He_n \cdots Br_2$ systems. Section III describes the techniques used to stabilize multiple ground-state conformers, paying particular attention to the effects of varying the expansion conditions on the relative abundances of the conformers. Also, the possible population-transfer mechanisms and the notion of the different ground-state conformers existing in a thermodynamic equilibrium are discussed. Section IV presents results of work aimed at characterizing the vibrational dissociation dynamics of several rare gas–dihalogen complexes and gives an overview of the ongoing work using multiple conformers to characterize the dissociation dynamics of these systems on different regions of the excited-state intermolecular PESs.

II. GEOMETRIES OF RARE GAS–DIHALOGEN COMPLEXES

A. Historical Overview of Triatomic Rare Gas–Dihalogen Interactions

Early experimental spectroscopic investigations on $Rg \cdots XY$ complexes resulted in contradictory information regarding the interactions within them and their preferred geometries. Rovibronic absorption and LIF spectra revealed T-shaped excited- and ground-state configurations, wherein the Rg atom is confined to a plane perpendicular to the X—Y bond [10, 19, 28–30]. While these results were supported by the prediction of T-shaped structures based on pairwise additive Lennard–Jones or Morse atom–atom potentials, they seemed to be at odds with results from microwave spectroscopy experiments that were consistent with linear ground-state geometries [31, 32]. Some attempts were made to justify the contradictory results of the microwave and optical spectroscopic studies, and

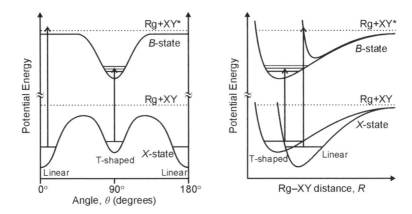

Figure 1. Schematic of the radial cuts of the ground- and excited-state potential energy surfaces at the linear and T-shaped orientations. Transitions of the ground-state, T-shaped complexes access the lowest lying, bound intermolecular level in the excited-state potential also with a rigid T-shaped geometry. Transitions of the linear conformer were previously believed to access the purely repulsive region of the excited-state potential and would thus give rise to a continuum signal. The results reviewed here indicate that transitions of the linear conformer can access bound excited-state levels with intermolecular vibrational excitation.

it became somewhat accepted that there were inherent limitations in the two techniques that precluded each from observing transitions of both conformers. One well-accepted thesis, which is illustrated in Fig. 1, was that the potential wells for the T-shaped conformer are similar in both the ground and excited electronic states resulting in favorable Franck–Condon factors for their optical transitions. The electronic excitation increases the antibonding character of the dihalogen, however, and the intermolecular interactions in the excited state will be weaker and perhaps purely repulsive. Thus, transitions of the linear ground-state configuration were believed to access purely repulsive regions of the excited-state potential [14].

In 1993, Gerry and co-workers published results of a microwave study on $Ar\cdots Cl_2$ that presented evidence of the T-shaped conformer and no indication of complexes with a linear geometry [33]. The authors speculated that the T-shaped conformer was energetically more stable than the linear conformer and due to the low temperatures achieved in the expansion, the linear complexes were not abundant and could not be spectroscopically observed. That same year, the Klemperer group reported optical transitions of both the linear and T-shaped $Ar\cdots I_2$ conformers in the I_2 $B–X$ spectral region [34]. Discrete features ascribed to transitions of the T-shaped complex were evident in the rovibronic spectrum, and the authors ascribed a continuum signal spanning the entire $B–X$ region to transitions of the linear conformer accessing a broad region on the excited-state potential. The compound $Ar\cdots I_2$ became the first system for which the experimental evidence confirmed the co-existence of the T-shaped and linear conformers in the expansion and for the next 10 years it would stand as the only example of such a system.

The absence of discrete features in the rovibronic spectra attributable to each of the different ground-state conformers of the $Rg \cdots XY$ complexes was problematic. During the latter part of the 1990s there were questions regarding the utility of these complexes as models for future studies of unimolecular reaction dynamics and intermolecular energy transfer. Specifically, the rotational contours of the T-shaped features are consistent with transitions accessing regions of the excited-state intermolecular potential associated with T-shaped or near T-shaped geometries. As a consequence, the experimental data characterizing the multidimensional PESs in the ground and excited electronic states of $Rg \cdots XY$ complexes were limited. These experimental shortcomings became particularly apparent as computational methods improved and became more powerful. Several high-level *ab initio* treatments of the $Rg \cdots XY$ ground-state intermolecular PESs indicated the existence of minima in the T-shaped and linear configurations, with the global minimum being in the linear configuration [35–37]. These theoretical results contradicted the early descriptions of the potentials based on Lennard–Jones and Morse pairwise-additive potentials, which supported the existence of the T-shaped conformer, but due to lack of observable linear rovibronic transitions they could not be experimentally verified.

B.   Rovibronic Spectroscopy of He $\cdots$ ICl and He $\cdots$ Br$_2$

While conducting research on the He $\cdots$ ICl complex, our group observed two previously unreported features in the ICl $B–X$, 3–0 LIF spectrum at 17,835 and 17,842 cm^{-1}, which are each ≈ 11 cm^{-1} to higher energy from the well-known T-shaped bands of He $\cdots$ I^{37}Cl and He $\cdots$ I^{35}Cl, respectively [38]. Spectra recorded by varying the backing pressure used to stabilize the complexes in the expansion indicated that all of the observed features could be associated with transitions of I^{35}Cl or I^{37}Cl molecules or He $\cdots$ ICl complexes with only one He atom. Spectra were also recorded in the ICl $B–X$, 2–0, and 1–0 regions and similar features were observed in both. An LIF spectrum recorded in the 2–0 region is shown in Fig. 2(a). Since the T-shaped He $\cdots$ I^{37}Cl band overlaps the very intense I^{35}Cl monomer band in this region, only the T-shaped He $\cdots$ I^{35}Cl and the two higher energy features are observed. In all of the LIF spectra, the intensity of the two newly observed bands is $\approx 3:1$ with the highest energy band being the more intense feature. This is precisely the ratio for the abundance of I^{35}Cl in comparison to I^{37}Cl, suggesting that the two bands contain I^{35}Cl and I^{37}Cl.

In addition to LIF spectra, action spectra of He $\cdots$ I^{35}Cl and He $\cdots$ I^{37}Cl were recorded to minimize the spectral congestion attributed to transitions of the different isotopomers and other molecular species. The action spectrum recorded in the ICl $B–X$, 2–0 region obtained with the probe laser fixed on the rotational

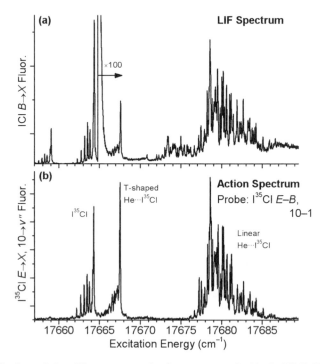

Figure 2. Laser-induced fluorescence and action spectra acquired in the ICl $B-X$, 2–0 spectral region. The LIF spectrum, panel (a), is dominated by the intense $I^{37}Cl$ and $I^{35}Cl$ transitions at lower energies. Features associated with transitions of the T-shaped $He \cdots I^{35}Cl(X, v'' = 0)$ and linear $He \cdots I^{37}Cl(X, v'' = 0)$ and $He \cdots I^{35}Cl(X, v'' = 0)$ conformers are observed to higher energy and are ≈ 100 times weaker than the monomer features. The action spectra, (b), was recorded by probing the $I^{35}Cl$ $E-B$, 10–1 transition, or the $\Delta v = -1$ vibrational predissociation channel.

band head of the $I^{35}Cl$ $E-B$, 10–1 is shown in Fig. 2(b). Similar spectra were also recorded with the probe laser fixed on $I^{37}Cl$ $E-B$ transitions. All of the features observed in the LIF spectrum are assignable to transitions of $I^{35}Cl$ or $I^{37}Cl$ molecules or to transitions of $I^{35}Cl$ or $I^{37}Cl$ containing complexes. The rotational structures of the higher energy features are broader and more complicated than the previously characterized T-shaped $He \cdots I^{35}Cl$ and $He \cdots I^{37}Cl$ bands. Simulations of the rotational contour of the highest energy band were performed assuming that this band is associated with transitions of $He \cdots I^{35}Cl$. While these rigid-rotor simulations consistently underestimated the observed number of rotational lines, a linear ground-state geometry, either $He \cdots I—Cl$ or $He—Cl—I$, seemed necessary to reproduce the overall shape of the band [38].

Ultimately, it was the strong agreement between $He \cdots I^{35}Cl$ excitation spectra calculated by McCoy and co-workers and experimental $He \cdots I^{35}Cl$ action spectra, such as shown in Fig. 2(b), that corroborated the proposed assignments of the newly observed $He \cdots ICl$ features in the spectra [39, 40].

The higher energy features can indeed be associated with transitions of $He \cdots ICl(X, v'' = 0)$ ground-state complexes with rigid $He \cdots I–Cl$ linear geometries. In contrast to the T-shaped band that is associated with transitions to the most strongly bound intermolecular vibrational level in the excited state without intermolecular vibrational excitation, $n' = 0$, the transitions of the linear conformer access numerous excited intermolecular vibrational levels, $n' \geq 1$. These levels are delocalized in the angular coordinate and resemble hindered rotor levels with the He atom delocalized about the $I^{35}Cl$ molecule.

Features associated with transitions of both T-shaped and linear conformers were also observed in rovibronic spectra of the $He \cdots Br_2$ complex [41]. The feature now attributed to transitions of the linear $He \cdots ^{79,79}Br_2(X, v'' = 0)$ conformer was observed previously in the Br_2 $B–X$, 8–0 region [42], but initial simulations of the feature using a rigid-rotor model could not satisfactorily reproduce the observed rotational contour. It was speculated that this feature could be associated with transitions of the ground-state, T-shaped complex to an excited-state level with bending excitation. Subsequent theoretical efforts aimed at accurately calculating the rovibronic spectrum of $He \cdots ^{79,79}Br_2$ complex revealed two ground-state conformers and showed that the higher energy feature is comprised of rovibrational lines associated with transitions of both the linear and T-shaped conformers [41, 43, 44].

The overall appearance of the linear $He \cdots ^{79,79}Br_2$ feature and the similar shift to higher transition energies from the monomer and the T-shaped band resemble those of the linear $He \cdots I^{35}Cl$ feature. Both LIF and action spectra of $He \cdots Br_2$ were recorded at varying downstream distances to determine if the features can be associated with transitions of different ground-state complexes [41]. The intensities of the two features did not scale together with cooling. A transfer of population from the T-shaped conformer to the species giving rise to the higher energy feature, presumably the linear $He \cdots ^{79,79}Br_2(X, v'' = 0)$ ground-state conformer, was observed as spectra were recorded further downstream, and thus in colder regions of the expansion. Calculations of the $He \cdots ^{79,79}Br_2$ excitation spectrum in the Br_2 $B–X$, 12–0 region and the strong agreement of those results with the experimental spectra provided more conclusive evidence for the proposed assignment. Just as with the case of $He \cdots ICl$, the calculations indicate that transitions of the linear $He \cdots Br_2(X, v'' = 0)$ conformer access multiple excited-state levels, $n' \geq 1$, that have the He atom delocalized in the angular coordinate relative to the dihalogen axis.

The angular-dependent adiabatic potential energy curves of these complexes obtained by averaging over the intermolecular distance coordinate at each orientation and the corresponding probability distributions for the bound intermolecular vibrational levels calculated by McCoy and co-workers provide valuable insights into the geometries of the complexes associated with the observed transitions. The $He + ICl(X, v'' = 0)$ and $He + ICl(B, v' = 3)$ adiabatic potentials are shown in Fig. 3 [39]. The abscissa represents the angle, θ,

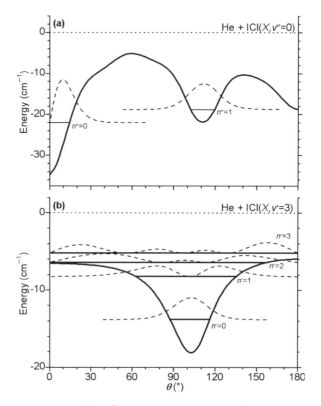

Figure 3. (a) The He + ICl($X,v'' = 0$) and (b) He + ICl($B,v' = 3$) intermolecular adiabatic potential energy curves calculated as a function of θ, the angle of He about the I–Cl axis. The probability densities for the two lowest He$\cdots$I^{35}Cl($X, v'' = 0$) intermolecular vibrational levels, $n'' = 0$ and 1, with experimental binding energies of $D_0''^{(L)} = 18.3$ and $D_0''^{(T)} = 15.2\,\text{cm}^{-1}$, respectively, are localized in the linear and T-shaped regions of the ground-state intermolecular potential. The probability densities of the $n' = 0–3$ intermolecular vibrational eigenstates are superimposed on the potential in (b). Adapted from Ref. [62].

between the intermolecular and I–Cl bonds, with $\theta = 0°$ corresponding to the linear He$\cdots$I–Cl geometry, and the ordinate is the energy relative to the dissociation asymptote of the complex. Plots of the $J = 0$ He$\cdots$I^{35}Cl probability amplitudes for the lowest energy intermolecular vibrational levels in the X-state, $n'' = 0$ and 1, and B-state, $n' = 0\text{-}4$, are superimposed on the potentials, with zero amplitude corresponding to the energy of that level. The ground-state probabilities strongly suggest that the $n'' = 0$ and 1 levels can be associated with the fairly rigid linear and T-shaped He$\cdots$ICl($X,v'' = 0$) conformers, respectively.

   The calculations of the rovibronic spectra verify that the T-shaped He$\cdots$ICl features in the LIF and action spectra can be associated with $n' = 0 \leftarrow n'' = 1$

transitions, in agreement with the previous assignment of transitions of a rigid
T-shaped complex to an excited-state level also with a rigid T-shaped geometry
[45–49]. The calculations also indicate that the higher energy features can be
associated with transitions of the linear, ground-state conformer to levels with
intermolecular vibrational excitation, $n' \geq 1 \leftarrow n'' = 0$. The calculated spectra
were obtained using an *ab initio* PES for the He + ICl$(X, v'' = 0)$ ground state
[50] and a simple pairwise additive potential for the He + ICl$(B, v' = 0)$
excited state [48]. Just by scaling the published He–I and He–Cl Morse
potential interactions each by a factor of 1.11 an even better agreement
between the calculated and experimental spectra was obtained so that
assignment of each of the observed lines could be made and the binding
energies of the excited-state levels could be determined [51]. An LIF spectrum
recorded in the ICl B–X, 3–0 region and the corresponding theoretical
excitation spectrum associated with the experimentally measured rotational
temperature of 0.2 K are shown in Fig. 4. The $R(0)$ rotor lines for the
$n' = 2, 3, 4 \leftarrow n'' = 0$ transitions are labeled to emphasize the spectral overlap

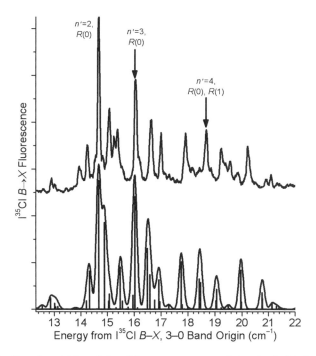

Figure 4. Experimental laser-induced fluorescence, upper plot, and calculated spectra, lower
plot, of the linear He $\cdots$ I^{35}Cl feature in the ICl B–X, 3–0 region. An I^{35}Cl$(X, v'' = 0)$ rotational
temperature of 0.19 K was measured for the experimental spectrum, and a temperature of 0.20 K was
used in the calculations. Adapted from Ref. [51].

within this feature and the energy spacings between the excited-state intermolecular vibrational levels.

The McCoy group also undertook calculations of the $He + Br_2$ interactions and excitation spectra. For this work, they generated a new $He + Br_2(X, v'' = 0)$ *ab initio* ground-state intermolecular PES at the CCSD(T) level of theory and used a diatomics in molecules PES for the $He + Br_2(B, v' = 12)$ excited-state interactions. Again the geometries of the lowest lying intermolecular vibrational levels are easily recognized when considering the angular-dependent adiabatic potential curve with the probability densities of the intermolecular vibrational levels superimposed, Fig. 5. The two lowest energy intermolecular levels in the

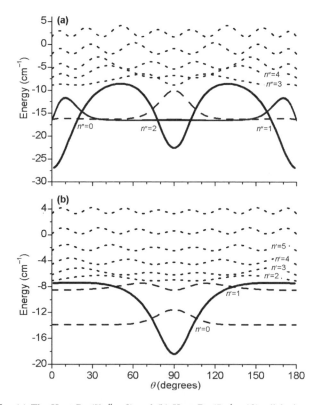

Figure 5. (a) The $He + Br_2(X, v'' = 0)$ and (b) $He + Br_2(B, v' = 12)$ adiabatic potential energy curves calculated as a function of θ, the angle of He about the Br–Br axis. The probability densities for the two lowest $He \cdots {}^{79,79}Br_2(X, v'' = 0)$ intermolecular vibrational levels, $n'' = 0, 1$ and $n'' = 2$, with experimental binding energies of $D_0''^{(L)} = 17.0$ and $D_0''^{(T)} = 16.6 \, cm^{-1}$, respectively, are localized in the linear and T-shaped regions of the ground-state intermolecular potential. The excited-state probability densities are also included, and in both plots the probability amplitudes localized in the linear wells are plotted with solid lines, those localized in the T-shaped minimum with dashed lines, and the free rotor states are plotted with dotted lines to differentiate between orientations of the levels. Adapted from Ref. [41].

ground-state, panel (a), represent the symmetric and antisymmetric linear combinations of the two levels that have the helium atom localized in the linear wells along the Br—Br axis near either of the Br atoms, $n'' = 0, 1$. The next lowest energy level, $n'' = 2$, corresponds to the helium atom being localized in a T-shaped configuration, relative to the Br—Br bond. Again, because of the localization of these probability densities, the $n'' = 0, 1$ and $n'' = 2$ levels can be considered different conformers of the ground-state $\text{He} \cdots \text{Br}_2(X, v'' = 0)$ complex. Similar to the $\text{He} \cdots \text{ICl}$ system, for $\text{He} \cdots \text{Br}_2$ it was found that the T-shaped feature consists of $n' = 0 \leftarrow n'' = 2$ transitions, while the higher energy linear feature consists of $n' \geq 1 \leftarrow n'' = 0, 1$ transitions.

The action spectrum recorded in the $^{79,79}\text{Br}_2$ B–X, 12–0 region and associated with a $^{79,79}\text{Br}_2(X, v'' = 0)$ rotational temperature of 0.28 K is plotted in Fig. 6. While the theoretical spectrum qualitatively reproduced the experimental spectrum, we chose to slightly shift the binding energies of the n' levels and use the resulting modified spectrum, lower plot in Fig. 6, to make assignments of the lines in the experimental spectrum. From these assignments, energies of the excited-state levels were directly obtained [41].

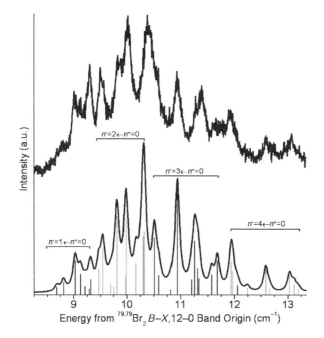

Figure 6. (a) Experimental laser-induced fluorescence, upper plot, and calculated spectra, lower plot, of the linear $\text{He} \cdots ^{79,79}\text{Br}_2$ feature in the Br_2 B–X, 12–0 region. A $^{79,79}\text{Br}_2(X, v'' = 0)$ rotational temperature of 0.28 K was measured for the experimental spectrum, and this temperature was used in the calculations. Adapted from Ref. [41].

TABLE I

Binding Energies of Rare Gas–Dihalogen Complexes for Which Linear Isomers Have Been Reported

	Vibronic State	Geometry	Binding Energy (Exp.)/cm^{-1}	Reference	Binding Energy (Theory)/cm^{-1}	Reference
He$\cdots$I^{35}Cl						
	$X(v''=0)$	Linear	22.0(2)	39	18.29	50
		T-shaped	16.6(3)	39	15.14	50
		Antilinear	$\cdots$		12.33	50
	$B(v'=2)$	T-shaped	16.3(6)	39		
	$B(v'=3)$	T-shaped	16.2(6)	51	13.3	49
He$\cdots^{79}$Br$_2$						
	$X(v''=0)$	Linear	17.0(8)	41	16.46	41
			16.6(8),	41		
		T-shaped	17.0(1.5)	52	15.81	41
	$B(v'=12)$	T-shaped	12.7(8),	41		
			13.1(1.5)	41	13.42	41
Ne$\cdots$I^{35}Cl						
	$X(v''=0)$	Linear	84(1)	55	76.19	50
		T-shaped	70(5)	55	62.59	50
		Antilinear	$\cdots$		53.73	50
	$B(v'=2)$	T-shaped	65(5)	55		
He$\cdots$I$_2$						
	$X(v''=0)$	Linear	16.3(6)	53	15.38	54
		T-shaped	16.6(6)	53	14.68	53
	$B(v'=23)$	T-shaped	12.8(6)	53		
Ne$\cdots$I$_2$						
	$X(v''=0)$	Linear	72(3)		38.8	15
		T-shaped	64.3–66.3,	56	58.454	15
			72.4–74.7			
	$B(v'=34)$	T-shaped	56.6–58.6,	56	58.363	15
			65.0–67.1			
Ar$\cdots$I$_2$						
	$X(v''=0)$	Linear	250(2)	57	237.8	58
		T-shaped	234.2–240.1	29	212.0	58
	$B(v'=30)$	T-shaped	220.0–226.3	29	223(3)	59

Our group has continued to examine several other rare Rg$\cdots$XY systems, and a list of the experimental and theoretical T-shaped and linear binding energies of these complexes is presented in Table I [52–55,57]. The general features and characteristics of the spectra for all of the complexes investigated are similar to those observed in the spectra for the He$\cdots$ICl and He$\cdots$Br$_2$ systems. The linear features are observed to higher transition energies than the T-shaped features. In contrast to the rather simple rotational structure of the T-shaped features, the linear features possess much broader and structured contours, often due to overlap of adjacent intermolecular bands. For all of the complexes, the results are consistent with the observation of transitions of the ground-state, T-shaped conformer to the lowest lying intermolecular level in the excited state, also with a rigid T-shaped geometry. The transitions of the ground-state, linear conformer access higher lying intermolecular levels that are often delocalized in the angular coordinate.

C. Spectroscopy and Geometries of Higher Order Rare Gas–Dihalogen Complexes: $Rg_n \cdots XY$, $n \geq 2$

The early work by Levy and co-workers indicated that the $He_n \cdots I_2$, $n = 2,3$ [28], $Ne_n \cdots I_2$, $n = 1$–6 [18], and $He_m Ar_n \cdots I_2$ [19] complexes have ground-state geometries in which all the rare gas atoms are localized in plane perpendicular to the dihalogen bond, in a so-called "toroidal" or "belt" configuration. Furthermore, they were able to establish useful guidelines for identifying rovibronic features in the visible excitation spectrum associated with higher order clusters [19]. First, there is a strong propensity for higher order $Rg_n \cdots XY$ complexes to dissociate into dihalogen products with n less vibrational quanta. This propensity breaks down for complexes containing argon, for which the large excited-state binding energies require multiple vibrational quanta from the dihalogen moiety to rupture a single van der Waals bond. Due to the anharmonicity of the dihalogen potential, it can also break down for complexes prepared in levels associated with high v' for which some of the product channels may not be energetically accessible. This trend, however, does hold quite well for $He_n \cdots XY$ and $Ne_n \cdots XY$ complexes. Second, the transition energies of the features in the excitation spectrum follow a band-shift rule [19]. This rule states that the transition energy of a $Rg_n \cdots XY$ complex is given by $E = E_0 + n\Delta E$, where E_0 is the band origin of the bare dihalogen molecule, and ΔE is a constant energy shift that represents the difference in the binding energies of the ground- and excited-state $Rg \cdots XY$ complexes.

Subsequent experimental and theoretical studies of other higher order clusters, such as $Ne_n \cdots Br_2$, $n = 2,3$ [20], $Ne_n \cdots Cl_2$, $n = 2,3$ [21], and $Ne_n \cdots ICl$, $n = 1$–5 [60], yielded similar results. The rotational contours ascribed to the higher order features in the $B–X$, v'–0 spectral region resemble those of the $Rg \cdots XY$ complexes, typically exhibiting an R-band head and a P-branch extending to lower energy. Such a contour has been shown to be consistent with the toroidal configuration. Analyses of rotationally resolved $B–X$ bands of $Ne_2 \cdots Cl_2$ and $Ne_3 \cdots Cl_2$ complexes using a rigid-rotor asymmetric top model revealed that Ne atoms localized within such a toroidal potential occupy equivalent positions [21]. Based on the band-shift rule the authors concluded that $Ne_2 \cdots Cl_2$ possesses a distorted tetrahedron geometry and the $Ne_3 \cdots Cl_2$ complex is planar, with all three neon atoms confined to a plane perpendicular to the Cl–Cl bond. In contrast, high-resolution studies of the

$He_2 \cdots Cl_2$ complex indicated that the weakly interacting He atoms are delocalized with a large amplitude motion around the toroidal potential, and the concept of an average geometry is no longer valid [22]. For the $Ar_3 \cdots Cl_2$ system, Janda and co-workers observed evidence for the existence of two ground-state conformers [61]. One conformer has a toroidal configuration, and the other has the argon atoms forming a triangle parallel to the Cl—Cl bond axis.

Transitions from ground-state $Rg_n \cdots XY$ intermolecular vibrational levels with toroidal geometries access the lowest lying level(s) in the excited state. Consequently, the predissociation dynamics are limited to this one spatial and energetic region on the intermolecular PES. Like the triatomic rare gas–dihalogen complexes, the capacity to access different regions on the excited-state intermolecular PES through transitions of multiple ground-state conformers may prove beneficial to more complete characterization of the photodissociation dynamics. Of particular interest, is the nature of energy partitioning into intermolecular modes possessing different symmetries and the efficiencies of the modes for energy redistribution within the complex. Also, due to the presence of multiple rare gas atoms and the multibody interactions in larger complexes, interesting dissociation dynamics may be observed when accessing highly delocalized intermolecular bending levels.

Motivated in part by our progress in stabilizing different conformers of the triatomic $Rg \cdots XY$ ground-state complexes, we have recently applied similar experimental techniques to interrogate the $He_n \cdots ICl$ and $He_n \cdots Br_2$, $n \geq 2$ complexes with varying ground-state geometries [62]. Action spectra of the $He_n \cdots I^{35}Cl$ and $He_n \cdots {}^{79,79}Br_2$ complexes were recorded with the probe laser fixed on transitions that would detect the nascent monomer population in the $\Delta v' = -1$, -2, and -3 dissociation channels. These spectra were collected using varying helium backing pressures and with the lasers intersecting the expansion at different downstream distances in order to preferentially interrogate clusters with varying sizes and perhaps geometries. For clarity, we will use the notation $(\#T, \#L)Rg_n \cdots XY$ to denote the geometry of the ground-state conformers. Here, #T denotes the number of rare gas atoms localized in the toroidal potential well lying in a plane perpendicular to the dihalogen bond, that is, in the T-shaped position, and #L indicates the number of Rg atoms localized in the linear well.

The three $He_n \cdots {}^{79,79}Br_2$ action spectra plotted in Fig. 7 were recorded with the probe laser fixed on the rotational band head of the ${}^{79,79}Br_2$ E—B, 1–11 transition [61]. In panel (a), the excitation laser is scanned through the Br_2 B—X, 12–0 spectral region. Since intermolecular vibrational levels in the He + Br_2 $(B, v' = 12)$ potential preferentially undergo VP via a $\Delta v' = -1$ mechanism forming He + $Br_2(B, v' = 11)$ products, this spectrum will be sensitive to transitions of $He \cdots {}^{79,79}Br_2$ complexes. As expected, the features associated with transitions of the ground-state, T-shaped and linear $He \cdots {}^{79,79}Br_2$

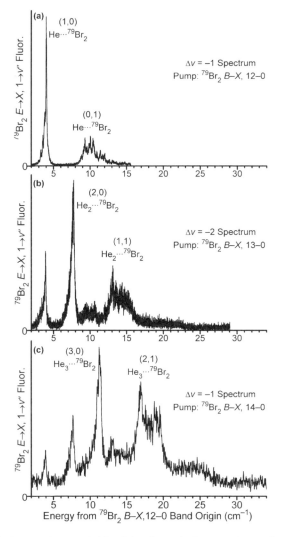

Figure 7. Action spectra recorded by fixing the probe laser on the rotational band head of the $^{79,79}Br_2$ $E-B$, $1-11$ transition and scanning the excitation laser through (a) the Br_2 $B-X$, $12-0$, (b) the $B-X$, $13-0$, and (c) the $B-X$, $14-0$ spectral regions. The (#T,#L) labels indicate where each of the helium atoms for the ground-state conformer associated with each observed feature. Adapted from Ref. [62].

$(X, v'' = 0)$ conformers, denoted as $(1,0)$He$\cdots$Br$_2$ and $(0,1)$He$\cdots$ Br$_2$, dominate the spectrum.

The spectrum plotted in Fig. 7(b) was recorded in the Br$_2$ $B-X$, $13-0$ region. This spectrum should preferentially detect transitions of He$_2 \cdots^{79,79}$ Br$_2$

complexes since $He_2 \cdots {}^{79,79}Br_2(B, v'=13)$ complexes tend to dissociate via a $\Delta v' = -2$ mechanism forming $2He + {}^{79,79}Br_2(B, v' = 11)$ products. Note that the $He_2 \cdots {}^{79,79}Br_2$ features are also present in this spectrum since collisions between He atoms and the ${}^{79,79}Br_2(B, v' = 12)$ products formed via VP of $He \cdots {}^{79,79}Br_2(B, v' = 13)$ complexes during the temporal delay between the pump and probe laser pulses, 20–50 ns, can result in vibrational relaxation down to the ${}^{79,79}Br_2(B, v' = 11)$ level. Furthermore, a fraction of the $He \cdots {}^{79,79}Br_2$ $(B, v' = 13)$ excited-state complexes can also undergo a $\Delta v = -2$ dissociation process forming $He + {}^{79,79}Br_2(B, v' = 11)$ products.

Nevertheless, the band-shift rule can be used to identify the geometries of each of the $He_2 \cdots {}^{79,79}Br_2(X, v'' = 0)$ conformers. Since the T-shaped $He \cdots {}^{79,79}Br_2$ feature associated with transitions of the $(1,0)He \cdots Br_2$ conformer is observed at $\approx +4\,\text{cm}^{-1}$ from the ${}^{79,79}Br_2$ monomer band origin, a feature associated with transitions of the $(2,0)He \cdots Br_2$ conformer, will appear at an additional approximately $+4\,\text{cm}^{-1}$ to higher transition energy. Thus, the feature at about $\approx +8\,\text{cm}^{-1}$ from the monomer band origin is assigned to transitions of the $(2,0)He \cdots Br_2$ conformer. Similarly, the addition of a He atom to the linear $(0,1)He \cdots Br_2$ conformer in the T-shaped well would produce a $(1,1)He \cdots Br_2$ complex, and transitions of it are observed at approximately $+4\,\text{cm}^{-1}$ from the linear band or at approximately $+14\,\text{cm}^{-1}$ from the monomer band origin.

The action spectrum plotted in Fig. 7c was recorded in the Br_2 B–X, 14–0 region. Features associated with transitions of $He_3 \cdots {}^{79,79}Br_2$ complexes are expected to become prominent in this spectrum since the $\Delta v' = -3$ product is being probed. Again following the band-shift rule the features associated with transitions of the $(3,0)He \cdots Br_2$ and $(2,1)He \cdots Br_2$ conformers are observed at about $+4\,\text{cm}^{-1}$ from the $(2,0)He \cdots Br_2$ and $(1,1)He \cdots Br_2$ features. Since the $(0,1)He \cdots Br_2$ feature is shifted by about $+10\,\text{cm}^{-1}$ from the monomer band origin, and the $(1,1)He \cdots Br_2$ feature is found at about $+14\,\text{cm}^{-1}$, the weak feature observed at about $+24\,\text{cm}^{-1}$ is tentatively attributed to transitions of the $(1,2)He \cdots Br_2$ ground-state conformer.

Similar action spectra have been recorded in the ICl B–X, v'–0 region to identify features associated with transitions of the different $(\#T,\#L)He \cdots ICl$ ground-state conformers [62]. Due to the limited number of bound $ICl(B, v')$ vibrational levels, $v' = 0$–3, the relatively weak ICl E–B, $v^\dagger$–v' transition strengths for low vibrational quanta in both states, and the spectral congestion in the region of the probe transitions, information concerning complexes larger than $He_n \cdots ICl$ is limited. Nevertheless, features associated with transitions of the higher order $(2,0)He \cdots ICl$, $(1,1)He \cdots ICl$, $(3,0)He \cdots ICl$, and $(2,1)$ $He \cdots ICl$ conformers were identified by acquiring actions spectra with the probe laser tuned to the appropriate ICl E–B, $v^\dagger$–v' transitions and using the band-shift rule.

III. Preferential Stabilization of Different Conformers of the Rare Gas–Dihalogen Complexes

While the concept of stereoisomers is commonly accepted for molecules, a corresponding definition at the quantum-mechanical level has not yet been formalized [63, 64]. A qualitative description of stereoisomers is often associated with a compound comprised of molecules that have the same number, kind, and connectivity of atoms, but differ in their spatial orientation. The quantum mechanical interpretation of this definition is that the PES characterizing a compound has multiple minima located at different structural orientations. If the wave function of the compound in a specific quantum state is localized in a minimum with negligible amplitude outside of that region, and if another state is localized in a different minimum, then the compound is said to have at least two stereoisomers. The concept of stereoisomers of van der Waals complexes is even harder to accept since the interactions are dominated by long-range electrostatic forces and, as a result, the moieties are typically weakly bound with significant zero-point motion. Consequently, van der Waals complexes are often found to be floppy with wave functions spanning wide regions of the PES, sampling varying intermolecular geometries and structures.

An examination of Figs. 3a and 5a reveals that the probability amplitude of the two lowest lying intermolecular vibrational levels, $n'' = 0$ and 1, in the $He + ICl(X, v'' = 0)$ potential and the three lowest levels, $n'' = 0$–2, in the $He + Br_2(X, v'' = 0)$ potential are well localized in different angular regions of the potential. The $He + I^{35}Cl(X, v'' = 0)$ and $He + {}^{79}Br_2(X, v'' = 0)$ potentials each have an appreciable barrier, ≈ 11 and $\approx 10 \, cm^{-1}$, respectively, between the T-shaped and linear wells, and in both cases $>99\%$ of the linear (T-shaped) probability amplitudes are found in the linear (T-shaped) wells. Thus, these results suggest that complexes localized in the T-shaped and linear regions of the intermolecular PES can be regarded as distinct stereoisomers, or conformers, of the complex. Since these two states represent distinct conformers, the conversion of T-shaped complexes into the linear form cannot be attributed to simple cooling of the intermolecular vibrational levels, and mechanisms must exist to bring about the isomerization of the complexes.

In 1999, the Klemperer group published experimental results from a dispersed fluorescence study of the linear and T-shaped $Ar \cdots I_2(X, v = 0)$ conformers to obtain information about the X- and B-state potentials of the complex and to gain insight into the photodissociation dynamics in the B electronic state [65]. In carrying out their analysis, they assumed that the linear and T-shaped conformers existed in a thermodynamic equilibrium, with the abundance of the linear conformer being three times that of the T-shaped conformer. In an effort to test the validity of the thermodynamic hypothesis Bastida, et al. [66], performed a molecular dynamics simulation of the isomerization of $Ar \cdots I_2(X, v'' = 0)$ complexes implementing the same expansion conditions used by the Klemperer group. The collision-induced population transfer between the linear and T-shaped $Ar \cdots I_2(X, v'' = 0)$ conformers was studied as a function of distance downstream from the nozzle orifice. The results revealed a strong agreement between the populations at each distance and those predicted using thermodynamics. The authors posited that kinetic effects do not play a significant role in the isomerization of the complex and that the populations of the conformers are under thermodynamic control. Despite the significant barrier between the two conformers, $\approx 100 \, cm^{-1}$, the simulation showed that isomerization could occur at collision energies of only a few wave numbers, and the authors proposed an exchange, or "swap cooling," mechanism to an explain these results. The concepts of a thermodynamic equilibrium and the collisional isomerization of the ground-state conformers are central to this chapter, and they will be addressed in greater detail in the next section.

A. Stabilization of Linear $Rg \cdots XY$ Triatomic Complexes: $He \cdots ICl(X, v'' = 0)$ and $He \cdots Br_2(X, v'' = 0)$

In initial work performed in our laboratory on $He \cdots ICl$, changes in the relative intensities of the T-shaped features and linear features were observed with varying backing pressures [38]. At lower helium backing pressures, the T-shaped features were much more intense than the higher energy features, but at much higher pressures, ≥ 65 psi, the higher energy features became more intense than the T-shaped bands. The localization of the T-shaped and linear probability amplitudes in the ground-state PES and the changes in the relative intensities of these features with varying gas density and collision frequency indicated that it was possible to manipulate the expansion conditions to control the relative abundances of the $He \cdots ICl$ ground-state conformers formed in the expansion. Furthermore, the relative populations of the conformers are not governed by the rates of formation of the different conformers, that is, under kinetic control, rather the system may proceed toward thermodynamic equilibrium. In order to address the notion of a thermodynamic equilibrium and to determine if the populations can be controlled, we carried out a detailed investigation of the $He \cdots ICl$ signal dependence on source conditions by systematically recording LIF spectra at several distances downstream from the nozzle orifice and using different He backing pressures [39, 67].

The relative intensities of the T-shaped and linear $He \cdots ICl$ features in the LIF and action spectra have a strong dependence on the downstream distance where the spectra are recorded, and thus on local temperature [39, 67], as is evident in Fig. 8. A shift by a factor of ≈ 20 in intensity from the T-shaped to the linear $He \cdots I^{35}Cl$ feature was observed as the downstream distance was varied when using a He backing pressure of 155 psi. While precise Franck–Condon

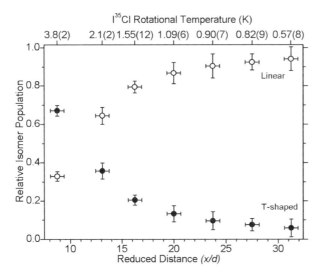

Figure 8. The relative peak intensities of the T-shaped and linear $He \cdots I^{35}Cl$ fluorescence excitation features at 17,831 and 17,842 cm^{-1}, respectively, are plotted as a function of reduced distance along the expansion. The $I^{35}Cl(X, v'' = 0)$ rotational temperature determined at each distance is shown on the top abscissa. Taken with permission from Ref. [67].

factors are not known for these transitions, it is clear that a significant change in the relative populations of the $He \cdots I^{35}Cl(X, v'' = 0)$ conformers occurs with cooling along the expansion. The observed shift in intensity also depends on backing pressure, spanning from ≈ 15 to ≈ 30 when increasing the He backing pressure from 95 to 215 psi. Since the rotational temperatures were nearly the same at a given downstream distance when using the different backing pressures, the pressure-dependent studies indicate that the relative populations of the conformers are not just dependent on local temperature, but also on the total number of collisions experienced by a complex.

Similar experiments performed on $He \cdots Br_2$ revealed the same trend as observed for the $He \cdots ICl$ complexes, with the intensity of the linear feature increasing relative to the T-shaped feature with increasing downstream distance and backing pressure [41]. Although the intensity changes were not as drastic as observed in the $He \cdots ICl$ spectra, the results showed that the $He \cdots ICl$ complex was not a special case and that stabilization of different conformers may be a more general attribute of $Rg \cdots XY$ complexes. For all of the $Rg \cdots XY$ systems studied thus far, see Table I, it was possible to manipulate the relative abundances of the T-shaped and linear conformers by manipulating the expansion conditions and by recording the spectra at different downstream distances.

These systematic studies not only demonstrated that the relative populations of different $Rg \cdots XY$ conformers are not kinetically trapped within the expansion, but they provide a simple method for determining whether spectral features may be associated with transitions of different conformers. Furthermore, monitoring the population shift as a function of local temperature in the expansion, perhaps at only two downstream distances, is a quick way to determine which of the ground-state conformers is more energetically stable. The population shift from the T-shaped to linear $He \cdots I^{35}Cl(X, v'' = 0)$ conformer [39, 67], and similarly for the $He \cdots^{79}Br_2(X, v'' = 0)$ conformers [41], with decreasing temperature indicated that the linear species is more strongly bound than the T-shaped for these species.

B. Stabilization of Multiple Conformers of Higher Order Complexes: $He_n \cdots ICl(X, v'' = 0)$ and $He_n \cdots Br_2(X, v'' = 0)$, $n \geq 2$

Action spectra were recorded at varying downstream distances and with different He backing pressures to investigate if the populations of the conformers of the higher order $Rg_n \cdots XY$ complexes could also be manipulated. The $\Delta v' = -2$ action spectra of the $He_n \cdots I^{35}Cl$ complexes recorded at two distances downstream from the nozzle orifice, $Z = 8.8$ and 19.1, corresponding to monomer rotational temperatures of 2.34(3) K and 1.09(10) K, respectively, are shown in Fig. 9a and b [62]. It is evident that the intensity of the feature

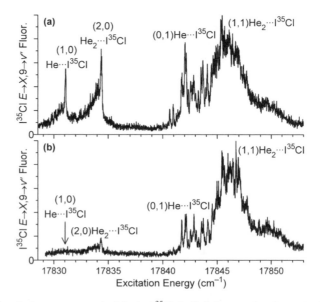

Figure 9. Action spectra acquired in the $I^{35}Cl$ B–X, 3–0 spectral region and with the probe laser tuned to the $I^{35}Cl$ E–B, 9–1 transition. Both spectra were recorded using the same source conditions, but with the lasers intersecting the expansion at $Z = 8.8$, (a), and $Z = 19.1$, (b). Monomer rotational temperatures of 2.34(3) K and 1.09(10) K were measured at the two distances [62].

affiliated with transitions of the toroidal $(2,0)He_2 \cdots I^{35}Cl(X, v'' = 0)$ conformer decreases drastically relative to the $(1,1)He_2 \cdots I^{35}Cl$ feature with decreasing temperature, thereby indicating that the $(1,1)He_2 \cdots I^{35}Cl(X, v'' = 0)$ conformer is more strongly bound. The $He_2 \cdots {}^{79}Br_2(X, v'' = 0)$ conformers exhibited the same trend [62], with the $(1,1)$ conformer being more strongly bound than the $(2,0)$ conformer, which is in agreement with recent theoretical calculations [68]. Unfortunately, the poor signal-to-noise levels in the $\Delta v' = -3$ action spectra precluded a definitive determination of the more stable $He_3 \cdots {}^{79}Br_2$ complex.

High level *ab initio* calculations of the $He_2 + ICl(X, v'' = 0)$ and $He_2 + Br_2(X, v'' = 0)$ interactions, as well as quantum mechanical calculations of the bound $He_2 \cdots ICl$ and $He_2 \cdots Br_2$ intermolecular vibrational levels offer valuable insights into the interactions of these higher order complexes [68, 69]. For the $He_2 \cdots ICl$ complex, even though there are five minima in the PES, the lowest five intermolecular vibrational levels are found to be localized in wells with the following geometries: (1) a police nightstick, or the $(1,1)He_2 \cdots ICl$ geometry using our notation; (2) a linear $He \cdots I—Cl \cdots He$, or $(0,2)He_2 \cdots ICl$ geometry; (3) and (4) a tetrahedral, or $(2,0)He_2 \cdots ICl$ geometry; and (5) a second police nightstick geometry with the He localized at the Cl end, or $(1,1)ICl \cdots He_2$ [69].

The probability distributions for the first three levels indicate that the He atoms are quite localized in each of the minima, and these should be thought of as distinct conformers of the $He_2 \cdots ICl$ complex based on similar arguments as described above for the lowest levels of the ground-state $Rg \cdots XY$ complexes. The most stable conformer, police nightstick conformer, is predicted to be bound by 33.51 cm^{-1} with the linear and tetrahedral conformers lying only 1.91 and 3.05 cm^{-1} to higher energy [69]. Recent work completed in our lab indicates that the $(1,1)He_2 \cdots ICl$ conformer is the most stable observed conformer with a binding energy of $38.4(9)–41.3(1.0)$ cm^{-1} [70]. The tetrahedral $(2,0)He_2 \cdots ICl$ is definitely less strongly bound, but we currently do not have an estimate of the relative binding energy of it compared to the $(1,1)He_2 \cdots ICl$ conformer. Note that we do not see any evidence of a linear $(0,2)He_2 \cdots ICl$ conformer.

The calculations for $He_2 \cdots Br_2$ predict that the linear $(0,2)He_2 \cdots Br_2$ conformer is the most strongly bound, followed by the police night stick $(1,1)He_2 \cdots Br_2$, and the tetrahedral $(2,0)He_2 \cdots Br_2$ conformers [68]. The binding energies for these conformers are similar to those of $He_2 \cdots ICl$ with energies of 32.24, 32.44, and 30.93 cm^{-1}, respectively. Recall for $He_2 \cdots Br_2$ we only see spectroscopic evidence of the $(1,1)He_2 \cdots Br_2$ and $(2,0)He_2 \cdots Br_2$ conformers.

The experimental results on $He_2 \cdots ICl$ and $He_2 \cdots Br_2$ demonstrate that by varying the expansion conditions it is possible to manipulate the relative abundances of the higher order complexes and drive the ground-state population to the more energetically stable configuration. The stabilization of multiple $Rg_n \cdots XY$ conformers suggests that the influence of the multibody interactions on the photodissociation dynamics can be interrogated with rare gas atoms localized in different regions around the dihalogen, and such investigations are not restricted to the case where all of the atoms are confined to the same region of space. Also, the similarity between the contours of the linear $\text{He} \cdots \text{XY}$ bands and the features of the nontoroidal conformers suggests that these complexes also access delocalized intermolecular vibrational levels in the excited-state. Thus, these dynamical studies could be performed with varying amounts of internal motion and angular momenta. A brief summary of the dissociation dynamics of $\text{He}_2 \cdots \text{I}^{35}\text{Cl}(B, v')$ complexes prepared with varying amounts of intermolecular vibrational excitation within the $\text{He} + \text{He} \cdots \text{I}^{35}\text{Cl}(B, v' = 3)$ potential is presented at the end of section IV.C [70].

C. Formation of $\text{Rg} \cdots \text{XY}$ van der Waals Complexes in Supersonic Expansions

The use of supersonic expansions to stabilize weakly bound complexes is well documented, and the physics within the expansions has been intensively investigated [27]. A typical supersonic expansion is established using a high pressure gas reservoir comprised of a small concentration of the molecule of interest entrained in a monatomic carrier gas and a small orifice that allows escape of the gas to vacuum. The hydrodynamic flow through the orifice converts the random motion in the high pressure region behind the orifice to a directed mass flow in the vacuum, resulting in a quick decrease in temperature, collision frequency, and density with distance from the orifice. Weakly bound van der Waals complexes are most efficiently stabilized just downstream of the orifice via three-body collisions, where a dihalogen molecule and rare gas atom are stabilized in the complex and the excess energy liberated when forming the complex is taken away by a another rare gas atom. In this region of the supersonic expansion, the collision frequency is sufficient for three-body collisions to occur while the local temperature is low enough that the collisions do not dissociate existing complexes.

The density of the jet decreases rapidly with increasing distance downstream from the nozzle orifice, and eventually the multibody collisions become much less probable than two-body collisions. Within this region of the expansion, the probability for forming new complexes falls off quickly, but two-body collisions serve to cool the complexes as internal energy of the complexes is transferred to kinetic energy of the colliding rare gas atoms. The gas density continues to drop with increasing distance and eventually the expansion will transition to a molecular flow that is essentially free of collisions. Within this collision-free regime, the cooling of the complexes should cease and their internal energy distributions should be frozen. However, if the evolution of the expansion into the collision-free regime is gradual the conditions favoring two-body collisions may be sustained for a time sufficient to allow the complexes to be cooled to the lowest intermolecular levels within the T-shaped and linear wells. Additionally, the collisions may induce isomerization converting one conformer into the other.

D. Isomerization Mechanisms

Simulations of the isomerization of the linear and T-shaped $Ar \cdots I_2(X, v'' = 0)$ conformers offer insights into two mechanisms by which collisions with rare gas atoms may convert population of one conformer to the other even at very low temperatures [66]. The first mechanism is a conventional isomerization reaction, depicted in Fig. 10a. A $He^{(a)} \cdots XY(X, v'' = 0)$ complex collides with a second helium atom, $He^{(b)}$. The colliding $He^{(b)}$ atom gains kinetic energy during the collision encounter because of the intermolecular attraction associated with the $He^{(b)} + He^{(a)} \cdots XY(X, v'' = 0)$ atom-complex van der Waals potential. This kinetic energy results in an increase in the internal energy of the collision complex, perhaps above the barrier along the angular coordinate, so that the $He^{(a)}$ atom can move from one minimum to the other. The colliding $He^{(b)}$ atom then takes away the excess energy as the complex relaxes into the other conformeric form.

The second mechanism, Fig. 10b, is an exchange, or "swap", mechanism. In this case, the colliding $He^{(b)}$ atom approaches an unoccupied minimum region of the intermolecular PES forming a quasistable $He^{(b)} \cdots He^{(a)} \cdots XY$

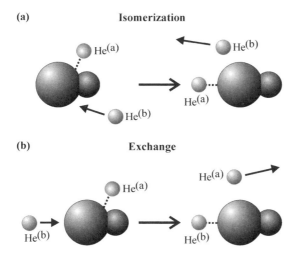

Figure 10. The transfer of population between the T-shaped and linear conformers may occur via two reaction mechanisms [66], isomerization (a) and exchange (b), even at the very low temperatures, $< 2\,K$, observed in the expansion at $x/d > 10$. Adapted from Ref. [67].

$(X, v'' = 0)$ collision complex. The internal energy of the collision complex may be greater than the $He^{(a)} + XY(X, v'' = 0)$ binding energy, and the ejection of the $He^{(a)}$ atom and the formation of a $He^{(b)} \cdots XY(X, v'' = 0)$ complex would result. The "swap" mechanism may be particularly important when the collisional energy decreases, and the $Ar \cdots I_2(X, v'' = 0)$ simulations indicate that in the regions of coldest collisions the exchange mechanism may account for 50% of the total isomerization [66].

E. Relative Binding Energies

The ability to detect discrete rovibronic spectral features attributed to transitions of two distinct conformers of the ground-state $Rg \cdots XY$ complexes and to monitor changing populations as the expansion conditions are manipulated offered an opportunity to evaluate the concept of a thermodynamic equilibrium between the conformers within a supersonic expansion. Since continued changes in the relative intensities of the T-shaped and linear features was observed up to at least $Z = 41$ [41], the populations of the conformers of the $He \cdots ICl$ and $He \cdots Br_2$ complexes are not kinetically trapped within a narrow region close to the nozzle orifice. We implemented a simple thermodynamic model that uses the ratios of the peak intensities of the conformer bands with changing temperature in the expansion to obtain experimental estimates of the relative binding energies of these complexes [39, 41].

The model [39] was developed using three assumptions: the conformers are in thermodynamic equilibrium, the peak intensities of the T-shaped and linear features are proportional to the populations of the T-shaped and linear ground-state conformers, and the internal energy of the complexes is adequately represented by the monomer rotational temperature. By using these assumptions, the temperature dependence of the ratio of the intensities of the features were equated to the ratio of the quantum mechanical partition functions for the T-shaped and linear conformers (Eq. (7) of Ref. [39]). The ratio of the $He \cdots I^{35}Cl$ T-shaped:linear intensity ratios were observed to decay single exponentially. Fits of the decays yielded an approximate ground-state binding energy difference, $\Delta E = D_0''^{(T)} - D_0''^{(L)}$, of 2.5(6) cm^{-1}. This value was in good agreement with the theoretical value, 3.14 cm^{-1}, calculated by McCoy and co-workers [40].

Based on the strong agreement between the calculated and experimental values for the ΔE values obtained by using this simplistic thermodynamic model, it was tantalizing to make the assertion that the conformers are in thermodynamic equilibrium within an expansion. However, several factors, including the use of peak intensities rather than integrated intensities, and ignoring possible variations of the Hönl–London factors and fluorescence quantum yields between the two conformers, could have introduced significant errors in the experimental values [39]. Furthermore, while the rotational temperatures of the T-shaped conformers, which are preferentially formed in the early regions of the expansion, seem to match the monomer values, we were not able to measure accurate rotational temperatures of the linear conformers. It is possible that the mechanisms that convert population from the T-shaped conformer to the lower energy linear conformer result in non-Boltzmann population distributions, thereby contradicting one of the assumptions of this thermodynamic model.

In part, to test whether the populations of the two conformers are in thermodynamic equilibrium, we recently undertook experiments aimed at directly measuring the relative binding energies of the T-shaped and linear $He \cdots I^{35}Cl(X, v'' = 0)$ conformers [70]. These experiments utilized resonant, two-photon transitions of the ground-state conformers to common intermolecular levels in the ion-pair states. The difference in the observed transition energies is a direct measure of the relative stability of the ground-state conformers, ΔE. The ΔE for the T-shaped and linear $He \cdots I^{35}Cl(X, v'' = 0)$ conformers determined in this manner is definitively $5.4(3)$ cm^{-1} with the linear conformer more strongly bound that the T-shaped conformer [70]. This difference is considerably larger than the $2.5(6)$-cm^{-1} value estimated using the thermodynamic model [39]. These results show that while the populations of the $Rg \cdots XY$ conformers are not kinetically trapped, they are also not in thermodynamic equilibrium. Nevertheless, observing the temperature dependence of the relative intensities of the vibronic features attributable to transitions of different ground-state conformers does reveal which conformer is energetically more stable.

As mentioned above, for all of the systems listed in Table I the relative intensities of the linear and T-shaped features were observed to change with downstream distance. While the most drastic intensity changes were observed for $He \cdots ICl$, sufficient changes were also observed in $He \cdots Br_2$ and $He \cdots I_2$ so that the thermodynamic model could be used to estimate the relative binding energies of the conformers for these systems. The linear $He \cdots Br_2(X, v'' = 0)$ conformer is $\approx 0.4(2)$ cm^{-1} more strongly bound than the T-shaped conformer [41]. The only system for which we found the T-shaped species to be the more strongly bound is $He \cdots I_2(X, v'' = 0)$ with $\Delta E \approx 0.3(2)$ cm^{-1} [53], determined using the resonant, two-photon scheme.

Results from our efforts and from earlier reports indicate that the differences in the binding energies of the T-shaped and linear $Ne \cdots ICl(X, v'' = 0)$, $Ne \cdots I_2(X, v'' = 0)$, and $Ar \cdots I_2(X, v'' = 0)$ conformers are significantly larger than of the $He \cdots XY$ systems. We expect that the large mismatch in the energies of the conformers would make it even more difficult for these systems to reach a thermodynamic equilibrium. Furthermore, these complexes were stabilized in mixed carrier gases, Ne in He or Ar in He, in order to reduce the propensity for forming higher order complexes. In so doing, the number of

Ne + Ne $\cdots$ XY and Ar + Ar $\cdots$ I$_2$ collisions within the expansion was greatly reduced and opportunities for isomerization events minimized [55]. Consequently, we made no attempt to apply the thermodynamic model to these heavier rare gas atom systems.

IV. INTERROGATING DYNAMICS ON VARYING REGIONS OF THE EXCITED-STATE SURFACE

The preceding discussion demonstrates that the conditions of supersonic expansions can be manipulated to stabilize multiple conformers of the ground-state, Rg $\cdots$ XY complexes. These discussions focused on the He$_n$ $\cdots$ I^{35}Cl $(X, v'' = 0)$ and He$_n$ $\cdots$ ^{79}Br$_2(X, v'' = 0)$ complexes with $n \geq 1$, but thus far there are no indications that other Rg$_n$ $\cdots$ XY systems will not exhibit the same general attributes. The Franck–Condon windows associated with transitions of the different ground-state conformers can provide access to multiple regions of the excited-state PESs so that a more thorough understanding of collision dynamics and energy-transfer mechanisms can be developed. Thus far, the electronically excited Rg $\cdots$ XY complexes have become the focus of dynamics investigations since depending on the amount of XY vibrational excitation, the electronic character of the XY molecule, and the size of the Rg atom, it is possible to study "simple" direct vibrational predissociation, dissociation through multiple, sequential intermediates, electronic predissociation, or statistical IVR. Thus, it was the early goal of investigations on the Rg $\cdots$ XY systems to build models and establish general propensities for intermolecular interactions of molecules, in general.

Part A of this section includes a brief overview of previous work undertaken to characterize the excited-state dynamics of the Rg $\cdots$ XY complexes induced from the T-shaped or toroidal geometry. While this material has recently been thoroughly reviewed [14], we include the results of frequency and time-domain experiments, including typical observables and expectations, and of theoretical efforts that provide general information on the underlying physical phenomena to illustrate what can be learned from these systems. To date, the linear He $\cdots$ I^{35}Cl$(X, v'' = 0)$ conformer has been the most extensively studied linear Rg $\cdots$ XY system, and the ongoing work interrogating the excited-state dynamics of this complex may provide additional insights into the role of intermolecular orientation and energy on dynamics and energy-transfer mechanisms. In part B, we describe recent dissociation dynamics results obtained by promoting the linear He $\cdots$ ICl conformer to varying regions on the excited-state PES. Specifically, we present results from VP studies that indicate the role of intermolecular vibrational excitation and the delocalization of the excited-state wave functions on the dynamics [51]. Transitions of the linear conformer also access the region of the excited-state PES near the dissociation limit, and both direct dissociation and rotational predissociation are observed [51]. Lastly, in part C, the dissociation dynamics of the conformers of the higher order complexes are briefly summarized.

A.   Photoinitiated Dynamics of T-shaped $Rg \cdots XY$ $(X, v'' = 0)$Complexes

1.   Vibrational to Translational Energy Transfer: Experimental Results

The rare gas–homonuclear dihalogen interactions are one of the simplest classes of van der Waals interactions, since they do not have permanent electrostatic interactions and are dominated by long-range van der Waals attraction and exchange repulsion [13]. Consequently, a majority of the experimental and theoretical efforts undertaken over the past 30 years have focused on characterizing the dynamics and interactions of these systems. The T-shaped $He \cdots I_2$ complex was the first triatomic van der Waals complex detected in LIF spectroscopy experiments [71]. The observed blue shifts of the T-shaped features from each of the I_2 $B-X$, $v' - 0$ band origins was approximately constant over the vibrational regions spanning from $v' = 3-29$. This constant shift in transition energy is indicative of the excited-state $He \cdots I_2(B, v')$ complex having nearly the same binding energy regardless of the amount of iodine vibrational excitation and of a weak interaction between the $He-I_2$ intermolecular vibrational mode and the $I-I$ molecular vibrational mode. An additional suggestion of this weak coupling is the narrow line widths of the T-shaped features in the fluorescence excitation spectrum. The line widths indicate that complexes prepared in the $n' = 0$ intermolecular vibrational level undergo predissociation on time scales decreasing monotonically from 221(20) ps at $v' = 12-37.9(1.3)$ ps at $v' = 26$ [72], solid circles in Fig. 11, corresponding to ≈ 1000 to ≈ 90 iodine vibrational periods. Since the v'-dependent decrease in the lifetime is not accompanied by a quenching of the fluorescence, the principal decay mechanism for $He \cdots I_2(B, v')$ is VP, not electronic predissociation. In partial support of this conclusion, the Levy group showed that the $\Delta v' = -1$ VP channel accounted for 98% of the observed fluorescence [74].

An investigation of the $Ne \cdots I_2$ complexes within the same spectral regions yielded similar results [18]; a nearly constant blue shift of the spectral features from the monomer band origins and excited-state lifetimes that are long compared to the molecular vibrational period, solid squares in Fig. 11. The vibrational state distributions measured following VP of $Ne \cdots I_2(B, v')$ also showed a propensity for forming products in the $\Delta v' = -1$ product channel. The major difference in the dynamics of the complexes through the $v' = 13-26$ region was that for a given v', the lifetime of the $Ne \cdots I_2$ complex is shorter than $He \cdots I_2$.

Zewail and co-workers performed a series of time-resolved experiments characterizing the vibrational predissociation dynamics of $He \cdots I_2(B, v')$,

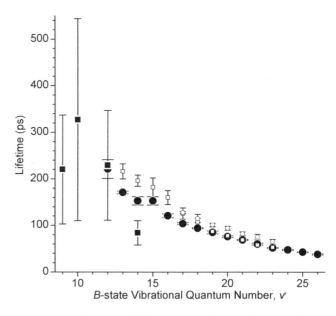

Figure 11. Experimental $He \cdots I_2(B,v')$ and $Ne \cdots I_2(B,v')$ excited-state lifetimes, circles and squares, plotted as a function of vibrational quanta, v'. The values taken from line width [72] and time-domain [73] measurements are shown as solid and open symbols, respectively.

$Ne \cdots I_2(B, v')$, and $Ne_n \cdots I_2(B, v')$, with $n = 2-4$ [23, 73, 75] and compared their results with the frequency-domain studies of the Levy group [72]. Using picosecond, pump–probe spectroscopy they monitored the rate of appearance of the $I_2(B,v)$ fragments in a specific $v < v'$ decay channel. The measured VP rates for $He \cdots I_2(B, v')$ and $Ne \cdots I_2(B, v')$ also exhibited nonlinear increase with increasing v', open circles and squares in Fig. 11, respectively. These rates are consistently lower than obtained from the line width measurements, which may include inhomogeneous broadening and thus represent upper limits. A particularly interesting result of the time-resolved experiments was the appearance of rotational coherences in the $He \cdots I_2$ and $Ne \cdots I_2$ transients corresponding to the formation of nascent $I_2(B, v' - 1)$ fragments. Although the VP processes in $He \cdots I_2$ and $Ne \cdots I_2$ have been shown to be dominated by $V-T$ energy transfer, the presence of the coherences underscored the importance of considering the effects of rotation on the VP dynamics.

2. Vibrational to Translational Energy Transfer: Theoretical Results

The $He \cdots I_2$ and $Ne \cdots I_2$ systems have become representative systems for investigating VP dynamics, wherein fragmentation proceeds via a direct coupling of the initially prepared quasistable state of the complex to the $\Delta v' = -1$ translational continuum. The nonergodic nature of VP makes these complexes very attractive to theoreticians, and they quickly became benchmark systems for testing theoretical methodologies for unimolecular dissociation [76]. Beswick and Jortner were the first to consider the VP of triatomic van der Waals complexes, developing an approach that regarded the predissociation process as a half-collision event involving vibrational-to-translational, V–T, energy transfer [77–79]. Treating this process as a pure V–T mechanism was justified since the small rotational constant of iodine makes vibrational to rotational energy transfer, V–R, improbable. This was initially substantiated by Beswick and Delgado–Barrio, who first incorporated the effects of rotation in $\mathrm{He} \cdots \mathrm{I}_2$ vibrational predissociation using a rotational decoupling scheme [80].

Within the context of scattering theory, Beswick and co-workers [77–79, 80, 82] employed a perturbative Golden Rule treatment that considers the decay of the complex into a manifold of coupled translational continua to determine the VP decay rates from the energy dependence of the scattering matrix. This approach resulted in the well-known energy [77, 79] and momentum gap laws [83] stating that the rate of VP is determined by the translational energy, E_{trans}, or momentum of the fragments after dissociation and, that the rates for VP decrease with increases in E_{trans}. Under the Golden Rule approximation, the decay rate is proportional to the overlap integral between the initially prepared, quasibound state of the complex and the continuum wave functions of the dissociation product states. At larger translational energies, the continuum wave function oscillates more rapidly, thereby resulting in a smaller overlap integral and a smaller rate of predissociation.

Given that the $\mathrm{He} \cdots \mathrm{I}_2$ and $\mathrm{Ne} \cdots \mathrm{I}_2$ excited-state intermolecular binding energies are approximately constant throughout the B-state, the anharmonicity of the $\mathrm{I}_2(B)$ potential results in a decrease in the energy available to the product translational degrees of freedom, and, as a result, a decrease in the VP rate with increasing v'. The observed excited-state lifetimes of the $\mathrm{He} \cdots \mathrm{I}_2(B, v')$ and $\mathrm{Ne} \cdots \mathrm{I}_2(B, v')$ complexes are in accord with the energy gap law prediction.

The experimental results on $\mathrm{He} \cdots \mathrm{I}_2$ provide targets for improving theoretical techniques including numerous classical and quasiclassical trajectory methods [84–88] and both time-dependent and time-independent quantum methods [76, 78, 81, 89–93]. While much of this methodology did not include rotational effects, it became apparent that even for inherent V–T processes introducing the rotational degrees of freedom could have a substantial effect on the decay widths, and that neglecting rotation results in an underestimation of the complex lifetimes [88]. The need for more thorough calculations was further clarified by Zhang and Zhang [93], who carried out two- and three-dimensional quantum wave packet calculations on the dissociation dynamics of $\mathrm{He} \cdots \mathrm{I}_2$.

The inclusion of the iodine rotations resulted in an effective centrifugal barrier in the potential that pushed the continuum wave function out of the bound-state region, thereby reducing the overlap integral and decreasing the VP rate.

3.   Vibrational to Rotational Energy Transfer

The frequency-domain line width and time-domain, pump–probe measurements provide information about the lifetimes of the complexes, but they tell us little about the nature of the dynamics incurred. Processes involving the partitioning of excess energy into rotational, as well as translational, degrees of freedom, $V–R$ and $V–T,R$, are more common than simple $V–T$ energy transfer, and, since the rotational excitation of the fragments is due to coupling with anisotropic terms of the intermolecular potential [94, 95], the ability to examine the disposal of energy into fragment rotation is vital to our understanding of the underlying forces governing the dynamics.

Lester and co-workers first demonstrated the use of a two-laser, pump–probe technique to record the product rotational state distributions of nascent ICl* fragments following photodissociation of Ne $\cdots$ ICl and He $\cdots$ ICl complexes from $n' = 0$ intermolecular vibrational levels in the ICl $B^3\Pi_{0+}$ and $A^3 \Pi_1$ electronic states [17, 46, 48, 96]. The VP dynamics of both complexes, which were found to be dominated by the $\Delta v' = -1$ decay channel and obey the energy-gap law, resulted in slightly different product rotational distributions. The measured $ICl(A, v', j')$ rotational distributions following excitation of the Ne $\cdots$ ICl complex to a range of vibrational levels in the A electronic state were inverted, with a local maxima at high j' values, quite narrow, and were found to scale with the square root of energy released into dissociation [97]. The rotational distributions measured for VP of Ne $\cdots$ ICl(B, v') exhibited a similar behavior [46].

In contrast, the rotational-state distribution measured with the excitation laser fixed on the band head of the T-shaped He $\cdots$ ICl feature in the ICl $B–X$, 3–0 region was distinctly bimodal [46, 48]. Our group repeated these measurements with slightly better frequency resolution so that single He $\cdots$ ICl rotational transitions could be excited and rotational averaging from the complex was minimized [51]. The $ICl(B, v', j')$ product state distribution is plotted in Fig. 12b and is only slightly more structured than observed previously. Lester and co-workers [46, 48] noted that such a distribution is consistent with an impulsive inelastic scattering picture of the VP dynamics. Furthermore, the features of the product rotational-state distribution are analogous to classical rotational rainbows observed in energy and angular distributions of differential cross-sections. By using semiclassical scattering theory, Waterland, et al. [48] qualitatively reproduced the rotational distributions and provided evidence of final-state interactions during fragmentation that arise from the angular forces, and thus the anisotropy of the PES, as the fragments separate. Even though the

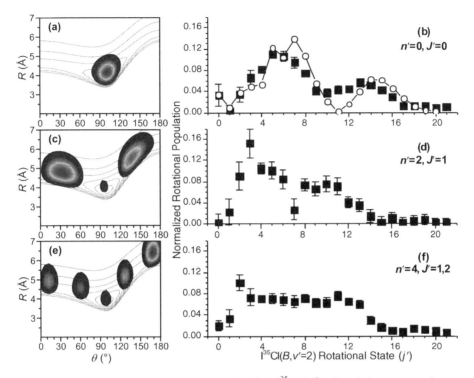

Figure 12. Potential energy contour plots for He + I^{35}Cl($B,v' = 3$) and the corresponding probability densities for the $n' = 0$, 2, and 4 intermolecular vibrational levels, (a), (c), and (e) plotted as a function of intermolecular angle, θ and distance, R. Modified with permission from Ref. 40. The I^{35}Cl($B,v' = 2,j'$) rotational product state distributions measured following excitation to $n' = 0$, 2, and 4 within the He + I^{35}Cl($B,v' = 3$) potential are plotted as black squares in (b), (d), and (f), respectively. The populations are normalized so that their sum is unity. The I^{35}Cl($B,v' = 2,j'$) rotational product state distributions calculated by Gray and Wozny [101] for the vibrational predissociation of He ··· I^{35}Cl($B,v' = 3,n' = 0,J' = 0$) complexes are shown as open circles in panel (b). Modified with permission from Ref. [51].

probability density for the $n' = 0$ level is localized in the T-shaped well, Fig. 12a, the potential in this region is very anisotropic and forces resulting in different angular momenta can result. Within a classical context, the observed maxima in the rotational distribution correspond to the recoiling He atom feeling these different forces and scattering off the PES near the I and Cl ends of the ICl molecule. Thus, the positions of the peaks observed in the rotational distributions can be directly related to the anisotropy of the PES sampled by the products. Lester and co-workers [48] further noted that the characteristics of the rotational distributions could also be attributed to the preparation of the initial state, for example, the zero-point energy of the complexes or the nature of the quasistable state coupling to the dissociative surface. However, full-dimensional quantum mechanical calculations on Ne $\cdots$ ICl [98] and He $\cdots$ ICl [49] showed that these factors contributed very little to the distributions and corroborated the general assertions of the classical calculations that the rotational excitation of the ICl fragments arises from final state interactions, that is, the torque imparted on ICl by the departing He atom.

It is interesting to compare the He $\cdots$ ICl results with those obtained for the He $\cdots$ Cl$_2$ and Ne $\cdots$ Cl$_2$ complexes. Although the Cl$_2(B,v',j')$ rotational distributions were also distinctly bimodal, they exhibited a few characteristics not observed in the fragment ICl rotational distributions of the He and Ne $\cdots$ ICl complexes. First, the Cl$_2(B,v)$ rotational distributions with $v < v'$ exhibited an alternating intensity between even and odd rotational levels [99, 100]. Cline, et al. [99], pointed out that the symmetry of the intermolecular potentials preserves the parity of the initial complex wave function in the parity of the Cl$_2(B,v)$ product rotational state, and the observed intensity alteration is due to the different weighting of the parity states imposed by spin statistics. Second, the Cl$_2(B,v)$ rotational distributions from both complexes were found to be relatively independent of the Cl$_2(B,v')$ vibrational quantum number. For example, all of the rotational distributions resulting from the dissociation of He $\cdots$ Cl$_2$ in the Cl$_2$ B–X, 6–0 through 10–0 spectral regions exhibited maxima at $j' = 4$ and $j' = 20$ [see Fig. 3 of Ref. (100)]. Third, the rotational distributions from Ne $\cdots$ Cl$_2(B,v')$ and He $\cdots$ Cl$_2(B,v')$ are somewhat invariant to the amount of energy released to the fragments. This is best illustrated in Fig. 9 of Ref. [98], which shows similar rotational distributions in the $\Delta v' = -1, -2$, and -3 channels following the dissociation of the He $\cdots$ Cl$_2(B,v' = 12)$ complex. Despite the factor of 3 increase in energy released to the products between the $\Delta v' = -1$ and -3 channels, all three plots have the same two maxima, a minimum at $j' = 6$ and the same j_{max} value. Of course, the parity conservation constraints are not expected in a complex containing a heteronuclear dihalogen moiety, such as He $\cdots$ ICl or Ne $\cdots$ ICl, but the last two attributes are not consistent with the impulsive half-collision predissociation mechanism used to describe the fragmentation of He $\cdots$ ICl and Ne $\cdots$ ICl.

The Franck–Condon model, which decomposes the initial quasistable state of the wave function in terms of free-rotor states and gives the product rotational distribution if no torques are present, was applied to Ne $\cdots$ Cl$_2$ and He $\cdots$ Cl$_2$ and correctly predicted the low j' behavior of the distribution, as well as the invariance of the distributions in the $\Delta v' = -1$ and -2 channels [99, 100]. However, this model could not accurately account for the behavior of the distributions at high j' nor could it account for the invariance of the minima, or nodes, in the distributions. It was proposed that a more sophisticated version of the Franck–Condon model might produce better overall agreement with the experimental data. Cline et al. [99], noted that the breakdown of the quasiclassical, impulsive model made it quite difficult to clearly interpret the origin of the bimodal distributions and proposed that both quantum interference and rotational rainbow effects are likely involved.

Gray and Wozny [101, 102] later disclosed the role of quantum interference in the vibrational predissociation of $He \cdots Cl_2(B, v', n' = 0)$ and $Ne \cdots Cl_2(B, v', n' = 0)$ using three-dimensional wave packet calculations. Their results revealed that the high j' tail for the VP product distribution of $Ne \cdots Cl_2(B, v')$ was consistent with the final-state interactions during predissociation of the complex, while the node at in the $He \cdots Cl_2(B, v') \Delta v' = -1$ rotational distribution could only be accounted for through interference effects. They also implemented this model in calculations of the VP from the T-shaped $He \cdots I^{35}Cl(B, v' = 3, n' = 0)$ intermolecular level forming $He + I^{35}Cl(B, v' = 2)$ products [101]. The calculated $I^{35}Cl(B, v' = 2, j')$ product state distribution remarkably resembles the distribution obtained by our group, open circles in Fig. 12(b).

Their proposed model [101, 102] states that the rotational product state distribution depends on the product of the probability density of the initially prepared quasibound state with a classical force induced by the $Rg + XY$ product–channel surface, and short-time dynamics along the exit channel. The first step involves the evolution of the initially prepared quasibound state of the complex onto the lower energy potential. This is the rate-limiting step and a superposition of states, or wave packets, that are localized in two different angular regions of the PES results. Each wave packet quickly propagates out the exit channel, forming rotational state distributions indicative of the forces sampled from that region of the potential. It is the interference between the two wave packets at large intermolecular distances that gives rise to the structure in the product rotational state distributions.

For $Ne \cdots Cl_2$, the two regions where the wave packets originate were confined to one angular region around the Cl_2 moiety, and the wave packets propagated separately from these regions yielding roughly the same rotational distributions [101]. No interference effects were observed. In contrast, two distinct angular origination regions were observed for $He \cdots Cl_2$ and propagating the wave packets separately resulted in two different rotational distributions, one peak at low j' and one peaked at high j'. However, the weighted sum of these two distributions did not correspond very well with the experimental results [101]. It was only by simultaneously propagating both origination regions that the structure of the Cl_2 rotational distribution was faithfully reproduced, with the quantum interference between the two wave packets giving rise to the distinctive node in the distribution. The agreement of the calculations with the experimental data, Fig. 12b, indicates similar dynamics for the vibrational predissociation of the T-shaped $He \cdots I^{35}Cl(B, v' = 3, n' = 0)$ complex [101].

4. Intramolecular Vibrational Energy Redistribution

The preceding discussion was limited mostly to VP processes occurring by direct coupling of the quasibound state of the complex to the dissociative continuum, which is the simplest and most commonly observed decay route for the complexes. However, these systems also serve as ideal venues for studying an array of more complicated dynamical processes, including IVR, and electronic predissociation. This brief section will focus on the former, underscoring some of the inherent dynamical differences between $Rg \cdots XY$ complexes by discussing the IVR behavior of a few systems.

Triatomic complexes containing He and Ne are generally characterized by small excited-state binding energies relative to the energy available from the dihalogen stretch and, in accord with energy gap laws, dissociation in the lower vibrational levels occurs predominantly by direct coupling to $\Delta v' = -1$ translational continuum. However, the anharmonicity of the dihalogen potential reduces the magnitude of the vibrational energy transferred to the inter-molecular mode with increasing v' perhaps causing the $\Delta v' = -1$ channel to close in well-defined energy regions. In these regions, the interplay between VP and IVR can become significant as decay may proceed via a direct coupling to the $\Delta v' = -2$ continuum or via an indirect coupling to the $\Delta v' = -2$ continuum through an intermediate, doorway state [103]. Because the doorway state corresponds to a vibrationally excited intermolecular mode in the $v' - 1$ manifold, the nodal pattern of the wave function is more complicated than in the initial state and the coupling of the intermediate state to the continuum generally produces highly structured and irregular rotational product state distributions. Also, a different doorway state may be accessed for each initial dihalogen vibrational level, resulting in an erratic dependence of the excited-state lifetimes and spectral shifts on v'. These signatures of IVR-mediated VP were observed for the $He \cdots Br_2$ complex in the highest Br_2 vibrational levels, $v' > 40$, making it the first triatomic complex containing He or Ne for which IVR was shown to play an important role [52, 104, 105].

Although energy conservation constraints dictate which VP channels are open, it is the nature of the intermolecular interactions, the density of states and the coupling strengths between the states that ultimately dictate the nature of the dynamics and the onset of IVR. These factors are dependent on the particular combinations of rare gas atom and dihalogen molecule species constituting the complex. For example, Cline et al. showed that, in contrast to $He \cdots Br_2$, $\Delta v' = -2$ VP in the $He \cdots Cl_2$ and $Ne \cdots Cl_2$ complexes proceeds via a direct coupling of the quasibound state to the $v' = -2$ continuum rather than by two sequential $\Delta v' = -1$ steps [99, 100]. Furthermore, drastically different intermolecular interactions and more complicated dissociation dynamics are found for $Ar \cdots Cl_2$. Not only does IVR take place in the $Ar \cdots Cl_2$ complex

[106, 107] but significant changes in the density of intermediate states and coupling strengths with v' give rise to several different IVR regimes depending on the Cl_2 vibrational level that is accessed during excitation [108].

The dissociation dynamics of $Ne \cdots Br_2(B, v')$ has been investigated using both frequency- and time-resolved spectroscopy by Cabrera, et al. [109]. They found that the excited-state lifetimes of the complexes deviate from those predicted with the energy-gap law, leveling off in the range of 10–20 ps for $v' > 20$. Interestingly, in double-resonance experiments monitoring the decay of the quasistable $Ne \cdots Br_2(B, v')$ complex, only an upper limit of 10 ps could be set for the complex in the $v' = 26$ level. The $Br_2(B, v' = 25)$ products, however, were found to form on time scales of 15 ps. Complimentary theoretical studies of the $Ne \cdots Br_2$ dissociation dynamics [110, 111] indicate that IVR may occur through intermediate resonances even though the prepared complex states are energetically well below the closing of the $\Delta v' = -1$ channel, which occurs at $v' = 28$. Therefore, rather than undergoing the typical IVR-mediated predissociation involving sequential $\Delta v' = -1$ steps, as described above, IVR in $Ne \cdots Br_2$ can occur within a single $\Delta v' = -1$ decay channel. Thus, $Ne \cdots Br_2(B, v' = 26)$ complexes prepared in the $n' = 0$ intermolecular level undergo IVR via an accidental resonance with a $Ne \cdots Br_2(B, v' = 25)$ level with significant intermolecular vibrational excitation lying above the $Ne + Br_2(B, v' = 25)$ dissociation limit. As discussed in Section IV. B, such a resonance was observed for $He \cdots I^{35}Cl$ when prepared just above the $He + ICl$ $(B, v' = 2)$ dissociation limit [39, 51].

B. Photoinitiated Dynamics of Linear $Rg \cdots XY(X, v'' = 0)$ Complexes

1. Vibrational Predissociation of Delocalized Levels

The influence of intermolecular bending excitation of the $Ne \cdots ICl(B, v' = 2)$ complex [97] and of intermolecular bending and rotational excitation of the $Ne \cdots Cl_2(B, v' = 11)$ complex [112] has been considered theoretically. To date, however, there are no complimentary experimental results reported to complement the predictions. The observation of the linear features in the excitation spectra of $Rg \cdots XY$ complexes enables dynamics to be investigated over broader regions of the excited state PES. Thus far, we have investigated the VP of $He \cdots I^{35}Cl(B, v' = 2, 3)$ dissociation with $n' \geq 2$ [51].

The probability distribution for the $n' = 2$ intermolecular level, Fig. 12c, indicates that this state resembles a bending level of the T-shaped complex with two nodes in the angular coordinate and maximum probability near the linear $He \cdots I–Cl$ and $He \cdots Cl–I$ ends of the molecule [40]. The measured $I^{35}Cl(B, v' = 2, j')$ rotational product state distribution observed following preparation of the $He \cdots I^{35}Cl(B, v' = 3, n' = 2, J' = 1)$ state is plotted in Fig. 12d. The distribution is distinctly bimodal and extends out to the rotational state, $j' = 21$, and is qualitatively similar to the distribution measured when exciting the T-shaped $n' = 0, j' = 0$ level, panel (b). As might be expected for a dissociation occurring from predominantly a linear geometry in comparison to that from a T-shaped geometry, the distribution is noticeably colder. An additional, subtle difference is that there is not a local minimum in the product population at the $j' = 1$ product state when exciting the $n' = 2, J' = 1$ state within the He + I^{35}Cl $(B,v' = 3)$ potential. We proposed that this local minimum results from the conservation of total angular momentum of the initial complex, and there should be a local maximum at $j' = 0$ and a minimum at $j' = 1$ with a symmetric complex wave function, such as that for He $\cdots I^{35}$Cl$(B, v' = 3, n' = 2, J' = 0)$, panel (b). In contrast, there should be a minimum at $j' = 0$ and a maximum at $j' = 1$ when the total wave function of the complex is asymmetric, such as for He $\cdots I^{35}$Cl$(B, v' = 3, n' = 2, J' = 1)$, panel (d).

The higher intermolecular vibrational levels, such as $n' = 4$, are even more delocalized in the angular coordinate, Fig. 12(e), and resemble free-rotor levels with the He atom rotating about the ICl molecule [40]. Excitation to the $n' = 4$, $J' = 0$, 1 states results in a distribution that is smoother and not distinctly bimodal, Fig. 12f. There is a minimum at $j' = 0$, but the population is not zero at this minimum, and again, the population distribution terminates at $j' = 21$. Following the model of Gray and Wozny [101], as the intermolecular probability distribution becomes delocalized in the angular coordinate, the nodal structure sequentially increases the number of regions on the repulsive potential associated with the product channels from which the wave packets originate significantly increases. Even though the wave packets still interfere as they evolve into the products, there is less structure in the product state distribution because of the increased number of contributing wave packets. Now that experimental data exist for the VP of levels with intermolecular excitation, it would be insightful to have complimentary calculations to further test the different theoretical methodologies.

2. Direct Dissociation above the Rg + XY(B,v′) Dissociation Limit

Continuum signals throughout the $B–X$ electronic region of the dihalogens have been observed when investigating the spectroscopy of the Rg $\cdots$ XY complexes. For example, a continuum signal observed in the LIF spectrum of Ar $\cdots I_2$ was observed and attributed to transitions of the linear conformer throughout the I_2 $B–X$ region [113]. Surprisingly, these continuum signals were even observed to extend above the $I_2(B)$ dissociation limit [114]. These observations led to several speculations regarding the excited-state dynamics and the properties of the excited-state Ar + $I_2(B)$ potential in the linear geometry. For example, upon electronic excitation it was believed that the antibonding character increases to the extent that this region of the intermolecular potential is fully repulsive. The excitation of the linear conformers in this region would promptly dissociate into separate $Ar + I_2(B,v')$ fragments. The molecular fragment would fluoresce and a continuum signal would be observed.

As we have reviewed here, the linear region is not fully repulsive, and transitions of the ground-state, linear conformer access vibrationally excited intermolecular levels that are delocalized in the angular coordinate. As depicted in Fig. 1, however, the internuclear distance is significantly longer in the excited state at the linear geometry. Consequently, there is favorable Franck–Condon overlap of the linear conformer with the inner-repulsive wall of the excited-state potential. It is therefore possible for the linear $Rg \cdots XY$ conformers to be promoted to the continuum of states just above each $Rg + XY$ (B,v') dissociation limit.

We have recorded action spectra for a number of the $Rg \cdots XY$ complexes to identify the precise transition energies at which the linear conformers access the continuum regions. While interesting for the investigation of dissociation dynamics, these transition energies also provide a very accurate measure of the binding energy of the ground-state, linear conformer; the transition energy of the beginning of the continuum signal less the corresponding monomer band origin transition energy is precisely the binding energy. Thus far, we have recorded such spectra and obtained linear conformer binding energies for $He \cdots$ ICl [39, 115], $Ne \cdots ICl$ [55], $He \cdots Br_2$ [41], $He \cdots I_2$ [53], $Ne \cdots I_2$, and $Ar \cdots I_2$ [57] with the values included in Table I.

An action spectrum for $He \cdots I^{35}Cl$ recorded with the probe laser fixed on the $I^{35}Cl$ E–B, 11–2 $R(0)$ rovibronic transition and scanning the excitation laser through the ICl B–X, 2–0 region is shown in Fig. 13a [51]. The continuum signal turns on ≈ 22 cm^{-1} above the $I^{35}Cl$ B–X, 2–0 band origin and extends for at least 140 cm^{-1} to higher transition energies. It was definitively shown that the continuum signal arises from transitions of the linear $He \cdots I^{35}Cl(X,v'' = 0)$ conformer, not from the T-shaped conformer, which would have extremely weak transition strength to the inner repulsive wall of the excited-state surface. The linear conformer when promoted to the repulsive, inner wall of the potential dissociates very quickly and a rotationally cold $I^{35}Cl(B,v',j')$ product state distribution results. The three product state spectra shown in Fig. 13 b–d were obtained by fixing the excitation laser at the transition energies labeled in panel (a) and scanning the probe laser across the $I^{35}Cl$ E–B, 11–2 band. With only three or four rotational rotational lines observed, the spectra illustrate that the three distributions are extremely cold, and cut off at the energetic limit based on the excitation energy and the $He + I^{35}Cl(B,v' = 2)$ asymptote and the amount of rotational energy within the ground-state complex.

When exciting to even higher transition energies, the product state distributions are still found to be quite cold, with the fraction of the available energy found in rotation actually decreasing as higher energy regions in the continuum are accessed. For example, the rotational distribution is characterized

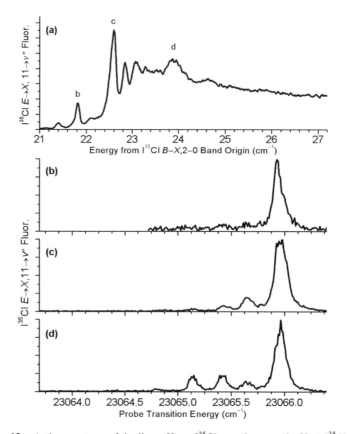

Figure 13. Action spectrum of the linear $He \cdots I^{35}Cl$ complex near the $He + I^{35}Cl(B,v' = 2)$ dissociation limit obtained by scanning the excitation laser through the ICl $B–X$, 2–0 region and monitoring the $I^{35}Cl$ $E{\rightarrow}X$ fluorescence induced by the temporally delayed probe laser, which was fixed on the $I^{35}Cl$ $E–B$, 11–2 band head, (a). The transition energy is plotted relative to the $I^{35}Cl$ $B–X$, 2–0 band origin, $17,664.08\,cm^{-1}$. Panels (b), (c), and (d) are the rotational product state spectra obtained when fixing the excitation laser on the lines denoted with the corresponding panel letter. The probe laser was scanned through the ICl $B–X$, 11–2 region. Modified with permission from Ref. [51].

by $\langle E_{rot} \rangle = 1.10\,cm^{-1}$ when the linear complexes are excited at $\approx 5\,cm^{-1}$ above the $He + I^{35}Cl(B,v' = 2,j')$ dissociation limit. This corresponds to $\approx 25\%$ of the total available energy to the $I^{35}Cl(B,v' = 2)$ product. When the complexes are prepared at $\approx 20\,cm^{-1}$ above the dissociation limit, the rotational energy is only $\approx 8\%$ of the available energy.

Our results on $Ar \cdots I_2$ [57] are particularly noteworthy because of the extensive previous work aimed at characterizing the continuum fluorescence signals that were found to extend above the $I_2(B)$ dissociation limit. It was proposed that these signals resulted from a one-atom caging mechanism, in which the linear $Ar \cdots I_2(X, v'' = 0)$ conformer is promoted to states where the I_2 is no longer bound. However, the proximity of the Ar atom cages or confines the dissociation path, recoiling one of the iodine atoms back toward the other and ejecting the Ar atom. The I_2 molecule is restabilized in a bound vibrational level from where it fluoresces. We verified that the continuum signals do arise from transitions of the linear $Ar \cdots I_2(X, v'' = 0)$ complex. However, our data provide no evidence that the excited-state complexes undergo a one-atom caging mechanism. Instead, we found that the continuum signals result from bound–free transitions of the linear $Ar \cdots I_2(X, v'' = 0)$ complex to the inner-repulsive walls of numerous $Ar + I_2(B,v')$ intermolecular potentials. The bound–free continuum signal associated with transitions to each $Ar + I_2(B,v')$ potential spans an energy region $> 700 \, cm^{-1}$. Thus, the continuum signal below and above the $I_2(B)$ dissociation limit is just the sum of all of the continuum signals that turn on above each $Ar + I_2(B,v')$ asymptote.

3.   Rotational Predissociation above the $Rg + XY(B,v')$ Dissociation Limit

The action spectrum recorded through the turn-on region of the continuum signal, Fig. 13a, contains rotational structure that is similar to that within the linear feature [39, 51]. The calculations of the $He \cdots I^{35}Cl$ intermolecular levels within the $He + ICl(B,v' = 2)$ potential indicate that the $n' = 7$ level should be within $\approx 1 \, cm^{-1}$ of the dissociation limit, and this level may give rise to the observed features [40]. In partial support of this assignment, action spectra were recorded at varying distances downstream, and the intensities of the individual lines change in a similar manner as observed within the linear feature [39, 51]. The discrete lines at 21.7 and 22.6 cm^{-1}, labeled (b) and (c) in Fig. 13a, gain intensity with cooling and are presumed to be the $P(1)$ and $R(0)$ lines line accessing the $n' = 7$ intermolecular vibrational level.

   The probability densities for the high lying intermolecular vibrational levels, similar to that shown in Fig. 12e, indicate that these levels can be considered free-rotor levels with the He atom being delocalized about the ICl molecule and having significant internal angular momentum. Such a level would most likely have a centrifugal barrier such that the complex could remain quasibound even if it is energetically above the $He + I^{35}Cl(B,v' = 2)$ dissociation limit. In $He + I^{35}Cl(B,v' = 2)$ scattering experiments this level would most likely give rise to a shape resonance at a very low internuclear translational energies. It is precisely a quasibound intermolecular vibrational level, referred to as a doorway state [103], that is proposed to give rise to the IVR-mediated VP observed in the $He \cdots Br_2$ system when prepared in levels with $v' > 40$ [52, 104, 105]. Of the systems investigated, Table I, the $He \cdots I^{35}Cl$ system at the $He + ICl(B,v' = 2)$ dissociation limit is the only one that such a resonance was definitively observed. There is some structure in the turn-on region of the $He \cdots ^{79,79}Br_2$ complex in the Br_2 $B-X$, 11–0 region [41], but the signal-to-noise level is not sufficient for further analysis.

C. Photoinitiated Dynamics of Higher Order $Rg_n \cdots XY(X, v'' = 0)$ Complexes

The ability to stabilize different conformers of the higher order, ground-state $Rg_n \cdots XY$ complex opens a new avenue for dynamics and energy-transfer studies. Since all of the observed $He_n \cdots ICl$ and $He_n \cdots Br_2$ features follow the band-shift rule the He atoms must not be strongly interacting in either the ground or excited electronic states. It therefore seems likely that transitions of the $(1,1)He_2 \cdots ICl$ complex will access excited-state levels that have one He atom localized in the T-shaped well and the other in bending or hindered-rotor levels. Our group is currently undertaking dynamical studies on $He_2 \cdots ICl$. As mentioned above, we have already measured the binding energy of the $(1,1)He_2 \cdots ICl$ conformer using action spectroscopy to find the bound-free continuum associated with the $He + He \cdots ICl(B, v' = 3)$ dissociation limit. It would also be insightful to perform time-resolved experiments on the different conformers of these systems to directly monitor the kinetics for forming the different products and intermediates as a function of the different excited-state levels prepared.

Acknowledgments

R.A.L. is indebted to the David and Lucile Packard Foundation for a Fellowship in Science and Engineering and to the National Science Foundation for a CAREER Award, CHE-0346745. The authors would also like to thank J.P. Darr and J. J. Glennon from Washington University and A. B. McCoy and S. E. Ray from The Ohio State University for valuable discussions and for fruitful collaborations.

References

1. P. Casavecchia, *Rep. Prog. Phys.* **63**, 355 (2000).

2. K. Liu, *J. Chem. Phys.* **125**, 132307 (2006).

3. J. I. Steinfeld, J. S. Fransisco, and W. L. Hase, *Chemical Kinetics and Reaction Dynamics*, 2nd ed., Simon & Schuster, Upper Saddle River, NJ, 1999.

4. R. D. Levine and R. B. Bernstein, *Molecular Reaction Dynamics and Chemical Reactivity*, Oxford University Press, New York, 1987.

5. S. Buelow, G. Radhakrishnan, J. Catanzarite, and C. Wittig, *J. Chem. Phys.* **83**, 444 (1985).

6. W. H. Breckenridge, C. Jouvet, and B. Soep, *J. Chem. Phys.* **84**, 1443 (1986).

7. G. Radhakrishnan, S. Buelow, and C. Wittig, *J. Chem. Phys.* **84**, 727 (1986).

8. C. Jouvet, M. Boivineau, M. C. Duval, and B. Soep, *J. Phys. Chem.* **91**, 5416 (1987).

9. C. Wittig, S. Sharpe, and R. A. Beaudet, *Acc. Chem. Res.* **21**, 341 (1988).

10. R. E. Smalley, L. Wharton, and D. H. Levy, *Acc. Chem. Res.* **10**, 139 (1977).

11. D. H. Levy, *Adv. Chem. Phys.* **47**, 323 (1981).

12. M. I. Lester, *Adv. Chem. Phys.* **96**, 51 (1996).

13. A. Rohrbacher, J. Williams, and K. C. Janda, *Phys. Chem. Chem. Phys.* **1**, 5263 (1999).

14. A. Rohrbacher, N. Halberstadt, and K. C. Janda, *Annu. Rev. Phys. Chem.* **51**, 405 (2000).

15. M. C. Heaven and A. A. Buchachenko, *J. Mol. Spectrosc.* **222**, 31 (2003).

16. S. E. Novick (2004).

17. J. C. Drobits, J. M. Skene, and M. I. Lester, *J. Chem. Phys.* **84**, 2896 (1986).

18. J. E. Kenny, K. E. Johnson, W. Sharfin, and D. H. Levy, *J. Chem. Phys.* **72**, 1109 (1980).

19. K. E. Johnson, W. Sharfin, and D. H. Levy, *J. Chem. Phys.* **74**, 163 (1981).

20. B. A. Swartz, D. E. Brinza, C. M. Western, and K. C. Janda, *J. Phys. Chem.* **88**, 6272 (1984).

21. S. R. Hair, J. I. Cline, C. R. Bieler, and K. C. Janda, *J. Chem. Phys.* **90**, 2935 (1989).

22. W. D. Sands, C. R. Bieler, and K. C. Janda, *J. Chem. Phys.* **95**, 729 (1991).

23. M. Gutmann, D. M. Willberg, and A. H. Zewail, *J. Chem. Phys.* **97**, 8048 (1992).

24. B. Miguel, A. Bastida, J. Zuniga, A. Requena, and N. Halberstadt, *Faraday Discuss.* **118**, 257 (2001).

25. G. C. Schatz, R. B. Gerber, and M. A. Ratner, *J. Chem. Phys.* **88**, 3709 (1988).

26. P. Villarreal, A. Varade, and G. Delgado-Barrio, *J. Chem. Phys.* **90**, 2684 (1989).

27. *Atomic and Molecular Beam Methods*, G. Scoles (ed.), Oxford University Press, New York, 1988, vol. I.

28. W. Sharfin, K. E. Johnson, L. Wharton, and D. H. Levy, *J. Chem. Phys.* **71**, 1292 (1979).

29. J. A. Blazy, B. M. DeKoven, T. D. Russell, and D. H. Levy, *J. Chem. Phys.* **72**, 2439 (1980).

30. R. E. Smalley, L. Wharton, and D. H. Levy, *J. Chem. Phys.* **68**, 671 (1978).

31. S. J. Harris, S. E. Novick, and W. Klemperer, *J. Chem. Phys.* **61**, 193 (1974).

32. S. E. Novick, S. J. Harris, K. C. Janda, and W. Klemperer, *Can. J. Phys.* **53**, 2007 (1975).

33. Y. Xu, W. Jäger, I. Ozier, and M. C. L. Gerry, *J. Chem. Phys.* **98**, 3726 (1993).

34. M. L. Burke and W. Klemperer, *J. Chem. Phys.* **98**, 6642 (1993).

35. C. F. Kunz, I. Burghardt, and B. A. Heß, *J. Chem. Phys.* **109**, 359 (1998).

36. R. Prosmiti, C. Cunha, P. Villarreal, and G. Delgado-Barrio, *J. Chem. Phys.* **119**, 4216 (2003).

37. R. Prosmiti, C. Cunha, P. Villarreal, and G. Delgado-Barrio, *J. Chem. Phys.* **116**, 9249 (2002).

38. M. D. Bradke and R. A. Loomis, *J. Chem. Phys.* **118**, 7233 (2003).

39. D. S. Boucher, J. P. Darr, M. D. Bradke, R. A. Loomis, and A. B. McCoy, *Phys. Chem. Chem. Phys.* **6**, 5275 (2004).

40. A. B. McCoy, J. P. Darr, D. S. Boucher, P. R. Winter, M. D. Bradke, and R. A. Loomis, *J. Chem. Phys.* **120**, 2677 (2004).

41. D. S. Boucher, D. B. Strasfeld, R. A. Loomis, J. M. Herbert, S. E. Ray, and A. B. McCoy, *J. Chem. Phys.* **123**, 104312 (2005).

42. D. G. Jahn, W. S. Barney, J. Cabalo, S. G. Clement, A. Rohrbacher, T. J. Slotterback, J. Williams, and K. C. Janda, *J. Chem. Phys.* **104**, 3501 (1996).

43. A. A. Buchachenko, R. Prosmiti, C. Cunha, G. Delgado-Barrio, and P. Villarreal, *J. Chem. Phys.* **117**, 6117 (2002).

44. A. Valdés, R. Prosmiti, P. Villarreal, and G. Delgado-Barrio, *Mol. Phys.* **102**, 2277 (2004).

45. J. M. Skene and M. I. Lester, *Chem. Phys. Lett.* **116**, 93 (1985).

46. J. M. Skene, J. C. Drobits, and M. I. Lester, *J. Chem. Phys.* **85**, 2329 (1986).

47. J. M. Skene, Ph. D. Dissertation, University of Pennsylvania, 1988.

48. R. L. Waterland, J. M. Skene, and M. I. Lester, *J. Chem. Phys.* **89**, 7277 (1988).

49. R. L. Waterland, M. I. Lester, and N. Halberstadt, *J. Chem. Phys.* **92**, 4261 (1990).

50. R. Prosmiti, C. Cunha, P. Villarreal, and G. Delgado-Barrio, *J. Chem. Phys.* **117**, 7017 (2002).

51. J. P. Darr, R. A. Loomis, and A. B. McCoy, *J. Chem. Phys.* **122**, 044318 (2005).

52. D. G. Jahn, S. G. Clement, and K. C. Janda, *J. Chem. Phys.* **101**, 283 (1994).

53. S. E. Ray, A. B. McCoy, J. J. Glennon, J. P. Darr, J. R. Lancaster, E. J. Fesser, and R. A. Loomis, *J. Chem. Phys.* (2006).

54. R. Prosmiti, A. Valdés, P. Villarreal, and G. Delgado-Barrio, *J. Phys. Chem. A* **108**, 6065 (2004).

55. D. B. Strasfeld, J. P. Darr, and R. A. Loomis, *Chem. Phys. Lett.* **397**, 116 (2004).

56. A. Burroughs, G. Kerenskaya, and M. C. Heaven, *J. Chem. Phys.* **115**, 784 (2001).

57. J. P. Darr, J. J. Glennon, and R. A. Loomis, *J. Chem. Phys.* **122**, 131101 (2005).

58. R. Prosmiti, P. Villarreal, and G. Delgado-Barrio, *Chem. Phys. Lett.* **359**, 473 (2002).

59. A. A. Buchachenko and N. F. Stepanov, *J. Chem. Phys.* **104**, 9913 (1996).

60. J. C. Drobits and M. I. Lester, *J. Chem. Phys.* **86**, 1662 (1987).

61. C. R. Bieler, D. D. Evard and K. C. Janda, *J. Phys. Chem.* **94**, 7452 (1990).

62. D. S. Boucher, J. P. Darr, D. B. Strasfeld, and R. A. Loomis, (to be submitted to *J. Phys. Chem. A*, 2008).

63. E. B. Wilson, *Int. J. Quantum Chem: Quant. Chem. Symp.* **13**, 5 (1979).

64. R. G. Woolley, *J. Math. Chem.* **23**, 3 (1998).

65. A. E. Stevens Miller, C.-C. Chuang, H. C. Fu, K. J. Higgins, and W. Klemperer, *J. Chem. Phys.* **111**, 7844 (1999).

66. A. Bastida, J. Zúñiga, A. Requena, B. Miguel, J. A. Beswick, J. Vigué, and N. Halberstadt, *J. Chem. Phys.* **116**, 1944 (2002).

67. D. S. Boucher, M. D. Bradke, J. P. Darr, and R. A. Loomis, *J. Phys. Chem. A* **107**, 6901 (2003).

68. A. Valdés, R. Prosmiti, P. Villarreal, and G. Delgado-Barrio, *J. Chem. Phys.* **122**, 044305 (2005).

69. A. Valdés, R. Prosmiti, P. Villarreal, and G. Delgado-Barrio, *J. Chem. Phys.* **125**, 014313 (2006).

70. J. P. Darr and R. A. Loomis, 2006.

71. R. E. Smalley, D. H. Levy, and L. Wharton, *J. Chem. Phys.* **64**, 3266 (1976).

72. K. E. Johnson, L. Wharton, and D. H. Levy, *J. Chem. Phys.* **69**, 2719 (1978).

73. M. Gutmann, D. M. Willberg, and A. H. Zewail, *J. Chem. Phys.* **97**, 8037 (1992).

74. M. S. Kim, R. E. Smalley, L. Wharton, and D. H. Levy, *J. Chem. Phys.* **65**, 1216 (1976).

75. D. M. Willberg, M. Gutmann, J. J. Breen, and A. H. Zewail, *J. Chem. Phys.* **96**, 198 (1992).

76. R. B. Gerber, V. Buch, and M. A. Ratner, *J. Chem. Phys.* **77**, 3022 (1982).

77. J. A. Beswick and J. Jortner, *Chem. Phys. Lett.* **49**, 13 (1977).

78. J. A. Beswick and J. Jortner, *J. Chem. Phys.* **68**, 2277 (1978).

79. J. A. Beswick and J. Jortner, *Adv. Chem. Phys.* **47**, 363 (1981).

80. J. A. Beswick and J. Jortner, *J. Chem. Phys.* **73**, 3653 (1980).

81. J. A. Beswick, G. Delgado-Barrio, and J. Jortner, *J. Chem. Phys.* **70**, 3895 (1979).

82. J. A. Beswick and J. Jortner, *J. Chem. Phys.* **69**, 512 (1978).

83. G. E. Ewing, *J. Chem. Phys.* **71**, 3143 (1979).

84. S. B. Woodruff and D. L. Thompson, *J. Chem. Phys.* **71**, 376 (1979).

85. S. K. Gray, S. A. Rice, and D. W. Noid, *J. Chem. Phys.* **84**, 3745 (1986).

86. M. J. Davis and S. K. Gray, *J. Chem. Phys.* **84**, 5389 (1986).

87. S. K. Gray, *J. Chem. Phys.* **87**, 2051 (1987).

88. G. Delgado-Barrio, P. Villarreal, P. Mareca, and G. Albelda, *J. Chem. Phys.* **78**, 280 (1983).

89. E. Segev and M. Shapiro, *J. Chem. Phys.* **78**, 4969 (1983).

90. R. H. Bisseling, R. Kosloff, R. B. G. M. A. Ratner, L. Gibson, and C. Cerjan, *J. Chem. Phys.* **87**, 2760 (1987).

91. S. Das and D. J. Tannor, *J. Chem. Phys.* **92**, 3403 (1990).

92. O. A. Sharafeddin, H. F. Bowen, D. J. Kouri, S. Das, D. J. Tannor, and D. K. Hoffman, *J. Chem. Phys.* **95**, 4727 (1991).

93. D. H. Zhang and J. Z. H. Zhang, *J. Phys. Chem.* **96**, 1575 (1992).

94. G. E. Ewing, *Chem. Phys.* **63**, 411 (1981).

95. G. E. Ewing, *Faraday Discuss. Chem. Soc.* **73**, 325 (1982).

96. J. C. Drobits and M. I. Lester, *J. Chem. Phys.* **88**, 120 (1988).

97. J. C. Drobits and M. I. Lester, *J. Chem. Phys.* **89**, 4716 (1988).

98. O. Roncero, J. A. Beswick, N. Halberstadt, P. Villarreal, and G. Delgado-Barrio, *J. Chem. Phys.* **92**, 3348 (1990).

99. J. I. Cline, B. P. Reid, D. D. Evard, N. Sivakumar, N. Halberstadt, and K. C. Janda, *J. Chem. Phys.* **89**, 3535 (1988).

100. J. I. Cline, N. Sivakumar, D. D. Evard, C. R. Bieler, B. P. Reid, N. Halberstadt, S. R. Hair, and K. C. Janda, *J. Chem. Phys.* **90**, 2605 (1989).

101. S. K. Gray and C. E. Wozny, *J. Chem. Phys.* **94**, 2817 (1991).

102. S. K. Gray and C. E. Wozny, *J. Chem. Phys.* **91**, 7671 (1989).

103. A. García-Vela and K. C. Janda, *J. Chem. Phys.* **124**, 034305 (2006).

104. T. Gonzalez-Lezana, M. I. Hernandez, G. Delgado-Barrio, A. A. Buchachenko, and P. Villarreal, *J. Chem. Phys.* **105**, 7454 (1996).

105. A. Rohrbacher, T. Ruchti, K. C. Janda, A. A. Buchachenko, M. I. Hernández, T. González-Lezana, P. Villarreal, and G. Delgado-Barrio, *J. Chem. Phys.* **110**, 256 (1999).

106. D. D. Evard, J. I. Cline, and K. C. Janda, *J. Chem. Phys.* **88**, 5433 (1988).

107. D. D. Evard, C. R. Bieler, J. I. Cline, N. Sivakumar, and K. C. Janda, *J. Chem. Phys.* **89**, 2829 (1988).

108. O. Roncero, D. Caloto, K. C. Janda, and N. Halberstadt, *J. Chem. Phys.* **107**, 1406 (1997).

109. J. A. Cabrera, C. R. Bieler, B. C. Olbricht, W. E. van der Veer, and K. C. Janda, *J. Chem. Phys.* **123**, 054311 (2005).

110. T. A. Stephenson and N. Halberstadt, *J. Chem. Phys.* **112**, 2265 (2000).

111. B. Miguel, A. Bastida, J. Zuniga, A. Requena, and N. Halberstadt, *J. Chem. Phys.* **113**, 10130 (2000).

112. N. Halberstadt, J. A. Beswick, and K. C. Janda, *J. Chem. Phys.* **87**, 3966 (1987).

113. M. L. Burke and W. Klemperer, *J. Chem. Phys.* **98**, 1797 (1993).

114. K. L. Saenger, G. M. McClelland, and D. R. Herschbach, *J. Phys. Chem.* **85**, 3333 (1981).

115. J. P. Darr, A. C. Crowther, and R. A. Loomis, *Chem. Phys. Lett.* **378**, 359 (2003).

AUTHOR INDEX

Advances in Chemical Physics, Volume 138, edited by Stuart A. Rice
Copyright © 2008 John Wiley & Sons, Inc.

SUBJECT INDEX

Advances in Chemical Physics, Volume 138, edited by Stuart A. Rice
Copyright © 2008 John Wiley & Sons, Inc.